Migration, Mining, and the African Diaspora

Migration, Mining, and the African Diaspora

Guyana in the Nineteenth and Twentieth Centuries

Barbara P. Josiah

palgrave
macmillan

First published in 2011 by
PALGRAVE MACMILLAN®
in the United States—a division of St. Martin's Press LLC,
175 Fifth Avenue, New York, NY 10010.

Where this book is distributed in the UK, Europe and the rest of the world, this is by Palgrave Macmillan, a division of Macmillan Publishers Limited, registered in England, company number 785998, of Houndmills, Basingstoke, Hampshire RG21 6XS.

Palgrave Macmillan is the global academic imprint of the above companies and has companies and representatives throughout the world.

Palgrave® and Macmillan® are registered trademarks in the United States, the United Kingdom, Europe and other countries.

ISBN: 978–0–230–11589–7

Library of Congress Cataloging-in-Publication Data

Josiah, Barbara P.
Migration, mining, and the African diaspora : Guyana in the nineteenth and twentieth centuries / Barbara P. Josiah.
p. cm.
ISBN 978–0–230–11589–7 (hardback)
1. Mineral industries—Social aspects—Guyana—History. 2. Mineral industries—Economic aspects—Guyana—History. 3. Mineral industries—Environmental aspects—Guyana—History. 4. Mines and mineral resources—Guyana—History. 5. Miners—Guyana—Economic conditions—19th century. 6. Miners—Guyana—Economic conditions—20th century. 7. Miners—Guyana—Social conditions—19th century. 8. Miners—Guyana—Social conditions—20th century. 9. Migrant labor—Guyana—History. 10. African diaspora. 11. Guyana—Economic conditions. 12. Guyana—Social conditions. I. Title.

HD9506.G98J67 2011
331.6′26609881—dc23 2011026975

A catalogue record of the book is available from the British Library.

Design by Newgen Imaging Systems (P) Ltd., Chennai, India.

First edition: November 2011

10 9 8 7 6 5 4 3 2 1

Printed and bound in Great Britain by
CPI Antony Rowe, Chippenham and Eastbourne

To my son
Raul I. Josiah

CONTENTS

List of Tables and Figures ix

Preface xi

Acknowledgments xvii

List of Abbreviations xix

Introduction 1

1 African Diaspora Migrant Miners and Guyana's El Dorado 9

2 Migration and Mining Strategies in a Colonial Society 23

3 Mining Factors in a Diversified Economy 39

4 The Perils of Labor in Mining: Migration and Mortality 57

5 Aspects of Infrastructure Development: Gold and Diamonds 71

6 Another Approach: Organizing Bauxite Production 93

7 Evolving Relations: Mining and Trade Unionism 109

8 Internal Migration and Village Dynamics: Families and Communities Coping 129

9 Knowledge Transfer and Cooperativism: Agriculture and Mining Eras 145

10 African Continuities, Jewels, and Economic Linkages to Mining 157

Conclusion 175

Appendix A 179

Appendix B 183

Appendix C 185

Notes 187

Selected Bibliography 249

Index 265

Tables and Figures

Tables

1.1	Some early African-Guyanese villages and prices	14
1.2	West Indians and Africans in Guyana, 1841–1891	16
2.1	Gold and diamond miners registered, 1888–1910	36
3.1	Value of selected exports, 1891–1900	42
3.2	Value of gold and miners registered, 1892–1900	43
3.3	Value of selected exports from Guyana, 1900–1910	44
3.4	Licenses issued to women, 1909–1910	45
3.5	Value of gold, diamonds, and bauxite, 1920–1956	55
4.1	Selected deaths in the gold mining regions, 1891–1895	61
4.2	Deaths in the goldfields, 1894–1901	64
A1	Purchasers of plantation New Orange Nassau (Buxton), Demerara, amounts paid ($)	179
A2	Persons for transports of lots, Winkle Village, 1890	181

Figures

3.1	Map of Guyana with gold, diamond, and bauxite locations	40
3.2	Men washing for diamonds and gold miners	48
5.1	River scenes	79
10.1	Clifford Hector, jeweler	165
10.2	Maurice July, jeweler and apprentice son	165

Preface

In the context of Guyana's (British Guiana's) social and economic history, this book on African diaspora migrant workers and their contributions to the establishment of the mining industry, offers unique insights into aspects of the country's development during the nineteenth and twentieth centuries. When gold mining began in earnest in the 1880s, it diversified the country's economy from an overreliance on the export of sugar and its by-products and boosted the revenue and financial benefits for government and private investors. Gold mining created a variety of employment opportunities for the male-dominated workforce mostly comprising of persons of African ancestry. Labor migrants from the Caribbean islands, French Guiana and Suriname, North and Central America, and Africa joined their local counterparts in establishing temporary and permanent communities in the mining regions. In 1890, the export of diamonds began. Soon, profits from gold and diamonds prompted legislators to enact laws allowing more access to government land. Dairy farming, rice, and timber production expanded. During World War I, bauxite exports began and further diversified the economy. The majority of bauxite employees were also African diaspora labor migrants. This book traces the mining of a precious metal and gems from labor intensive origins to the eve of gold's emergence at capital intensive heights after the 1950s. It aligns bauxite with gold and diamonds to tell the story of mining in Guyana.

Mining workers functioned within a capitalist economy and understood the utility of material possessions and physical comforts. In earlier decades, their ancestors had transferred some knowledge of African practices that were applicable to their experiences in New World environments to empower themselves and survive. At the turn of the nineteenth century, many of the challenges workers in Guyana faced in the mining era had already confronted them when agriculture was dominant. This, in fact, raises an important question—why did these workers pursue mining in the face of the environmental perils associated with gold and diamond prospecting. The research reveals that mining enabled some workers to cope with under- and unemployment while others utilized the occupation as an alternative to agricultural work in a plantation society plagued by conflict and adversities. They had long participated in farming and related activities, established their own financial institutions, bought real estate, generated employment, organized labor action, and challenged the power structure's authority to impose laws to control their lives and work. In the changing environment of a diversified economy, mining offered an opportunity for the African-descended to continue those activities and to ensure their offspring functioned at higher and better socioeconomic levels.

While the vestiges of continuous adversarial relations between the power structure and the working class often formed the backdrop of socioeconomic relations in the colony, this study digresses and includes the workers' notion of "order and progress" and modernization from the agricultural through the mining era. Consequently, the study delineates the connection between knowledge transfer and African continuities blending with Western models for progress and development. Primary and secondary sources such as administrative reports, mining ordinances and regulations, death and financial records as well as contemporary newspapers, gazettes, and oral accounts generated by the economic elites and the working class unfold the history of mining in a colonial society.

In this study, pendulum or seasonal migration to and from the coast to the interior gold and diamond areas and movement to and settlement in bauxite producing regions are examined in relation to the country's diverse environments and infrastructure development. In gold and diamond mining, aspects of colonial policies, which intertwined with export-oriented production, are discussed in the context of an ongoing need to import labor migrants. The organization of bauxite mining offers a counterpoint to the establishment of the gold and diamond sectors. When bauxite production began, North American managerial policies differed somewhat from that of the British colonial bureaucrats who had established gold and diamond mining. While the production of the three commodities emerged at strategically critical times in international affairs—during a global economic depression and a World War—gold and diamond mining was speculative in orientation, but that of bauxite was deliberate. Overall, mining benefited the participants including the government, private investors, business persons, and workers. A critical contention of this study is that mining workers were not simply exploited by and for the benefit of others, but they also ensured some rewards for themselves from their labor.

Evidence reveals that workers understood the value of their labor and the utility of self-discipline in coping with their circumstances to achieve improvements. Archival sources provided the facts on the economic and environmental situations within which the workers functioned. Colonial administration, policies, strategies, laws, amount and value of exports, and control and exploitation of workers, provided the basis for a number of scholarly studies on Guyana's mining industry. A critical divergence in the current analysis is its focus on the concepts and actions of mining workers who envisioned specific outcomes from their earnings and provided for the future.

The labor migrants engaged their circumstances to improve their conditions. Although most of the workers were agricultural in orientation, with the coming of mining they adapted to and integrated the new activity into their existing economic enterprises. The discussion opens a window on people who were not simply reactionaries to the doings of others but also initiated activities for their own well being. The study affords a prism to analyze connections between mining and economic linkages to other enterprises such as the jewelry industry, pioneered by skilled persons of African ancestry in the country. Beyond mining, the workers had identities, aspirations, and initiatives that are given analytical space here.

For centuries, before the emergence of mining, development in Guyana followed a trajectory similar to other British colonies in the Americas where the labor of persons of African descent was important for production. As the nineteenth century ensued, global economic developments, due in part to capitalism, industrialization, and free trade caused changes in many societies. Minerals and ores were in demand to satisfy the increasing needs of industrialization. In Guyana, colonial administrators and economic elites recognized the importance of sufficient labor for mining and the production of a variety of other commodities. Legislators were urged to recruit additional immigrants and to enact more stringent labor laws. The belief existed that labor control ensured profits for investors and local and metropolitan governments. Workers functioned in the context of capitalism within which the acquisition of financial resources determined survival. Consequently, the study explores the relationship between the initiatives and actions of persons who grappled, not always unsuccessfully, with colonial realities including hostile environments in an era characterized by "order and progress" and modernization.

At one level, colonial elites and administrators favorably viewed a diversified economy as a way to export more products, and as a means of accruing profits. At another level, while the working class subscribed to the tenures of modern development such as education and property-holding, they often challenged the ways in which the power structure sought to impose labor and taxation laws and other restrictions on them. The research reveals that during this period continuous conflict between the working class and the plantocracy gave mining the appearance of an attractive alternative employment avenue to sugar estate work. However, gold and diamond mining occurred in the country's virgin and often dangerous hinterland regions. Cost minimizing policies of both government bureaucrats and officials of private companies limited infrastructure development in the interior and prolonged frontier conditions. As a result, many gold and diamond miners suffered ill health, were incapacitated, died untimely or endured tremendous adversities. Most workers frequently journeyed between their homes and mining sites due to the lack of amenities for the settlement of families. Despite this they remained an important factor in the expansion of Guyana's mining frontiers. Equally noteworthy, were the contributions of bauxite workers whose production was highly valuable to the munitions, automobile, and aircraft industries during World Wars I and II.

When bauxite, the third member of Guyana's mining family emerged, life for those miners achieved semblances of predictability and stability. Unlike workers in gold and diamond mining, those in bauxite did not have to contend with the vagaries of the weather. Hazards of deaths by drowning and fatal encounters with wild animals, reptiles, and insects, pervasive in gold and diamond mining, were almost unheard of in bauxite production. Centers of bauxite mining were mostly within easy traveling distance to urban areas, compared with those of gold and diamond-producing areas, which were far distant. Management of bauxite operations early provided infrastructure to facilitate permanent residences for workers and their families near production areas. In the gold and diamond mining regions, government and private attempts to provide residential settlements were piecemeal and never achieved the sophistication of the bauxite areas. Notwithstanding, for decades, prospectors journeyed hundreds of miles inland to excavate gold and diamonds. Many miners survived the human, geographic, and ecologically hostile environments, but myriad others died searching for gold and diamonds in Guyana's El Dorado.

The country's gold and diamond miners literally "walked between the raindrops," traversed many dangerous waterways, and faced deadly situations as they developed the industry. Unlike many bauxite workers who relocated their entire households permanently, miners of gold and diamonds invariably left their families at home for lengthy periods as they prospected hundreds of miles away. To no avail, at the urgings of the mineocracy, colonial legislators enacted laws, ostensibly to discourage the operations of small-scale independent groups of gold and diamond miners. These self-financed miners were mostly of African ancestry. Meanwhile, women never shirked their responsibilities as bread winners in the temporary absence of the male heads of households. By the early twentieth century, mining women further diversified the workforce. Defying the gender roles of the times in which they lived, women became claim and prospecting license holders and miners.

Gold and diamond miners also worked in the sugar, rice, and timber industries or in their own fishing and farming enterprises on the coast when it was not propitious to trek to the interior. Vagaries of the weather, environmental impediments, investment limitations, manipulation of markets by mining magnates, legal difficulties of miners, wartime conditions, uncertain food supplies, inadequate transportation to and from the interior, and work opportunities in bauxite sometimes hindered gold and diamond mining. In essence, the dynamics of this study of gold, diamonds, and bauxite intertwine migration and mining, to juxtapose colonialism from the vantage points of colonial elites and the working class.

The book underscores that profit-maximizing techniques were adopted so that administrators, private investors, and their allies reaped optimum benefits. The paths miners walked were not always "paved with gold" or "studded with diamonds" and establishing and maintaining the bauxite industry involved arduous labor and capital intensive removal of soil overburden in virgin terrain. For many of the migrating gold and diamond workers, mining often meant dying in attempts to make a living or struggling and surviving against all odds. Ironically, as the export of gold, diamonds, and bauxite revitalized the economy, development and underdevelopment existed side by side. Still, the labor of the workers was as pivotal as land and capital to diversify the country's economy. The workers' efforts, which made mining economically viable, remain the untold part of Guyana's history. The centerpiece of this book is the long overdue assessment of the importance of mining in the country's socioeconomic development.

In the 1880s, output of gold and diamonds increased revenue, impacted national development, and diversified the country's exports. When bauxite production began in 1916, the new activity had a similar effect as its precursors in the mining industry. With the proliferation of gold and diamonds, the jewelry industry expanded. An often overlooked aspect of the country's development is that this industry initially relied on African-descended craftspersons. The inclusion of the roles of knowledge transfer and African continuities and their connections to economic enterprises with linkages to the mining industry displays inherent features of human development. Labor migrants not only diversified the country's economy, but some of them achieved a modicum of material success to enhance the future of succeeding generations. The challenge here is to understand mining workers as subjects of their history and to provide a meaningful interpretation of their story in a national and international context. Mining workers' contributions are central to the migration, labor, economic, and social history of Guyana.

Acknowledgments

The research for this publication was triangular in nature due to its local, national, and international scope. Documents housed at the Guyana National Library, Guyana National Archives, the Caribbean Research Centre, University of Guyana, and the Geological Survey and Gold Commission were utilized. The holdings of Howard University Libraries, Moorland Spingarn Research Center, the Library of Congress, and the National Archives were invaluable. Documents in the United Kingdom at the National Archives, the British Library, and the London School of Economics concluded an effort to analyze the contributions of people of African origin to development in Guyana.

This book resulted from the unique environment of the Department of History and Geography, Howard University, Washington, DC where I spent eight years as a graduate student and three years as a faculty member. I thank the faculty and staff of the department, especially Drs. Emory J. Tolbert, the then chairperson, Selwyn H. H. Carrington, my dissertation advisor, Joseph P. Reidy and Aziz Batran who along with Ralph Gomes of the Department of Anthropology and Sociology were my committee members. Dr. Walter B. Hope, economic geographer, deserves my special gratitude because his reviews and valuable insights enhanced the manuscript in ways too numerous to mention here. Dr. Arthur Burt, Ms. Alma Kemp, former administrative assistant of the department, and Mrs. Bessie Hill, her successor deserve recognition for their help over the years. I must express deep appreciation to Dr. David Deleon for his encouragement, support, and valuable feedback on my work. As Dr. Edna Greene Medford's former graduate assistant and student, I am forever indebted to her.

The staff of the Moorland Spingarn Research Center, Founders, the Undergraduate, and School of Business Libraries at Howard University deserves many thanks. I am grateful to the former and current staff of the Graduate School of Arts and Sciences—Drs. Phyllis Galloway and Marlene Sherrill, Ms. Betty Goodwin, and Gloria Lloyd. Dr. Avis Pointer deserves special honor and thanks for the excellent mentorship she provided voluntarily to Howard University and its graduate students. Dr. Roger Davidson and I began "the long walk" at Howard University together and we endured. Dr. Basil Wilson, former provost, Dr, Gerald Markowitz, Dept. of History, Lennox Weithers, computer technician, and former student Amilcar Herbert, John Jay College/CUNY, and Dr. Michael A. Gomez of New York University and a founder of the Association for the Study of the Worldwide

African Diaspora (ASWAD) can never be thanked enough for their help over the years. I am grateful to Dr. James Rose and the staff of the Caribbean Research Centre, University of Guyana; the General Registrar Office, the Geological Survey and Gold Commission, and the Guyana National Archives. The comments of the anonymous reviewers improved the work.

Financial support from the Sasakawa (Nippon) Peace Foundation Fellowship, the Albert J. Beveridge Grant of the American Historical Association, the Department of History, and the Graduate School of Arts and Sciences, Howard University enabled me to carry out extensive field and archival work. John Jay College of the City University of New York, my undergraduate alma mater, and where I am a faculty member in the Department of History, aided the completion of this work with research time-off. I thank the editor for permission to utilize in revised form my article: "Creating Worlds: A Study of Mutuality and Financing among African Guyanese, 1800s to 1950s," *The Journal of Caribbean History,* 38:1 (2004) I am deeply indebted to the staff of Palgrave Macmillan, Chris Chappell, Sarah Whalen, and Richard Bellis; and at Newgen, Rohini Krishnan.

I do owe a lot to Keith Easton, Lance and Marjorie McCaskey, and Glenda Mingo Stewart. I will always remember Elaine Matthews (mother), Herman Grafton Headley (father), and my grandmother Miriam Linton Maraketa Small (1889–1982). "Aiee" as she was called and who nurtured me, was my first history professor. My dear son, Raul I. Josiah, Howard University Class of 1993, is the sunshine of my life as well as his lovely wife Beverley Carla and offspring Amanda Michelle and Marcus I. At all times, Raul's presence and attitude personifies how wonderful life is.

Abbreviations

BG	British Guiana
BGLU	British Guiana Labour Union
BGMWU	British Guiana Mine Workers' Union
BPP	British Parliamentary Papers, Colonies West Indies
DEMBA	Demerara Bauxite Company Ltd.
FSD	Foreign Service Despatch
CO	Colonial Office
COOC	Colonial Office Original Correspondence
GMWU	Guyana Mine Workers' Union
GNA	Guyana National Archives
HU	Howard University Libraries
HMSO	His/Her Majesty Stationery Office
MPCA	Man Power Citizen's Association
NACP	National Archives College Park
NA	National Archives, Washington, DC
PRO	Public Record Office
RT	Records of the Treasury
TNA	The National Archives, UK
WID	West Indian Department

Introduction

Guyana on South America's Caribbean coast was part of the European colonial plantation complex from the sixteenth to the twentieth centuries. Initially, enslaved Africans dug massive drainage canals on the coast for settlement and production. Subsequently, they labored mostly on plantations producing agricultural staples including sugar, cocoa, coconuts, cotton, and coffee. The enslaved also worked in forestry undertakings such as lumber and charcoal production. Similar to the quests for El Dorado elsewhere in the Americas, in Guyana, Africans were part of the early gold prospecting parties of the sixteenth century.[1] The 1833 Abolition Act, partially ended enslavement in the British West Indies, of which Guyana was administratively and is culturally a part. An Apprenticeship scheme, lasting from 1834 to 1838 followed. The arrangement enabled ex-slave owners to pay the semi-freed workers only after they worked gratis for a stipulated time. Prior to their paid apprenticeships, the enslaved had been allowed to conduct economic activities, and they saved money. After emancipation in 1838, many of them bought real estate and established villages.

Some freed persons remained in agricultural work as self-employed farmers or as full- or part-time employees on sugar plantations. Quite a number of them moved to urban areas and became hucksters, shopkeepers, or businesspersons, or worked in skilled or unskilled capacities. By the 1850s, some of them were part of a gold rush to Venezuela. Three decades later, Guyanese, West Indians, and other labor migrants of African ancestry were instrumental in establishing the gold and diamond industries in the country's interior.[2] The bauxite industry, started in 1916, depended on the labor of this segment of the workforce. This study examines migratory movement from agriculture to gold and diamonds in mining environments that sharply contrasts with bauxite production. The story of migration, mining, and its workers, vital threads in the fabric of the country's development, is told here.

Abiding interests in labor control and exploitation; resistance and continuous conflict; and the seemingly uneven development of the physical environment, which the working class occupied, have resulted in a paucity of studies on how migrant miners grappled with challenges in this colonial society. The body of literature on Guyana during the nineteenth and twentieth centuries mostly focuses on control/resistance modes and the continuous conflicts that ensued because of laws, which were enacted to control the exploited working class. General studies focusing on the struggles between the colonial hierarchy and the laboring class include, Ashton

Chase's, *A History of Trade Unionism in Guyana, 1900–1964* (1964); Walter Rodney's, *A History of the Guyanese Working-People, 1881–1905* (1982); Brian Moore's, *Race, Power, and Social Segmentation in Colonial Society: Guyana After Slavery, 1839–1891* (1987) and *Cultural Power, Resistance, and Pluralism: Colonial Guyana, 1838–1900* (1995); Maurice St. Pierre's, *Anatomy of Resistance: AntiColonialism in Guyana, 1823–1966* (1999); and Kean Gibson's, *The Cycle of Racial Oppression in Guyana* (2003). Following the same genre is Odida T. Quamina's, *The Mineworkers of Guyana: The Making of A Working-Class* (1987), a study of the bauxite mining complex of Mackenzie-Wismar-Christianburg (Linden). Sherwood A. J. Lowe's, *Historical Analysis of Mining Policy, Law & Administration in Guyana: 1831–2002* (2003) focuses on gold, diamonds, and bauxite with colonial administrators, locally and in Great Britain, North American officials, businesspersons, mining elites, and investors at center stage.[3] African diaspora labor migrants and the situations they faced and coped with during the development of Guyana's mining industries—a neglected theme in the country's history is the centerpiece of this book.

Guyana has long been a sparsely populated country with most of its inhabitants living in coastal areas. The country of 83,000 square miles is interspersed with many long, wide, and dangerous rivers and creeks. Initially, access to the hinterland gold and diamond fields involved perilous overland journeys and movement on waterways with numerous dangerous falls and rapids presenting formidable barriers to traveling. Many workers died by drowning, unattended illnesses, and other mishaps on their way to or from interior mining sites. Although a core of workers persevered in internal and seasonal migrations and sustained mining, an alarmingly high mortality rate was one factor inhibiting wide-scale expansion of the gold and diamond industries and interior production, generally. The effects of the ravages on the workforce, addressed in this study, have long remained uninvestigated.

The study fills important gaps in Caribbean historiography. Focusing on the interplay between gold, diamonds, and bauxite, it analyzes aspects of the lives of those associated with the development of the mining industry. It is the first study to introduce women workers into the activity of mining in the country. Until now, the environmental causation of migratory movement between gold, diamonds, and bauxite mining was unexplored. Moving beyond a mere study of the development of mining from production to processing and utility to the using of profits—beneficial to the power structure—it incorporates the ways in which workers utilized their earnings to better their conditions. Although sociopolitical activism is an important aspect of the groups' history, this study differs from the ubiquitous control/resistance/conflict theme in histories of the country, and engages pivotal questions. They include: why these workers migrated into mining; what were their contributions to economies; what benefits they gained from the occupation; and how African Guyanese with nurtured skills as craftspersons were instrumental in establishing the jewelry industry. The study seeks answers beyond the power of colonial bureaucrats, the plantocracy, and the mineocracy. It offers insights on the workers in relation to their successes, failures, and determination to overcome obstacles through self-reliance. Mining workers are analyzed through their programs, plans, and accomplishments; and socioeconomic developments are discussed in the context of modernization.

From the early 1800s, industrialization and modernization accelerated. To satisfy the insatiable need for raw materials by overseas industrial complexes, many leaders in Latin America were authoritarian and resorted to their brand of "order and progress." Regionally, continuous conflict ensued. In Guyana, tumult was intermittent, but this study contends that members of the working class utilized their concept of "order and progress" to their advantage. Key was the ingenious ways they applied knowledge transfer in their financial activities during the agricultural and mining eras that enabled the thrifty to improve their situations. Inherently, Africans and their descendants in capitalist Guyana had the discipline (order) to save money and to progress.[4] The colonial hierarchy's misunderstanding of this community's actions coupled with their overbearing attitudes to the working class generally detracted them from focusing on urgent infrastructure developmental issues.[5] Consequently, tragedies that many gold and diamond miners encountered in hostile interior environments also affected family members and friends living on the coast. Chaotic as their experiences were, many of the workers persevered in their mining quests. The perpetual movement from familiar urban, rural, and overseas settings to the strangeness of hinterland gold and diamond prospecting and the stability of riverain bauxite mining towns had their own uniqueness. In this setting, despite the ongoing adverse effects of labor constraints along with the perils of the travel and work, mining migrants expanded frontiers, produced commodities that diversified and contributed to economies, and provided for their future.

By the late 1880s, laws were periodically enacted, which expanded prospecting opportunities in interior areas, but the mining hierarchy's disinterest in relocating and residing in the mineral-rich regions had a deleterious effect. For example, piecemeal infrastructure development such as road and railroad construction provided limited access to the interior mining regions. Environmental and geographical obstacles persisted and inhibited the potential for greater profits from gold and diamonds. Still, overseas investors and multinational companies were given special concessions as encouragement to prospect in the country.[6] Consequently, employment opportunities as well as royalties, revenues, and profits from mining increased. These developments, along with reforms in land utilization policies, played crucial roles in the furtherance of the colony's export-oriented economy and the diversification of its production.

The production of gold and diamonds was dictated by the two rainy and two dry seasons of the country. Seasonal rains caused rivers to overflow their banks and made interior travel to and from the mining sites, and work dangerous. Drought conditions also prevented access to sites because the waterways' levels affected travel. Life and work in gold and diamond mining were by no means utopian, but the occupation afforded economic opportunities for local workers and African diaspora labor migrants who traveled to prospect. Imagine the horrors of coastal plantation conditions influencing workers to regard interior gold and diamond mining as an option. Long after enslavement ended conditions on and off coastal plantations remained precarious. Four factors accounted for this: first, plantation conditions scarcely changed from the days of slavery, and restrictive Vagrancy and Labor Laws were enacted and caused continuous conflict between workers and the plantation hierarchy; second, the introduction of indentured immigrants complemented or replaced the nascent African work-force; third, planters' hostile attitudes and contradictory behavior in times of "boom and bust" which resulted from the imposition or removal of preferential treatment on the British market; and fourth, economic fluctuations in the sugar industry itself due to international competition contributed to the tenuous nature of prevailing situations and influenced many agricultural workers to undertake mining. Although insufficiently acknowledged, the emergence of mining diversified the country's economy, added to the value of its exports and earnings, and contributed to local and national development. Neglect of the importance of miners to domestic and metropolitan prosperity persisted because many studies focused on their agricultural counterparts on Caribbean plantations.[7]

Parallel and consecutive changes precipitated workers' decisions to move from a hostile coastal plantation environment to ecologically and geographically challenging interior mining. During the pre-mining period, social and economic conditions in the country fluctuated. By 1884, adversity prevailed in many sectors of the still largely coastal agricultural economy. Simultaneously, gold production in the interior regions increased and many local men and a few women along with workers from overseas decided to engage in mining. Migratory movements had long been a feature of development in the Americas. Until the late nineteenth century, most of the movements to and in Guyana were targeted for agricultural production. With the advent of mining, the labor, contributions, sacrifices, sufferings, and strivings of African diaspora migrant workers in situations of daunting physical, environmental, and technological obstacles in the gold and diamond regions; and the work of their bauxite counterparts were pivotal in the country's further economic development.

As the twentieth century unfolded, based on market situations, workers made decisions to mine temporarily for gold or diamonds because both commodities can occur in propinquity. The miners who criss-crossed the country for decades preferred this type of work-life to any on the coast. When bauxite mining began some gold and diamond miners switched permanently to this new area of production. Similarities and differences in the three industries, regarding miners' travel, living, labor conditions, and modes of production are scrutinized here. Comparatively, gold and diamond mining are juxtaposed with bauxite to emphasize the workforce's flexibility and the long-term impact of versatile workers on economies. Mining historiography is augmented by this novel way of intertwining these commodities. In the face of tremendous odds, these workers were largely responsible for the country's mining operations, but production was often treated as though the factors were only land and capital, without the indispensable input of labor. These workers constituted what became one of the largest mining undertakings in the Caribbean and Latin America.

This undertaking illustrates the relationship between the contributions of pioneering mining migrants and the value of the commodities they produced. The nexus of the gold, diamond, and bauxite industries to the general economic growth of the country is shown. The workers influenced the gradual construction of roads and buildings and their maintenance; the intermittent clearing of rivers; and sporadic placement of medical, administrative, technical, security, and other personnel to facilitate more profits from mining. In fact, gold and diamond miners helped to open and to extend the interior frontier, expanding Guyana's economic enterprises similar to that of neighboring Venezuela, Brazil, and Suriname.

Uniquely, the study will show that in addition to the small number of women who participated in mining, many others took care of various responsibilities when the male gold and diamond miners departed for the interior. Women looked after elderly and dependant relatives, their children, and themselves. They worked in agriculture, fishing, and other enterprises, and dealt with tax and other business requirements. Women domiciled in the bauxite producing areas were similarly occupied and were also employed as teachers, nurses, seamstresses, and domestics. Some women were itinerant hucksters and others sold products at local markets and shops. Mining influenced various kinds of economic undertakings and linkages associated with increased employment in diverse sectors of the economy.

This study, identifying mining workers who became entrepreneurs, is a departure from the traditional view of the working class as largely oppressed and exploited under colonialism as was the case in Southern Africa. Unlike Ghana, Guyana's mining industry has no pre-colonial roots. Studies on gold mining and craftspersons in Ghana emphasize their ancient and pre-colonial origins. Ghanaians provided labor and expertise as jewelers and functioned at high and intermediate levels of the mining industry from pre-through post-colonial times.[8] Hence, the significance of the establishment of the local jewelery industry is manifold due to the importance of the miner's expertise in redeeming gold and diamonds and on those of African ancestry engaged in traditional practices of jewelery, coin, and ornament making. Continuities in mining and technological skills enabled some incredible achievements and ushered in the activities of millionaires and wealthier gold and diamond miners.

The quest for El Dorado, the fabled city of gold, propelled more European explorers to the Americas after their much heralded 1490s, arrival. Almost four centuries later, following long and arduous searches, Eldorado in Guyana was found. The discovery of gold in commercial quantities was timely because revenue and profits from sugar, the country's main export, were decreasing. Recruiting labor migrants of African ancestry for potentially profitable ventures in gold mining and other economic undertakings continued apace. Initially, Europeans were the main employers in the industry with a mostly African-descended work force. By the mid-1890s, small-scale and self-financed groups of prospectors increased. Modified mining regulations enabled some employees to become entrepreneurs and employers who created employment opportunities.[9]

Widespread use of only the interpretations and views in bureaucratic and companies' records foster historical interpretations from a "top-down" perspective. On the contrary, the approach here is to reassess conventional depictions of workers based on their material conditions and aspirations, as well as evidence that they generated including real estate, taxes, business, and other financial transactions. Oral accounts, death, census, and legal information in repositories in Guyana, Great Britain, and the United States, represent workers as "subjects of their history," how they saw themselves, and provided for the future educational, economic, and social achievements of their offspring. Simultaneously and sequentially, this perspective is an analysis of the power structure and the laboring class attitudes, and highlights workers as quintessential history-makers whose labor is core to the wealth of nations.

The challenges undertaken here and discussed in the chapters that follow are: to evaluate what gold, diamond, and bauxite mining meant to the economy and the miners themselves—and to profile aspects of working class life other than continuous conflict. The abilities of this class should not be minimized or ignored because they established and sustained their villages soon after enslavement ended and were able to embark on a variety of economic activities rather than only sugar plantation work. As the nineteenth century ended, although the colonial hierarchy was determined to maintain their influence on the country's social and economic direction, the expanding mining industry provided opportunities for workers to cope with the vicissitudes of life.

Empire building was particularly important to Joseph Chamberlain, the British colonial secretary from 1895 to 1903. For Guyana, Chamberlain's policies included encouraging investors in the gold industry to contribute to local development. Unfortunately, cost-minimizing, the export and utilizing of most of the products and wealth overseas, and manipulations by international cartels fostered the country's underdevelopment.[10] Events related to World War I affected sections of the mining industry in opposite ways. While the export of gold and diamonds was severely limited, bauxite capitalized on the opportunity to demonstrate its utility in the automobile, ammunitions, and aircraft industries. From that period onward bauxite became one of Guyana's leading export-earners.

Before World War I, the 1891 constitutional reforms had extended the franchise to a wider section of the population. Soon, Andrew Benjamin Brown of Den Amstel Village, West Coast Demerara, an African Guyanese lawyer, was elected to the legislature. In keeping with the diverse bent of many of his compatriots, as an entrepreneur, Brown was involved in agricultural and mining enterprises. After the war ended, miners were actively involved in trade union activities. In fact, one of the first trade unions started in the Caribbean region was the British Guiana Labour Union, whose founder was Hubert Nathaniel Critchlow, a one-time gold miner and waterfront worker. Initially, the union Critchlow founded served the interests of the general workforce, but eventually miners formed their own labor organizations. Throughout the twentieth century, African Guyanese educated abroad, some at Historically Black Colleges and Universities, including Shaw, Howard, and Tuskegee, returned home and provided a variety of services in urban and rural areas.[11]

The precious metals and gems frontier had expanded mostly without any planned residential communities. Lack of educational and recreational facilities militated against the relocation and settlement of families and fostered prolonged and intermittent absences of the miners' from their primary residences. Stability and proximity to family members were important to the workers, but migration and mining provided a needed alternative or a supplemental employment avenue to agriculture. Sometimes, mining as a means of making a living seemed justified by fortunate endings and economic benefits. Often, however, try as they might, miners could never make up for the emotional damages their absences caused family members. Unfortunately, at other times, the absentee male heads of household, died untimely and left their families destitute. The human and familial sides of mining are discussed in this volume with the focus on miners of African ancestry because they bore the brunt of the occupational hazards involved in reclaiming gold and diamonds and establishing the bauxite industry. Perspectives on internal and overseas migration solely for mining work converged. Consequently, several forms of migration are encapsulated here. Internal, overseas, pendulum, seasonal, cyclical, chain, rural-to-rural, urban-to-rural, urban-to-urban, temporary, and permanent migrations are all features of the discussion.[12] These workers offer a wide panorama on the effects of migratory movement and divert attention away from external migration that has been well-documented in Caribbean history.[13]

As a group, miners were not focused on conflict and resistance, in the conventional sense, but were more interested in forging their lives in situations where they could exert some control over their labor—although "the proliferation of departures" was "conditioned by and within a changing global culture and political economy,"[14] Migration to mining provided human capital to areas of scarce populations; fostered employment; the sending of monetary remittances to workers' homes, the acquiring of skills, and furtherance of investments and infrastructure development. Eventually, many gold, diamond, and bauxite miners who originally hailed from overseas—Afro-Caribbeans and South Americans, Africans, and African Americans—made Guyana their permanent home. By the 1950s, the magnet of El Dorado in Guyana had lost some luster and miners increasingly joined the exodus of Caribbean migrants to Europe and North America; creating a reverse migration.

Mining led to profits for local, British, and North American interests. Profits from mining, which were invested abroad contributed to overseas development. Ultimately, the reverse migration trend was a reflection of endeavors to enjoy the benefits of living in "more-developed" countries. Some funds were invested in selective local development and infrastructure works; albeit to increase production. Yet the research reveals that part of the society's colonial experience included a marked indifference to the welfare of the workforce, thus spawning the coming of additional community, ethnic, and labor organizations to empower their constituents. Another facet of the society, often left unexplored—the different ways in which some miners ensured benefits for themselves and continued the accomplishments achieved by their mostly agricultural ancestors—is underscored. Miners saved their money, acquired and maintained properties, paid taxes, educated their children, and created economic linkages to other businesses such as transportation, construction, manufacturing, retail, and wholesale. The abilities of the African descended enabled them to face the realities of colonialism, modernization, and human and geographically hostile environments to ensure a better life for their offspring.

Chapter 1 overviews the evolution of the African diaspora labor force in gold and diamond mining and the multiracial population from 1838 to 1890. Modes of travel to interior locations are detailed while the topography, climate, flora, and fauna are described to delineate the precarious environment in which workers undertook mining activities. Chapter 2 examines the roles of entrepreneurs, legislators, sugar planters, the mineocracy, and personnel attached to multinational companies in organizing the gold industry. The colonial frame of mind of the economic elites focused on importing excess workers to depress wages and to produce more profits. The ability of workers of African ancestry to perform hard work made them a target for recruitment. At a time when pseudoscientific and stereotypical racial views were dominant in the writings of count de Gobineau, Houston Stewart Chamberlain, and others who considered the achievements of certain types of Europeans as the hallmark of civilization, discussions on labor recruitment and migration strategies for mining in Guyana resembled those views.

Chapter 3 asserts the role of gold and diamonds in diversifying the country's economy and the worth of the miners in relation to the value of the commodities they produced. Gold, diamonds, sugar, timber, and balata are juxtaposed to show the value of economic diversity. The bauxite and jewelry industries—added features of the twentieth century economy—are showcased. Women who defied gender roles to become part of the country's mining industry are visible here. In chapter 4, previously unutilized death registers outline the perils of labor in gold and diamond mining. The mortalities of the African diaspora labor migrants are graphic, macabre, and wrenching aspects of the occupational hazard of mining. The state acting as the legal guardian of private companies at the expense of workers' rights added to the perils symptomatic of the lives of workers in colonial Guyana.

Infrastructure developments are examined in Chapters 5 and 6, to show that the slow pace of development in the gold and diamond regions was a contrast to bauxite's because the organizers of the latter learned from the mistakes of the former. Chapter 7 examines trade unions in the mining industry and the impact of working environments on labor relations and unionization. Aspects of "order and progress" form the backdrop of the last three chapters. A unique facet of Chapter 8 is the coping mechanisms of the working class and the roles of women who remained at home. The penultimate chapter analyzes how knowledge transfer and cooperativism among the self-disciplined workers during a monoculture of agriculture and the advent of mining ensured the longevity of their habitats. Chapter 10 solidifies the utility of African continuities in occupational skills. The chapter is enhanced and enriched by tracing the history of the jewelry industry. Mining is moved beyond the activity itself to the creation of finery from gold and diamonds. Economic linkages to the mining industry that created employment and generated income distribution through other entrepreneurial ventures in the interior and on the coast bolster a critical contention of the book. While authoritarian modernization created conflict in this colonial society, a segment of the working class was self-disciplined, recognized the utility of "order and progress," incorporated their traditions with other practices, empowered themselves, and improved their circumstances. An important and long neglected aspect of Guyana's historiography, which is that people of African descent contributed tremendously to the country's socioeconomic development, is recognized in this publication.

CHAPTER 1

African Diaspora Migrant Miners and Guyana's El Dorado

In the Americas, long after enslavement ended in the nineteenth century, the labor of Africans and their descendants continued to propel the economies of many nations. In Guyana, from the 1880s, a local workforce spawned by an agricultural economy along with Caribbean and other African diaspora labor migrants crossed boundaries to work in the gold, diamond, and bauxite industries. Other races in the country—Aboriginal Indians, Europeans, East Indians, and Chinese—played varying roles in its mining development. Still, it was mostly those of African ancestry who braved the uncertainties of virgin terrain and raging rivers to establish the interior gold and diamond industries. Bauxite, although mined under different environmental conditions from those of gold and diamonds, also relied overwhelmingly on the labor of these workers. Propitiously, in the twentieth century when the output of precious metal and gems waxed and waned, much wealth was obtained from increased bauxite production. This chapter identifies the origin of the mining workforce during the nineteenth and early twentieth centuries.

The Cooperative Republic of Guyana, bordering Brazil, Venezuela, and Suriname, consists of a coastal plain, a hilly sand and clay belt, and mountainous and savanna areas. The temperature of the country ranges from 70 to 90 degrees Fahrenheit. Annual rainfall varies from 120 inches along the Atlantic coast to 200 inches in the interior mountainous regions, and less than 60 inches in the savannas. Demerara, Berbice, and Essequibo form the three counties of a country with numerous creeks and rivers, many of them impeded by rapids, cataracts, rocks, and waterfalls. The largest rivers of Guyana are the Essequibo and some of its tributaries including the Mazaruni, Potaro, and Cuyuni. Other large rivers are the Pomeroon, Waini, Barima, Barama, Demerara, and Berbice. The Corentyne River forms the boundary between Surinam and Guyana.[1] Until almost the mid-twentieth century and the advent of more motorized water craft, frequent air transportation, and better roads, most persons traveling to the interior had to navigate the above-named rivers and others in rudimentary crafts. Working in gold and diamonds proved no easy feat even for workers long accustomed to enduring the rigors of plantation conditions and hardships associated with colonial environs. With the coming of bauxite mining many former or potential gold and diamond miners flocked to its production. Bauxite-producing areas were within easy reach of established communities, and travel to work sites involved little if any of the environmental dangers associated with gold and diamond mining.

Early Mining: Dutch and British Administrations

Over a protracted period, from the sixteenth through the nineteenth centuries, mining expeditions into the interior of Guyana by government and private companies led to the discovery of gold in profitable amounts during the 1880s. By 1890, increasing quantities of diamonds were found. From the early 1880s, the migratory movement of African diaspora workers to the mining industry was the continuation of an evolutionary process, engineered in part, by the laboring class to improve their lives from enslavement to freedom. The late nineteenth-century movement was linked to earlier interests in mining, a feature of the country's development dating back to Dutch colonization.[2]

In May 1738, Laurens Storm van's Gravesande was appointed secretary-bookkeeper of the Dutch West India Company that administered Essequibo colony. Gravesande became commander of the colony in 1743, and his tenure until 1772 is often regarded as one of the most important periods in Guyana's history. Although interested in coastal plantation cultivation, Gravesande was also focused on prospecting for minerals in the hinterland.[3] He had certainly heard of the reputedly fabled city of gold that was the quest of early explorers of the "Wild" or "Arabian Coast" of Guyana. Sir Walter Raleigh's 1595 journey is, perhaps, the most celebrated. Raleigh embarked on a search for "that mighty, rich and beautiful Empire of Guiana, and that great Golden City, which the Spaniards call Eldorado."[4] In 1720, Simon Abrahams, of Jewish origins, was permitted to search for gold and silver in Berbice colony.[5] By 1721, Dutch settlers were also prospecting in areas of the Upper Essequibo River for silver. Enslaved and freed Africans were used as "porters and diggers" in the early mining expeditions.[6] Although most of the early expeditions produced only a "few specimens of... ores and stones,"[7] forays into the interior for gold and silver continued under the Dutch.[8]

The British seized the Dutch colonies of Essequibo, Berbice, and Demerara in 1781. The colonies, taken over by France in 1782, were returned to their Dutch administrators the next year. By 1796, Great Britain regained the colonies, only to have them restored to the Dutch in 1802. After the British recaptured the colonies from the Dutch in 1803 they were ceded to the British by 1815. In 1831 the three colonies were united as British Guiana and renamed Guyana in 1966.[9] From 1836 to 1845, at the behest of the British government, Robert Schomburgk carried out a series of exploratory mineral expeditions. By mid-1857, rumors began circulating "of the discovery of gold" in what was later called the Caratal District.[10] At this time, Venezuela claimed the western portion of Guyana, but private explorations continued in the colony.[11]

Between 1860 and 1870, British officials organized geological surveys of their Caribbean colonies and Guyana under James G. Sawkins and Charles Barrington

Brown.[12] In 1863, several mining groups sought permission to prospect in the Cuyuni area of Wariri, which was in the disputed territory claimed by Venezuela.[13] By October 31, 1863, one of them, the British Guiana Gold Company published a prospectus. In 1867, the company ceased operations because of poor management, loss of capital, and the border dispute.[14] Although, an 1875 survey report and the failures of fledgling companies were not encouraging for the future of prospecting in Guyana, mining forays continued.[15]

On October 5, 1879, Henri Ledoux, a French vice-consul, Jules Camon, an African-descended gold miner from Cayenne, French Guiana, and Raymond Didier began prospecting for gold in Guyana. Camon and Didier found considerable amounts of gold in the Potaro area of Akaiwanna Creek, sparking further interests in mining in the country.[16] A *Gold Mining Ordinance #16 of 1880* regulated gold and silver mining, with provisions for land grants, fees for miners' certificates, registering claims, and a royalty tax of 2 percent on gold produced[17] (See Appendix B). In 1882, 40 ounces of gold produced in the colony and valued at $900 were exported. Two years later, in 1884, mining operations in areas of the Essequibo and Cuyuni rivers led the way in producing 250 ounces. News circulated of the increasing gold output, and more prospectors were encouraged.[18] Many local men and a few women from coastal and rural communities ventured into gold mining. They were joined by other African diaspora prospectors from the Caribbean islands, North and South America, and Africa.

In 1887, a local miner named Gilkes made news when he found a large number of gemstones in Putareng Creek, a branch of the Mazaruni River. By 1900, the output of diamonds increased, and in subsequent years, diamond production sometimes outstripped that of gold.[19] As the importance of mining to the colony's economy gained recognition, the government enacted legislation in 1886 and 1887 to solidify its control and administration of the industry, and to increase its revenue.[20] In 1886, Great Britain had demarcated British Guiana's borders with Venezuela and Brazil. The next year, news of gold discoveries along the colony's boundary with Venezuela, in the vicinity of the Barima River, Essequibo, increased prospecting in the British-claimed area.[21] Shortly after, a border dispute arose between the British and Venezuelans over other areas in Essequibo. By 1888, prospecting had spread to areas of the Puruni, Potaro, Kanwarook, and Omai River, Essequibo County, as well as to various sites in Demerara County.[22] The following year, British colonial officials affirmed their jurisdiction of the entire colony.[23] A permanent Commissary was established in the North West District, police stations were constructed in Pomeroon and Barima, and surveying continued. The actions reaffirmed the resolve of the British administrators to disregard Venezuela's claim to any portion of minerals and ores-rich Essequibo.[24] Gold and diamonds were sparse in Demerara County, more pronounced in Berbice, and most profuse in Essequibo.[25]

The economic depression evident in the colony by 1888 had much to do with decreased earnings from the major export, sugar. Unemployment in the country increased, but almost simultaneously, increasing outputs of gold helped to stabilize the economy by providing employment opportunities, avenues for investments, and additional revenues for government. Gold's potential for increased profits caused some bureaucrats to focus on aspects of the colony's infrastructure, especially in interior mining areas.[26] British colonial administrators considered measures to facilitate exploration and expand economic enterprises in the hinterland. Ultimately, however, most efforts were designed only to improve access to gold and diamond mining regions; and the pace of interior infrastructure development was slow, selective, and limited. Government services were scarce and housing for mining workers was practically nonexistent, with temporary camps usually providing shelter.[27]

The prevailing high mortality rate of interior workers and labor demands in other sectors caused periodic shortages of gold and diamond miners. Often, the remedy for the short supply of workers was the perennial solution of increasing migration and enacting more stringent laws to control labor.[28] In fact, during the period of mining development covered by this study, gold, diamonds, and bauxite production were all labor-intensive undertakings. Bauxite production also utilized capital-intensive mechanisms. The workforce originated from diverse sources and mostly comprised Africans and their descendants.

Origins of African Diaspora Mining Workers

The origins of the late nineteenth-century African-descended migratory movement into mining in Guyana are embedded in the earlier Atlantic Ocean slave trade. Enslaved Africans in the country had originated from locations in Guinea, Angola, the Congo, and other areas of West Africa.[29] Another segment of the enslaved population originated from the Caribbean islands after the British slave trade ended in 1807. British Caribbean planters conducted an inter-colony and mostly clandestine slave trade from Trinidad and Tobago and the Bahamas that guaranteed a labor supply for British Guiana. From 1808 to 1821 more than 8,000 enslaved persons were illegally transported from the islands to the colony.[30] On the eve of the Apprenticeship period, the colony had roughly 85,000 enslaved persons, several thousands freed people, and a significant "coloured" population. Provisions of the law governing apprenticeship included the formerly enslaved registering to work for the planters. Former field slaves were required to work for six years while those who had been house-slaves had to work for four years. The apprentices worked gratis for seven and a half hours daily and received food, clothing, and housing. The working day could be extended, and apprentices were paid below market value wages for the time in excess of the stipulated hours they worked. After serving the required four or six-year term, apprentices would be fully freed.[31]

Many of the workers were dissatisfied with this arrangement. Hence, the Apprenticeship period was as tumultuous as the enslavement era it had followed. The ongoing conflict between the workers and the hierarchy led to a shortened Apprenticeship period. Instead of ending in 1840, by August 1, 1838, the mayhem that the apprentices caused led to an aborted Apprenticeship period. By September 7, 1838, a Vagrancy Law was passed to force the freed people to continue working on plantations, but some of them had other plans. As pointed out earlier, some of the workers were in no mood to continue living under circumscribed plantation conditions. They increasingly challenged the plantocracy's authority and took action in accordance with their new status.[32] From the late 1830s, several plantation owners were ending operations and selling off property. Some recently freed individuals and others bought house lots and land while many of their compatriots pooled their financial resources, bought former plantations, and formed communities nearby that enabled them to continue estate work or to pursue independent economic activities.[33] Real estate ownership empowered property owners, and assured them of survival under less rancorous conditions, which was central to their socioeconomic advancement.

The world occupied by Africans and their descendants was changing and the acquisition of financial and material resources was important in asserting their freedom. For a variety of acquisitions, including real estate, Africans in diasporic communities continued processes of financial accumulation by utilizing traditional practices of communal saving and pooling resources. As enslaved persons, Africans had an aim when they participated in independent economic activities. Although enslaved, their actions suggested that they never abandoned hope that their predicament would eventually end, and progress to other levels ensue. Some of them were self-disciplined, saved money and bought their freedom and that of family members and friends. A Savings Bank was established for the enslaved by colonial administrators in 1828[34] (See Chapter 9). The money they had acquired and saved through their own enterprises while enslaved, and the wages they earned during the Apprenticeship period, enabled them to become property-holders.[35]

From 1839 to the 1850s, African-descended people established villages including Victoria, Buxton, Friendship, Den Amstel, Fellowship, Bagotville, and Mocha-Arcadia in Demerara County. The 141 persons who bought Plantation Orange Nassau and renamed it Buxton contributed sums of money varying from $1,000, $362, $350, to $36. They originated from Plantations Lusignan, Good Hope, Non Pariel, and Enterprise (see Table A1). Similar efforts were replicated in Berbice and Essequibo Counties.[36] From August 10, 1838, former apprenticed laborers in Berbice County began receiving the Transports for land they had purchased. Initial portions of Plantation Ithaca, West Coast Berbice, were purchased by 120 adults and three children on April 20, 1844.[37] Liverpool, Gibraltar, Sandvoort, Litchfield, and Hopetown villages in the county had similar origins.[38] In Essequibo County, as early as February 28, 1839, individuals such as Yasuw Paterson and John Rose (December 17, 1839) purchased land at Dagaraad, Westfield, and Mocha. By 1841, the purchasers of what became known as Queenstown Village expended $22,000 for additional land. Other villages such as Danielstown, Perseverance, and Zorg were similarly formed.[39] Essequibo River has several large populated islands including Wakenaam, Leguan, and Hogg. Maria Johanna on Wakenaam had an evolution similar to other African-Guyanese villages of the post-emancipation era. Stophel Wolff, William Chance, and Abram Sam were headmen representing 103 persons who purchased the former plantation and established the village.[40] From the 1880s the villages supplied large numbers of mining workers (See Table 1.1).

The ability to save money and acquire property in those times must be placed in perspective because enslavement was institutionalized for centuries and few foresaw its end. Enslaved Africans participated in economic activities that prepared them for slavery's demise and the dynamics of freedom.[41] The judicious actions of the ancestors suggest an understanding of the capitalist economy and a willingness to participate in it. The manner in which they contrived ways to deal with changes

Table 1.1 Some early African-Guyanese villages and prices

Village	Price $
Victoria	10,000
Buxton	50,000
Friendship	80,000
Beterverwagting	22,000
Good Hope	50,000
Fellowship	6,000
Bagotville	22,500
Golden Grove	5,200
Anns Grove-Two Friends	17,800

Source: CO111/253. Original Correspondence, British Guiana, 1848. Despatch 139, 3 May 3, 1848. Enclosures: 48–67; J. Graham Cruickshank, "The Beginnings of Our Villages," *Timehri 7* (1921), 65–76; and Alan Young, *The Approaches to Local Self-Government in British Guiana* (London: Longman and Green, 1958), 10–15.

emerging from a new social status and different relationships with the power structure were preludes to their long-lasting progressive efforts despite the authoritarian regimes of the era.

The deputation from the African Association of British Guiana who presented a petition to the governor on May 31, 1842, were harbingers of the collective advocacy "for the improvement of the colonial population against" legislators and "old slave owners." Outlining their concept of "order and progress," the petitioners pointed out that the colonial hierarchy was mishandling the colony's financial assets to the detriment of the general population.[42] The complexities of life had already become even more profound when freed workers decided to forge a life for themselves and their families through their labor and acquisitions.[43] Thus, the assertion of Robin D. G. Kelly, "I looked in vain for glimmers of a new society in the 'liberated zones'... in Guyana's tragically, short-lived nineteenth-century communal villages," is vacuous.[44] The villages still survive in Guyana and elsewhere in the diaspora.[45] Acquiring and holding on to their real estate showed the abilities of working people to create permanent living spaces that fostered economic and social stability as well as continuity beyond the period of immediate emancipation.

Planters eventually realized the futility of trying to exercise too much control over workers in the years after emancipation. Yesterday's enslaved chattel laborers had become today's freed people who retained memories of their work, which had created gain for others but caused pain for themselves. After all, enslavement had made the plantation a hated place for the workers and freedom did not change it. Thus, buying land, forming villages, and distancing themselves in any way from the plantation were efforts of the former enslaved to exercise their freedom.[46] Planters had accepted as adequate the compensation for their enslaved property negotiated by their agents and lobbyists in the British Parliament. Yet planters could not initiate new management techniques and thus, after emancipation, they insisted on treating workers as if they were still enslaved. The unchanged social nature of the plantation environment and the reasons for its existence dictated that workers were handled like exploitable commodities. Planters' mentalities would not allow them to adjust to an economic order based on "free labor" and a social setting in which the nonwhite population would not accept unquestioningly the determination of economic elites to dominate them.[47] In an attempt to evade discord, many workers tried other means of social and economic survival rather than plantation residence and work.[48] Workers' ability to abandon plantation work caused a dilemma for planters who needed to replenish their workforce periodically.

Legislation and regulations urged by the agricultural and commercial interests to control the lives and labor of the working class and implemented by government had unintended results. For instance, experienced workers abandoned plantation work altogether and created space for other workers.[49] Planters were forced to import labor migrants under an Indentureship Scheme which accelerated from the 1840s.[50] As contract workers, the imported were bound to serve a contractor, usually on a sugar plantation, for a prescribed period, under stipulated conditions, at agreed wages. Indentured workers included Portuguese and Madeirans, East Indians, Chinese, and free and liberated Africans. African Americans and West Indians were also encouraged to become labor migrants.[51] The migration of liberated and free Africans lasted from the early 1800s to the late 1860s; 14,020 of them entered Guyana.[52] Indentured workers and other labor migrants of African ancestry added another component to the eventual mining workforce.

When the migrants and the nascent African-descended population met on plantations, many found common ground. They labored, mingled, and developed relationships which were transferred to the villages. Evolving village life assumed added complexities with the coming of the new Africans. In the ensuing years, the new immigrants, the emancipated people, and their descendants lived in villages—thus weaving intricate webs of socioeconomic relationships. The fusion of this segment of the population is reflected in the country's birth registers from 1869. In 1868, *Ordinance #1, the Registration of Births and Deaths Ordinance*, was enacted. The law that authorized the registering of all births and deaths in Guyana came into effect one year later.[53] By 1869, the British slave trade had been abolished for sixty-one years, complete emancipation and the employment of free labor had existed in the British Caribbean for thirty-one years, and liberated and free African migration to the region had officially terminated a few years previously. Sierra Leoneans, Angolans, Kongos, Krus, Akus, Cape Verdeans, and West Indians including Creole Natives of Anguilla, Barbados, St. Kitts, and British Guiana had formed various types of relationships. Many of the births recorded in the registers indicate parents who were from regional or African ethnic groups.[54]

African Americans were welcomed by members of *The Voluntary Subscription Immigration Society* and supporters of labor migration into the colony. In the 1840s, many of them including Garraway and Rachel Powell and their eight-year-old daughter Sarah; Don Carlos and Susan Hall and their children aged two and four years old, respectively; and the brothers Edward and Richard B. Chew were among the skilled and unskilled migrants from Baltimore, Maryland. Several Guyanese communities, both urban and rural, had branches of the African Methodist Episcopal

Church (AME) established by African Americans.[55] Taken together, the milieu of Africans and their descendants in Guyana after emancipation were among the many thousands of internal migrants who mined gold and diamonds, and later produced bauxite.

In the nineteenth century there were constant streams of African Caribbean labor migrants to the colony. Death Registers, extensively utilized in Chapter 4, confirm the composition of gold and diamond miners who died in interior regions had originated from Africa and its diasporas. *Reports of the Institute of Mines and Forests of British Guiana* (herein after, the Institute of Mines), the *Department of Mines*, and *The Gold Mining Gazette* record mining migrants from neighboring Suriname (Dutch Guiana) and French Guiana. Although some labor migrants worked in agriculture, many of them were recruited solely for or gravitated to gold, diamond, and later, bauxite mining.[56]

Table 1.2 shows West Indians and Africans in Guyana from 1841 to 1891. Guyana's censuses covering the period 1841 to 1891 indicate that large numbers of persons of African ancestry had contributed to the upswing in gold and diamond production. By 1891, the number of Creole Africans was 112,155 or just over 40 percent of the entire population. The various groups of Africans in the country eventually became the dominant workforce in gold, diamond, and bauxite mining. In the 1890s, a sponsored immigration of Africans and their descendants was geared to many needs, and undertaken by quasi-government and government enterprises as well as private companies and shipping agencies. As was the case of earlier immigrants, those who came during this period and after integrated countrywide into the communities of the African descended.[57]

By the late nineteenth century, the Industrial Age had further enveloped the Western hemisphere, and the exploration and excavation of minerals and ores in the Americas required increasing numbers of migrant workers.[58] Locally, although interior mining would entail as many if not more challenges than plantation and coastal work, a large number of mostly agricultural employees migrated seasonally to the gold and diamond fields. On the coast, an array of hostilities including attitudes of the colonial hierarchy, inhospitable working conditions, unreliable employment, and increased cost-of-living confronted workers.[59] While precarious environmental

Table 1.2 West Indians and Africans in Guyana, 1841–1891

Year	West Indians	Percentage of Population	Africans	Percentage of Population
1841	8,092	8.2	91	—
1851	9,278	6.8	14,251	10.4
1861	8,309	5.3	9,299	5.9
1871	13,885	6.7	7,541	3.9
1881	18,318	7.5	5,077	2.0
1891	21,025	8.1	3,433	1.2

Source: *Census Reports of British Guiana*, 1861–1891; E. D. Rowland, "The Census of British Guiana, 1891," *Timehri 4* (1892), 40–68.

conditions characterized life and work in the interior, independent-minded groups of prospectors operated under mutual agreements without an overarching managerial presence.

The coming of the mining industry highlighted vestiges of the legacies of the institution of slavery. Disruption from familiar home surroundings, arduous labor, exploitation, diseases, and untimely deaths still plagued the lives of Africans and their descendants. Yet against all odds, this group viewed self-reliance as a means of survival, and one of their greatest accomplishments was the establishing of villages and their ability to keep and maintain their habitats. Thus, all of their efforts to effect changes and to make progress relied on their availability and flexibility as workers when the mining industry emerged. In other words, the inhabitants of the nineteenth-century villages and their compatriots recognized that both economic and social progress depended on their own initiatives. Consequently, although mining took them to the limits of human endurance, they were willing to provide the crucial labor the industry needed at a strategic time in Guyana's development.

On several occasions, colonial administrators and their commercial counterparts recommended highly the recruitment of African-descended migrants similar to villagers. However, government officials and the economic elites exhibited contradictory behavior. While asserting that most of these workers were self-motivated and dependable, they were recommending more stringent laws to control them. This was a setting in which the workers were pushing beyond familiar coastal confines and work. By the advent of the mining industry, these workers had previously established and worked in most of the other main commercial undertakings of the colony, including rice cultivation. In addition, they remained active in sugar, timber, balata, stone quarrying, and other enterprises.[60] They became valued underpinnings of the mining industry.

Mining and Other Segments of Guyana's Population

The multiethnic population of Guyana includes indigenous people or Amerindians, Europeans, East Indians, and Chinese. Of all the racial groups in the colony during the period of this study, the enumeration of the indigenous population is most problematic because they are both sedentary and nomadic. Over the centuries, many members of the Amerindian population crossed the boundaries of Guyana, Brazil, Venezuela, and Suriname at will. Amerindians traversed areas that are not always easily accessible to census enumerators and other chroniclers. Amerindians also worked in the country's mining industry.[61] Nevertheless, statistics of this segment of the Guyanese population vary.[62]

Portuguese and Madeirans first arrived in the colony during the 1830s, and continued to do so until 1881, when state-aided immigration of this group ended. In that year, 11,926 of them lived in Guyana with the figure increasing to 12,166 a decade later.[63] British residents in Guyana numbered 2,776 in 1841 and 4,558 by the census of 1891. Europeans (including Portuguese and Madeirians), became involved in the precious metals and gems industries mostly in an entrepreneurial capacity.[64] Access to credit from private financial sources and investors and funding from the government bureaucracy itself enabled them to function as entrepreneurs. Europeans were investors and businesspersons, as well as agents and managers involved in the transportation of people, goods, and services between the coast and hinterland. Quite a number of them became employers in forestry and in the gold and diamond business.[65]

In the 1850s, the first group of Chinese indentured immigrants arrived. By 1891, a total of 13,434 Chinese immigrants had come to Guyana, but owing to factors including high mortality rates, only 3,714 of them were present that year. Chinese immigrants became associated with businesses in the interior of the colony. They were in "shop keeping, wood cutting, and entrepreneurship in the gold fields."[66] By 1913, many Chinese were investors in gold and diamond mining, held gold mining grants, concessions, and prospecting licenses, or were claim-holders. Although few Chinese were gold mining laborers in the initial years of the industry, by the 1930s an increasing number of them such as the Quan brothers of Windsor Forest, West Coast Demerara, were prospecting in Cuyuni, Siparuni, and other areas of Essequibo.[67]

East Indians were imported to Guyana from the 1830s. The 1851 census showed 7,682 East Indians were in the colony as a result of an indentured immigration scheme. Forty years later, this group increased to 105,463 persons.[68] In spite of the rapid numerical growth of East Indians in the country, their seeming lack of involvement in the mining industry commensurate with their numbers has never been addressed. African-descended persons are always correctly portrayed as the predominant workforce in the mining sector. East Indians are noted as agriculturalists whose so-called "lack of suitable physical strength" was often deviously cited by white colonials, as the specific reason, for their alleged invisibility in the gold and diamond fields. A more detailed discussion of this is given in the following chapter. Constraints on their labor by planters, legal measures, colonizing schemes, and land grants served to restrict the majority of East Indian workers to coastal sugar plantations and other agricultural work, even after their indentureship ended. As a result, they were never considered a significant part of the mining workforce. Until 1928, no official efforts were made to encourage the movement of East Indians into mining.[69] In fact, at the 1890 sitting of a governor-convened commission to organize the gold industry and to recruit labor, it was decided to import specifically African-descended workers, but no Asians, for mining. Governor Sweetenham himself stated "that the gold industry was dependent on the negro."[70]

East Indians were among gold and diamond mining workers although their numbers were in no way comparable to that of the African-descended component. For example, the census report of 1891 identified only eighteen East Indian gold-diggers compared with thousands of their African-descended counterparts. Similarly, the *Reports of the Institute of Mines*, *The Gold Mining Gazette*, and *Death Registers* for the gold and diamond regions show that for many years there was a paucity of East Indians in mining. Over the decades, and as bauxite production got under way in the twentieth century, an increasing number of East Indians were involved in the mining industry at all levels.[71]

Early Migratory Movement to the Gold and Diamond Fields

From the mid-1880s, there were many thousands of recruits working for private mining companies. Initially the prospecting companies endeavored to obtain their own labor supplies, but ultimately, the workers contracted and registered through the Institute of Mines. Later, this late nineteenth-century migratory movement of African diaspora workers included increasing numbers of small groups of self-financed persons who worked independent of the larger mining concerns. Mining employees and independent claim-holders required prospecting licenses and registration to work at sites. In very many instances largely due to frontier conditions, it was impossible to quantify the number of such persons.[72] Nonetheless, for decades migrating miners traveled interminable miles over land and water to reach the interior gold and diamond fields. Workers traversed terrain characterized by dangerous waterways, forests, and animals to prospect for the commodities as the topography, climate, flora, and fauna allowed. Most of the mining sites were located in interior Essequibo County or on the upper reaches of the Demerara and Berbice rivers. For many mining workers, the capital city, Georgetown, served as an embarkation point to the gold and diamond fields.

Workers who originated from a coastal village on the East Coast Demerara followed a particular itinerary. They traveled by the East Coast Demerara railway or along the East Coast Demerara road. Both rail and road took the mining workers to the Georgetown Ferry Terminus on the Demerara River. Here, they crossed the river by ferry to Vreed-en-Hoop Village, West Coast Demerara. Prior to 1900, miners traveled by road or water to Parika Village, East Bank Essequibo. Some of them also traveled on the West Demerara railway company's train from Vreed-en-Hoop to Leonora, West Coast Demerara; then they continued the journey by road to Parika. After 1899, the railroad extended to Greenwich Park where commuters journeyed to Parika Village by road. By 1912, the West Demerara railway terminated at Parika. From there, miners utilized boats or steamers to Bartica, a critical point en route to mining operations and a town at the confluence of the rivers Essequibo, Mazaruni, Cuyuni, and Potaro.[73] In 1841, the area's population comprised of mostly African-descended persons. *Ordinance #5 of 1887* created an appointed Board of Commissioners to organize the layout of the proposed town. As the gold industry expanded in the 1880s so too did "the development and growth of Bartica."[74] Although other routes to the gold and diamond fields were taken, based on the worker's origin, in general, the initial stages of the journey were directly to Bartica. Most persons journeying to Essequibo mining districts regard the entrepot town, Bartica, as the place where interior Guyana begins. Conversely, when they left the mining fields, that town was viewed as the place where the interior ended.

Well into the twentieth century, the method of traveling to the interior remained mostly an overland and across water experience, although trips became less hazardous on rivers with the advent of boats with higher horsepower capacity engines.[75] Prospecting parties had to contend with climatic conditions because rainfall could average as much as 90 inches a year. Two wet and two dry seasons annually were followed by a long wet season from April to August and a dry season until November.

From mid-November to mid-February and mid-February to mid-April, a short rainy season was followed by a short dry season.[76] The nature of the climate and frontier conditions which persisted for a long time in the mining region caused the majority of workers to travel back and forth—giving this migratory movement its pendulum features. Although the vicissitudes of mining entailed workers risking their lives trying to recover the precious metal and gems, in a sense, the quest for gold and the related benefits have motivated the search for this commodity by humans worldwide from time immemorial. It was the same in Guyana. Notwithstanding, this pendulum migratory movement of mining workers did not simply evolve because of a lust for the precious metal and gems or the love of adventure. The question is-why did the workers choose mining as an occupation?

Impetus for Mining: Push and Pull Factors

In the case of the local workers, continuous conflict, the harsh realities of the plantation environment, and the attitudes of the colonial hierarchy caused many of them to consider mining as a viable employment avenue. While the probability of earning better wages was a "pull factor" to the gold and diamond fields, plantation conditions served as "push factors" out of the sugar fields. In addition, legislative measures and regulations were utilized as means to stymie the independence of the working class and to limit their initiatives and abilities on the coast. Workers functioned within "colonialism as a system of oppression" that "involved the political subjugation and economic exploitation of the colonized."[77] "It is within this context that the most pernicious . . . experiences of the African diaspora" people "must be assessed." They "existed . . . in hostile, socially oppressive situations and conditions."[78] One can make a case that Guyana's colonial hierarchy often impulsively reacted to the actions of the working class. For example, after the recently freed began to purchase property, some planters and their allies inflated land prices. Soon after, legislators stipulated the number of persons who could purchase communal property. This reaction served to constrain the actions of would-be individual or group purchasers of real estate, and thus restricted them from providing self-employment in areas not directly related to the plantation.[79]

Arbitrary "Rules and Regulations" and various tax increases enacted by the government in the 1840s, 1850s, and later, seemed contrived to keep this segment of the population in a state of perpetual economic dependence on the sugar plantation as wage laborers. The equipment these self-employed workers utilized carried the highest taxes while those for the plantations' operations were either exempted or carried little taxation.[80] Although inhospitable socioeconomic conditions could limit opportunities for both urbanites and rural dwellers, their initiatives and self-determination were remarkable. Some of them migrated to nearby Suriname or Venezuela seeking mining and other employment. In April 1861 when the US Civil War began, some African-Guyanese enlisted and served in the Union Navy. Among them were George Burton, John Abrigo,, Joseph Massey, George Peters, John Baptiste, and William DeHaarte, sailors who had enlisted in New York, Boston, and Baltimore.[81] In the nineteenth century, other community members made their way to Brazil and were involved in railroad construction. By the turn of the century, some of them ended up working in the United Kingdom and South and West Africa.[82] These various types of actions as well as prospecting were endeavors by the working class to make a living away from the plantations' continuous conflicts.

In the pre-mining era planters, legislators, colonial bureaucrats, and their allies continued in their efforts to exercise various economic pressures on the laboring population. The "imposition of discriminatory taxation continued and caused considerable financial and economic distress among" African-descended persons.[83] Taxation, imposed on them, was not limited to consumer items and work equipment only. Porters, hucksters, and shopkeepers also had to obtain licenses to ply their trade. Some of them who were interested in providing transportation services as boat-operators or by mule- or donkey-carts had to purchase licenses for their work. Similar contraptions used by the plantation hierarchy required no such licenses to operate. Evidence reveals that colonial administrators implemented policies that benefited Portuguese immigrants and discriminated against those of African origins.[84] In the 1880s, during a time of economic depression, gold was increasingly recovered in the interior.[85] A downturn in the sugar exports continued with severe financial consequences for the country. Unemployment and socioeconomic problems occurred on the coast, culminating with riots in 1889 and 1891.[86] Simultaneously, interior gold and diamond output stimulated migration to the fields and created employment opportunities for local and international workers.[87]

Throughout the nineteenth century, relations between two groups—planters and workers—remained as rancorous as ever. In this environment, the constraining effects of the planter-inspired Vagrancy and Labor laws along with intermittent economic crises influenced the migratory movement into mining. A critical factor—the relative absence of an overarching and constraining presence of the mineocracy—largely absent from the fields—served as an impetus to attract workers into gold and diamond mining. Unlike many plantation officials who resided nearby, the mining hierarchy resided on the coast or abroad—far away from prospecting operations. In addition, after the first decade of operations, large conglomerates could not monopolize gold and diamond mining. Serving as pull-factors for some was the mushrooming of independent small-scale operators of African ancestry who made a living from mining, shouldered their responsibilities, provided for themselves and offspring, created employment, and contributed to the export economy.[88]

Mining, whether in gold, diamonds, or bauxite offered alternative employment opportunities away from agriculture for the laboring class. Prevailing social and economic conditions in the country partially explain and offer a backdrop for the exodus of migrants to the interior. Viewed through more expansive lens, local workers of African ancestry were never granted options like those afforded other labor migrants. For example, bureaucrats attempted to ameliorate the grievances of East Indians in the country, and this section of the working class always had India as a point of reference. In fact, the Indian government and other representatives were able to make supplications on behalf of these workers and to mitigate unfair and harsh aspects of their social and economic conditions. Mechanisms were in place for East Indian workers' repatriation to their homeland or to obtain land free or at very cheap prices.[89] In contrast, those of African origin in Guyana whose labor built and often sustained various economic undertakings hardly ever had these concessions extended to them. The slave trade and enslavement meant so much chaos and confusion for Africans that almost all of their descendants were never able to refer to a specific place in their ancestral homeland where they or their ancestors originated. Enslaved Africans were unpaid and they and their descendants received no reparations. Even after the 1890 Combined Court's decision in favor of granting fifty-seven former enslaved persons or their descendants' ownership to land in Winkle Village, Berbice County, no financial help was extended to them for land development[90] (See Appendix A 2). Still, members of this segment of the population progressed and sometimes turned "adversity to prosperity" while making tremendous contributions to the nation.[91]

Conclusion

All colonized peoples suffered in situations not of their own making. The effects of government policies on a majority of the colony's population were no different from those that European colonists devised and implemented worldwide on people they colonized and viewed as inferior.[92] The coming of the gold and diamond industries paved the way for the diversification of the country's economy and forever changed the lives of many persons. After emancipation in the 1830s, life for many persons of African ancestry in the intervening decades up to the 1880s and the advent of gold and diamonds, was challenging and sometimes devastating. In the agricultural and mining eras environmental conditions on the coast and interior were "push" and "pull" factors influencing the migratory movement of African diaspora workers. Half a century after their enslavement ended, exercising an option to work in mining offered no instant transformation in the quality of their lives, but it provided another employment avenue. The initial composition of workers involved in the mining industry also reflected migration trends in the plantation society in the years after British slave trading ended. Africans, West Indians, North and South Americans of African descent joined the nascent workforce in the quest for El Dorado and diamonds in Guyana. As the following chapter outlines, their exemplary work performance influenced labor recruitment policies and migration strategies to increase the colony's mining output.

Chapter 2

Migration and Mining Strategies in a Colonial Society

As the nineteenth century drew to an end, gold and diamond production increased in colonial Guyana, but labor was problematic. Interior mining was attracting many workers, mostly of African ancestry, away from coastal and agricultural enterprises. Meanwhile, significant numbers of them were injured or died in the hinterland causing a seeming labor shortage countrywide.[1] At a time when a costly government-sponsored Indentureship system importing mostly Asians operated alongside incoming migrant workers from the Caribbean and elsewhere, it was claimed that overall production was hindered because of labor shortages. As was often the case with insufficient labor, colonial officials and businesspersons recruited more migrants, and legislators enacted or enforced stringent laws to control the workforce. The expeditious solution of recruiting migrants to satisfy labor needs precluded the hierarchy from implementing innovative ways to maintain an experienced workforce and to scale back immigration costs.

Mining seemed to have entered a marriage of convenience with agriculture as the mineocracy, similar to the plantocracy, contrived to control and exploit workers. Policies favored by those charged with organizing mining were mere continuations of previous policy-decisions. To the nascent workforce, it seemed as though the struggles which their ancestors had waged against hierarchical control were no different in this new undertaking. "Order and Progress" suffered setbacks because the strategies of the organizers of mining could not accommodate input from the workers.[2] The bureaucracy's persistent resort to conventional policies and strategies inhibited the potential lucrativeness of gold and diamond mining and hindered the country's overall growth. Efforts to develop these two sectors of the industry by mining, administrative, and private officials were predicated by a focus on quick profits on limited investments. Colonial bureaucrats and their allies were unable to move beyond conventions applicable to the slave trading, enslavement, and post-emancipation eras in terms of their attitude to labor recruitment, management, and conserving the workforce. The local situation was unfolding decades after the construction of US government-financed transcontinental railways and roadways, at the behest of business interests, to circumvent dangerous and circuitous travel to gold fields in California, US.[3] Guyana's colonial status influenced migration and mining strategies and circumscribed wide-scale development.

During the heyday of colonialism, European control, exploitation, and exportation of the human and natural resources of Africa, Asia, Oceania, and the Americas were geared to produce wealth for overseas coffers. Beneficiaries of the system tended to laud colonialism as a virtual good and a harbinger of growth and development in the colonies. This is no surprise because by the late-nineteenth century racist views of the superiority of certain types of Europeans and the inferiority of other races had circulated and influenced the mind set, writings, and actions of colonial administrators and their business allies.[4] Skeptics examined contemporary situations and considered colonialism a scourge of underdevelopment and the cause of continuous conflict in former colonies.[5] In the Guyanese context, mining historiography also identified local and international collusions by entrepreneurs, businesspersons, mining magnates, and colonial governments that ensured profits for themselves and everyone else; except the workers.[6] While these perspectives may have validity, this chapter investigates the prevailing frame of mind of the colonial power structure who organized Guyana's mining industry and their recruitment and migration strategies from the last decade of the nineteenth century onward.

A Colonial Frame of Mind: Organizing Migration and Mining

For most of the 1880s, yields from gold mining were profitable. By the 1890s, government–convened commissions established entities to increase profits from the industry. Colonial officials, businesspersons, and interested parties understood that for interior economic undertakings to develop profitably, labor, in particular, needed organizing. A group of such persons collaborated and became the nucleus of the Institute of Mines and Forests of British Guiana. *Ordinance #9 of 1890,* legalized the quasi-government entity. Among the main concerns of the body were the recruiting, registering, and monitoring of gold, diamond, and forestry workers.[7] The headquarters of the Institute of Mines was located in Georgetown, the capital city. Branch offices opened in New Amsterdam, Berbice County and at Bartica, Potaro, Baramanni, and a few other locations in Essequibo County.[8]

A mandate of the agency was the targeting of workers of African ancestry for employment. However, the contentious issue of autocratic control of workers who labored under extremely hazardous conditions underscored the colonial hierarchy's inflexibility. The officials' consistent disregard of views that did not support their ideas on the organizing of labor and the mining industry led to a seemingly shorthanded workforce; ongoing costly migration schemes; and stringent laws to control the workforce.[9] Following the trend in the enslavement era, with the enactment of the Abolition Act of 1833, came a series of Apprenticeship, Contract, Labor, Immigration, and Vagrancy Laws all designed to circumscribe the initiatives of the laboring class. The reliance on laws and punishment revealed that the economic survival of the colonial power structure depended on their arbitrary control of labor.

Included in the Apprenticeship Order-in-Council was the stipulation that any apprentice found five miles away from a sugar plantation where he/she worked or lived could be arrested. On September 7, 1838, soon after the Apprenticeship period ended, an Order-in-Council legalized another Vagrancy Law. As the Indentureship system and immigration accelerated, laws were passed from the 1840s, to control workers movements. As early as 1842, the African Association of British Guiana aired views concerning the collusion of legislators and former slave owners to implement policies inimical to the interest of the entire colony. In 1890, as meetings to organize the mining industry and other economic enterprises proceeded, constant calls were made, especially by members of the Institute of Mines, to ensure the application of existing Vagrancy and Labor Laws and to enact even stricter ones to control workers.[10]

The members of the Institute of Mines were simultaneously involved in mining, agriculture, forestry, commercial, and even legislative activities. They made decisions narrowly based on their experiences, world views, and self-interests. Legislator B. Howell-Jones who represented the West Demerara constituency, had introduced the bill to the British Guiana Court of Policy to incorporate the Institute of Mines. Howell-Jones owned a sugar estate and was a director of the Barima Gold Mining Company and the Arakaka Placer and Mining Company.[11] He supported recruiting more labor migrants after enforcement of the existing labor laws on the workforce. But, he paid a price because in 1891, the country's constitution was reformed that led to a wider enfranchisement of the population. At the next election, Howell-Jones lost his legislative seat. Many among the newly enfranchised working class understood that he and his cronies aimed to continue their dominance politically, socially, and economically. Utilizing the ballot to challenge that dominance was a counterpoint to the ongoing disempowerment workers faced at the behest of administrative and economic elites in the colonial society. Previously, their leadership had had little, if any direct legislative influence and the working class struggled to ensure their progress and development. At this juncture the emergence of Andrew Benjamin Brown is noteworthy.[12]

In July 1890, Brown was called to the bar of the Middle Temple, England. Shortly afterward, he returned home to British Guiana and opened a law firm in Georgetown. Brown, an entrepreneur and investor in gold mining, financed prospecting crews to work in Mazaruni and Cuyuni. He joined the Progressive Association. As one of that political party's candidates in the 1896 general election, Brown won the West Coast Demerara constituency, unseating the incumbent Howell-Jones. Brown became the first "pure" African Guyanese to gain a seat in the Court of Policy, the premier chamber of the legislature. He was unopposed and reelected for three further successive terms; serving in that capacity for a quarter century. Brown was sometimes embroiled in sensitive legislative issues concerning migration strategies. Together with members of other political organizations, the Negro Progress Convention, labor unions, and newspaper journalists, Brown waged a protracted battle for immigration and labor reforms as the mining industry developed.[13]

Meanwhile from the 1890s through the early 1900s, members of the Institute of Mines and their supporters tacitly ensured that much of the labor for interior mining and other enterprises funneled through their organization. Mining workers who required an annual prospecting license costing fifty-cents monthly, had to register at any of the Institute's offices. Agents would encourage family members or friends of current recruits to circulate information about vacancies in mining companies, which the agency represented. Advertisements were placed by the Institute of Mines in local newspapers to recruit workers. On December 2, 1893, *The Echo* advertised a notice on behalf of the Institute of Mines based on an order issued by the Governor-in-Council. The notice explained that the agency had been granted government assistance to import Barbadian laborers to work in the gold and forestry industries. *The Gold Mining Gazette*, a publication of the Institute of Mines, also advertised for employees.[14] Agents, some of them attached to commercial shipping, or gold and diamond companies, also recruited labor migrants from the Caribbean and elsewhere overseas for employment in mining.

These agents played a multiplicity of roles in the development of the country's mining industry. On behalf of mining concerns, they negotiated the purchase and sale of prospecting claims as well as the acquisition of shares in various companies. Sometimes, companies' agents were owners of other businesses and functioned as managers. They included George Garnett, H. G. Williams, M. J. Morpungo, A. G. Fraser, James F. Webber, Ernest and Percival Forbes, and James Winter. British Guiana Gold Mines, British Guiana Dredging Syndicate, the Berbice Company, and the Guiana Company were among the private firms operating offices in rural and urban areas.[15]

Agent/owner, George Garnett was involved in shipping, commerce, and mining; and also served as a legislator. Garnett, an elected member of the Institute of Mines became its president in 1893. Both Garnett and Howell-Jones were directors of the Barima Gold Mining Company.[16] The two men and their colleagues represented government and private interests that intertwined to monopolize vital areas of the country's political and economic activities. In addition to ensuring that prospective employees registered and signed contracts, agents made arrangements for the purchase and supply of food and equipment needed in the interior. Agents established credit at stores in the city to facilitate the sending of goods to the interior. They also liaised with boat-captains and their assistants to transport employees to and from work sites.[17] In one way or the other, the Institute of Mines and its members played monopolizing roles in recruiting and furnishing workers for mining and maintaining production.

The Institute of Mines collected and supplied data on the mining industry to the government. Members of the organization recommended "legislation as may seem most advantageous to the interests" of both mining and forestry. Permission was granted for the body "to act as arbiter in disputes that arose through dealings, or relations with mining."[18] Paradoxically, members of the Institute of Mines appeared as witnesses before a Board of Commissioners, appointed by the governor of the colony in 1890, to investigate existing labor shortages in vital industries. Among the main issues addressed by the governor's appointees was the impact of the migratory movement of workers from coastal enterprises to interior mining and other economic activities. Chronic labor shortages adversely affected sugar production and it was envisioned that the contemplated banana and fruit industries and extension of the railway system would likewise encounter similar problems. The commission's appointees had been instructed to import more agricultural labor migrants from places other than Asia, but while the stipulation seemed straightforward, closer scrutiny revealed that mining development was central to the migration strategy.

Predetermined views on labor had followed a symmetric pattern as was evident from the arrangements in effect when the 1890 commissioners met. Preambles to the discussions clarified the governor's directive and self-serving letters from bureaucrats and businesspersons; members of the Royal Agricultural and Commercial Society, the Chamber of Commerce of the City of Georgetown, and the British Guiana Planters Association supported the convening of the commission. Some correspondence was even received from members of the same commission or witnesses scheduled to testify.[19] A critical concern was that the labor scarcity had forced some employers to pay increased wages to attract workers. As a remedy to the ongoing situation, recruiting workers of African ancestry from the Caribbean and Cape Verde islands, and elsewhere was favored. West Indians, Africans, and San Antones (Africans or mixed race Portuguese), and Cape Verdeans were often characterized as "the best laborers."[20]

At the 1890 hearings, the manager of Reliance Sugar Estate, Essequibo who preferred West Indian workers (especially Barbadians), stated that "over 300 people had gone within the last three months . . . 500 from last year . . . and about" 150 "persons whose names were on his books had left for the goldfields."[21] James Andrew, a member of the Institute of Mines and manager/attorney of Leonora Sugar Estate, West Coast Demerara, noted that 700 to 800 workers had recently left the plantation. Blairmont Sugar Plantation, West Coast Berbice, suffered a similar fate to that of Leonora because gold mining motivated between 700 and 1,000 workers to quit. Other sugar estates also experienced labor shortages due to the same reason and their representatives supported recruiting labor migrants of African ancestry.[22]

The work abilities of people of African origins as well as racism or pseudoscientific notions of their physical stamina influenced the conclusion that their labor was the one of choice for recruiting purposes. This was by no means an expression of consideration for their humanity because value was based simply on the utility of their skilled and exploitable labor that planters feared they were losing to mining and which the mineocracy desired to obtain.[23] Agricultural and mining interests were amenable to recruiting and organizing workers of a particular profile and to exerting as much control over them as possible. The evidence and findings of the commission underscored the value of the labor of Africans to Guyana's existence and prosperity. Although the labor migrants who were recruited seemed destined for plantations, it was a ruse to acclimatize them to local conditions in coastal agriculture production before sending them to work in interior gold and diamond fields. Migration and mining accelerated by 1916, when bauxite production began as Asian indentured immigration was ending. By the 1920s, as labor migration from the Caribbean Islands and overseas continued; the stratagem to disguise the ultimate intended use of recruits was ascertained. *The Report of the Institute of Mines* verified that newly recruited labor migrants facilitated seasoned workers "inured to the climate to work in the mining industry."[24]

Sentiments expressed about the unsuitability and repeated denigrations of Asian labor by several witnesses to the labor commissioners were contradictions at best, and at worse, insidious utterances. At the onset of the migratory movement from the coast to the interior, planters and their allies were concerned about the loss of labor from sugar production. Hence, the expressions about the unsuitability of East Indians for gold and diamond mining belied the true intentions of keeping this segment of the workforce permanently in coastal sugar and other agricultural production. Ironically, while the labor of Asians was decried as unsuitable for one sector of production—mining—it was touted as valuable for another—agriculture. Laws were enacted and strategies such as colonizing and land development schemes were devised to keep East Indians on the coast and in agriculture. Myths die hard and often take on a life of their own. Thus, colonial expatriates appeared before the *Royal Commission of Enquiry, 1939,* and aired stereotypical views on the unsuitability of East Indians for hard work and the senseless and spendthrift ways of Africans.[25]

By 1890, East Indians were already a significant segment of the country's population. However, one may conclude from the questions raised by the 1890 commissioners and the responses given to them that a perception existed that Asian labor was not the one to rely on for Guyana's future economic development. More than likely, the perception had much to do with the challenges then being mounted against the oppressive and exploitative system of Indentureship.[26] Activists and their supporters in Guyana, Great Britain, and India were dissatisfied with ongoing East Indian immigration because of the harshness and brutality of the indenture system with its similarities to enslavement.[27] Discomfort existed in other circles, regarding Chinese indentured migration and their treatment in the Americas, particularly in Mexico, Cuba, Peru, and later, Panama.[28] Discussants at the 1890 Labor Commission were well aware that from the perspectives of the activists, the time was not the most opportune to increase the migration of Asians.

George Garnett, the owner of the company bearing his name and who had vested interest in the Royal Mail Steam Packet Company, sometime legislator, president of the Institute of Mines, and balata and gold investor contended that Africans, Krumen or Congas, were very reliable workers. Garnett opined that those African workers were similar to villagers from East Coast Demerara, Mahaicony, and Berbice who were noted for their work skills and aptitude. Garnett categorically stated his dislike for West Indian migrants in general, except for men from St. Lucia and Jamaica, who were accustomed to work in Panama. He indicated that East Indians were unsuitable for mining because they "are not quite strong enough." Garnett added that only a few of them worked in the gold industry that employed four to five thousand workers.[29]

Supporting Garnett for not wanting East Indians in gold mining was James Andrew from Leonora Sugar Estate who emphasized that departing East Indian canecutters were never missed because of their poor laboring abilities.[30] J. D. Patterson, openly disliked workers who were not of African descent; and his selective embellishments and dire predictions were instructive. Patterson predicted that although the timber industry, which fell under the jurisdiction of the Institute of Mines, paid high wages, the labor shortage would worsen in that enterprise. He wanted only good African men, especially "Krumen from South Africa" (*sic*). He viewed forestry work as the hardest for which only Africans and Amerindians were suitable. Patterson declared that "coolies are no good at all" and admitted that he had "only seen one coolie' worker; he wanted "no Chinamen" because they "would be dead in 2 days" and concluded that it was the labor of those of African descent, which had disproportionately contributed to the colony's prosperity.[31]

The condemnation of Asians workers, the praise of those of African descent, and the ambivalence to the laboring ability of Amerindians, must be placed in the perspective of the prevailing times. Recruitment of Asians and their abject laboring conditions in South and East Africa, Fiji, Peru, and Cuba, for example, was attracting the attention of various humanitarian groups as well as that of their home governments in China and India. As already stated, some concerned local "parties" were advocating the end of East Indian immigration and the implementing of relevant reforms to ameliorate the harsh conditions of Indentureship. It was in this setting that the colonial hierarchy was organizing mining development. Entrepreneurs tended to focus on labor recruits who were the least protected by home governments and the most susceptible to recruitment and migration overtures. Potential employees' bleak economic conditions and inadequate earnings for themselves and their families would cause them to view labor migration as a means of social and economic opportunity.[32] Economic activities in the Americas relied on the labor and skills of the enslaved and their offspring. Garnett and the other witnesses' consistent advocacy for recruiting workers of African ancestry was indicative of their awareness of global capitalists' sentiments on migration issues. Considering that unfettered Asian indentured migration was not popular, it was no surprise that efforts to recruit cheap and exploitable labor should again focus on those of African ancestry as in the era of slavery.

The groups targeted for recruitment were vulnerable to labor exploitation as humanitarian efforts were directed mostly at other workers. Events in Liberia and Sierra Leone encouraged Krus and other African ethnics displaced by simmering unrest, to consider contract work elsewhere. At this time, considerable numbers of Cape Verdeans were moving to the United States, Argentina, and Brazil. In addition to the ongoing inter-migratory movements of peoples from the Guianas, West Indians had long been labor migrants in the circum-Caribbean loci. Therefore, it was not accidental that this group was the focus of the local recruiting plans. As the evidence from the 1890 meetings revealed, the colonial mindset on display was attuned to social and economic conditions worldwide which inhibited or was conducive to migration.[33]

The origins of workers had long played roles in recruiting and migration policies. Many enslaved persons in the Americas came from regions of Africa where gold and diamond mining and jewelery- and coin-making were traditional practices. Colonial Spaniards, who owned gold mines in New Granada, (present-day Colombia that once included Panama), Ecuador, and Venezuela, expressed preferences for buying persons from areas of Africa noted for gold mining. In the Carolinas and Georgia where rice flourished, the choice of Africans for enslavement from the Rice Coast of Africa was not random. In Guyana, Africans were always involved in early gold mining expeditions, and goldsmiths were identified in the Slave Registers.[34]

In 1890, an estimated "3,000 to 4,000" or "as many as possible" Africans and their diaspora counterparts were favorably viewed as recruits to work in the local gold and diamond fields. It was surmised that they would be valuable contributors to the country's economy because of similarities to their resident counterparts who had penchants for spending money on items such as "grog and dresses," which contributed to the revenue and enhanced trade.[35] T. J. Wakefield, a mine owner emphasized that persons working in the industry were well paid and spent more money, and helped to increase the colony's revenue. Wakefield asserted that "the governor and Court of Policy evidently" considered revenue gains "when they increased the duties on goods consumed in the (interior) bush," where many workers spent considerable time. Gold mining made an "enormous difference to the revenue both directly and indirectly" and obtaining those workers was optimum.[36]

Paradoxically, references to revenue gains when higher taxes were imposed on items purchased by those of African ancestry harkened to policies implemented earlier in the nineteenth century. In the 1840s and 1850s, higher taxation was solely levied on the items this segment of the population purchased, ostensibly to pay for indentured migrants who eventually replaced many of them as plantation workers. Invariably, references were made to the "higher" wages these workers earned to justify increasing their taxes, but the proponents did not recommend similar action on their own financial benefits.[37] Migration strategies of the colonial elites consistently ensured that the financial and human costs for the schemes fell on the African segment of the working class.

Howell-Jones and some others at the commission's hearings were opposed to carte blanche labor importation. They blamed government officials for not enforcing existing Labor and Vagrancy Laws to control the laboring class and accused workers, who disregarded the laws by refusing to work, of contributing to the labor scarcity. Noting the many idle able-bodied persons who congregated in urban and rural areas, Howell-Jones and his colleagues influenced decisions to enforce existing laws, enactment of more stringent ones, and inflicting stiffer penalties for offences in order to control workers more effectively. They advocated importing more labor migrants after the existing workforce was under strict control.[38]

David Gibson, a merchant, conceded that the existing Vagrancy/Labor Laws should be strictly enforced. He agreed that revamped labor recruitment policies could possibly guarantee more profits from gold mining, but was less sanguine about the effects and offered some cautionary advice. Gibson noted that more West Indian migrants were needed to open up the country's interior because infrastructure work on roads, railroads, telegraphs, and canals was necessary for expanding mining. He reminded his colleagues that although the gold industry had stimulated business in the commercial sectors, the entire economic development of the country's interior needed addressing. Gibson concluded that focusing unduly on the mining sector and enacting more stringent laws, could have a deleterious effect on labor and other economic enterprises.[39]

Implementing harsher measures to compel persons to work against their will under arduous conditions also had unintended consequences. The colonial hierarchy's ongoing uncompromising mindsets and reactions to workers' self-determination was causing continuous conflict throughout the Americas as the laboring class had mounted consistent challenges against the arbitrary and intemperate actions of the officials, legislators, planters, and economic elites, as stated so well in Demetrius L. Eudell's, *The Political Languages of Emancipation in the Caribbean and U.S. South*. Many workers knew what was in their best interests and how to take care of themselves without plantation employment; and signing contracts that made their situations untenable.[40] Attorneys, managers, planters, and others at the hearings pointed out that many prospective workers refused to bind themselves to an employer and objected to signing contracts. In fact, some workers had left the country rather than work under the existing conditions; and thereby added to the so-called labor shortage.[41]

Policies and inept management needed reform; while adequate funds for interior development were necessary to accelerate gold and diamond mining.[42] "Profits over people" and the cost-minimizing bent of mining management that had made workers reluctant to sign work contracts needed to be addressed. Finding ways to increase profits, neglecting to improve working conditions, forcing the laboring class to work, and scarcely paying attention to addressing their points of view on employment practices added to the uneven development of the gold and diamond industries. The premise here is: the setting of the 1890s hearing could have served as a counterpoint to changing the conventional frame of mind of the elites and that of the working class. However, as was generally the case in industrial relations worldwide, workers' points of view were often overlooked or never given the serious consideration they deserved by the power structure until wide-scale mayhem erupted. Eventually, the strengthening of workers' organization would ensure some reforms in industrial relations, but in Guyana of the1890s, the moment was lost to organize mining with semblances of satisfaction to both the employers and employees. The state of affairs continued indefinitely and this aspect of mining history serves as a prism to view long-standing antagonisms between management and the laboring class, which is outlined in Chapter 7. In a sense, while the former focused on short-term monetary gains from mining, the latter retained memories of the enduring pain associated with control by others and strove to improve their lot.

The 1890 hearings revealed that workers felt very strongly that signing work contracts "was like going back to the old slave story," and sugar plantation work "and would not do it... not even with a bounty" or advance because they suspected and were afraid that would-be employers intended "to bind them" and to exploit and oppress them as in "the days of slavery."[43] Many workers' long-standing traditions of growing crops and rearing sheep, goats, and pigs for domestic and commercial trade had made them self-reliant. Independent economic activities enabled them to control their labor, thus making them skeptical of overtures towards mining. Creole Africans could not be induced to work on the estates for more than three and a half days weekly because of their own enterprises. Farmers from West Bank Demerara crossed the river in their own boats and sold their produce at the Stabroek Market, Georgetown. Similarly villagers crossed the Corentyne River to work in Nikerie, Suriname where many of them earned higher wages than in Guyana. Increased wages in other enterprises and the concerted efforts of the colonial hierarchy to control workers made many of them reluctant to work in sugar, much less to sign contracts to work in mining.[44] Workers' decisions not to sign mining contracts, anticipated the coming of independent self-financed African diaspora mining groups by the mid-1890s.

Emphasizing that the workers' attitudes were based on their ongoing adverse experiences with the colonial hierarchy, two clergymen then identified suitable places for overseas recruitment and suggested migration strategies to induce more labor migrants into the colony. The first, Reverend D. J. Reynolds of the Wesleyan Mission and a native of Jamaica was well acquainted with labor situations in several places. Reynolds identified Manchester and St. Thomas-in-the-East as two places in Jamaica with suitable prospective labor migrants because persons from those areas had worked in Panama. He cautioned that those workers were not enamored with Demerara (Guyana), but he suggested offering high wages and the payment of an annual bounty as inducements. Reynolds admonished that "you have rather a bad name and they" Jamaicans "are rather chary of certain places" to work. He noted Jamaicans were already acclimatized to local conditions because their work in Panama had well-prepared them for gold mining. As a staunch supporter of Caribbean migration, Reynolds suggested that agents of steamers should be encouraged to transport an oversupply of workers from St. Vincent and rural areas of Jamaica. He knew that recruiters had done this to keep wages low in Panama. This ploy, the minister vouchsafed, was a well-known ruse in the Caribbean and North America to depress wages and to exploit workers because of their economic needs.[45]

Unlike Reynolds, native-born Reverend F. C. Glasgow, a Congregationalist of the London Missionary Society, did not view contracts or inducements as guarantees for reliable labor supplies. Glasgow had ministered for twenty-five years in Canal #1, West Bank Demerara and Buxton, East Coast Demerara and knew of the reluctance of villagers to sign contracts because they connected the arrangement to enslavement and the control of their labor. Glasgow felt that although the contracts were simple, the workers did not understand their obligations or the role market conditions played in increasing or decreasing their wages. He advocated educating the workers and "to treat them as children." Glasgow surmised that "unlike the British peasantry" the local laboring class needed more education, and added that "contracts in the British Isles may be very easy things, but there slavery did not exist."[46]

Paternalism is a possible explanation of the clergymen's constraints in crediting the workers with understanding which situations were in their best interests. The lingering legacies of enslavement still plagued the laboring class and the colonial power structure's attitudes to them did little to improve socioeconomic relations. Still, the clergymen's testimonies revealed a willingness to work within a socioeconomic system, which needed revamping. Glasgow even encouraged his own son to engage in contract work and was not averse to importing labor from Africa, Cape Verde, and the Azores.[47]

Migration and strategies for mining in this colonial society revealed festering issues relating to enslavement, labor conditions in post-emancipation times, the prevailing attitudes of employers, and workers who were committed to self-determination. Countrywide, more than fifty years after enslavement had ended; workers still deeply distrusted colonial bureaucrats and their allies. Similar to that of the economic elites, the frame of mind of African diaspora workers remained relatively unchanged on the question of the control of their labor. As such, the newly diversified economy of the colony was characterized by continuities and changes in both agriculture and mining. The attitudes of the economically powerful were virtually unchanged as they organized mining similar to the manner in which they had approached agriculture. The working class viewed the changed social and economic conditions since the days of enslavement and the advent of free labor as a means for them to exercise options. Workers realized that conditions related to gold and diamond mining could be deadly, and were hardly different from the exacting laboring and living conditions generally existing in the colony. As mining workers they wanted to exercise some control over their labor and how they lived and/or died.

Reasons for the colony's perceived inadequate labor supply included questionable management, adversarial socioeconomic relations, the tendency to view laws and their enforcement as sole solutions to labor shortages, manipulation of wages to the detriment of the workers, and sometimes unusually long rainy seasons, which intermittingly prolonged workers' illnesses. To address the labor shortfalls, a more efficient method of labor recruitment from places other than India and making arrangements to bring 3,000 contract workers from the West Indian islands became priorities. Beyond the present exigencies, plans were envisioned to recruit laborers from Africa, Cape Verde, and the Azores to prevent the stagnation of current industries and to ensure progress and development.[48]

A recurring sentiment in migration strategies was the potential profits that would accrue from utilizing the labor of more workers of African origins. These workers represented a profile fit for exploitation, that is, they were cheap and expendable. Economic elites and their colleagues simply did not care or were too blissfully unaware of situations that bedeviled miners and imperiled their lives. They overlooked the fact that the constant need to replenish labor for gold and diamond mining was caused by the workers' untimely deaths in the hinterland. When gold mining began, infrastructure schemes were proposed to encourage hinterland settlement. Establishing communities and ensuring the labor migratory trend were offered as reasons to develop the interior well into the 1950s. At the 1897 *West India Royal Commission*—convened seven years after the 1890 sitting of the Labor Commission, to address a severe Caribbean-wide economic crisis, many of the same issues were again aired. Similarly points of view on several economic issues as well as ideas and remedies to foster development were again largely ignored by the colonial hierarchy.

In 1897, Reverend Reynolds appeared before the *West India Royal Commission* and made recommendations to lessen the plight of the native-born African descent component of the population. Reynolds opined that government should offer land to this group instead of only to foreigners, as was the case with Asian indentures. He was referring to the government land colonization scheme whereby mostly foreign-born East Indians were granted free or cheap land at the end of their contracts as an inducement to remain in the colony instead of returning to India. Similar land development schemes, Reynolds posited, would allow those of African descent to relocate from expensive urban areas and settle where the cost of living was cheaper. Reynolds, an advocate for hinterland land settlement schemes suggested the importation of more Barbadian labor migrants to develop the gold and other industries.[49] For decades, Barbados like Jamaica and unlike Guyana had had surplus populations.

At the Royal Commission of 1897, the hazardous conditions causing high mortality rates in the interior were identified as a cause of labor shortages, but colonial bureaucrats again delayed implementing the recommendations to address the problems and stimulate additional internal migration. One year after the colony-wide riots in 1906, a "Monster Petition" signed by over 5,750 persons protesting the neglect of the recommendations of the 1897 Commission was sent to the colony's governor for onward transmission to the Colonial Office. Another such petition soon followed.[50] Census, death, and other records revealed that the perennial solution to alleviating the short supply of workers locally was to recruit more labor migrants of African ancestry. Parenthetically, this may have given validity to the lyrics by Black Stalin, a Calypsonian from Trinidad and Tobago, when he sang "the more Africans they kill, more come, more come."[51] The fixation on only immigration as a means of filling labor vacuums indicated that Africans and their descendants were regarded as expendable.

The number of miners who registered in six mining districts from January 1 to February 25, 1891, included 2,778 workers who were classified as Black. They were overwhelmingly native-born, Caribbean, South American and other African-descended persons including those of mixed (African) ancestry. A few East Indians were included in the column headed "Black." Seventeen Europeans were identified under a column headed "white." [52] Over the decades, the registration of workers was undertaken at the Institute of Mines, the Department of Mines, the Department of Lands and Mines, the Department of Labor, the Labor Bureau, and even the mining companies themselves. Overlapping registrations resulted in conflicting and inaccurate tallies of mining workers. Thus even approximate labor needs were nebulous.

At a time when the Department of Mines was already registering mining workers, the 1890 mandate of the Institute of Mines to recruit and to control labor, and oversee the operations of the gold and diamond industries became effective. When the Department of Lands and the Department of Mines merged in 1902, to form the Department of Lands and Mines, functions of the new department and the Institute of Mines overlapped. By July 1914, "New Enactments and Amendments in Laws and Regulations Affecting Crown Lands and Mines" were implemented under *Ordinance #20 of 1914. The Employers and Labourers' Amendment Ordinance* transferred the recruiting and registering of miners, the signing of their contracts, and payment of wages to the Department of Lands and Mines from the Institute of Mines, effective at end of 1914.[53]

The intertwining economic interests of bureaucrats and private business personnel continued under the new arrangements. James Winter, secretary and registering officer of the Institute of Mines, became an employee of the Department of Lands and Mines to oversee the transition. Registering and related duties formerly undertaken at branches of the Institute of Mines in other parts of the colony were taken over by district officers of the Department of Lands and Mines. With the emergence of the bauxite industry by 1914–1915, these agencies as well as the bauxite company recruited and registered mining workers. When the Labour Bureau was established in 1932, officials from the Institute of Mines were still involved in several aspects of gold, diamond, and bauxite mining, and their employers and employees. Although the Institute of Mines remained moderately involved in mining, it no longer wielded its previous authority. It fully yielded its responsibilities in 1936.[54] The Labour Bureau's headquarters were located at the Hadfield Street office of James Winter, secretary of the Institute of Mines. Nepotism survived in the labor recruitment agencies. Reverend Glasgow, on retiring from the colonial government in 1932, began working as a clerk at the Institute of Mines. When James Winter went on leave in 1933, his nephew, Alexander Winter took over his duties.[55]

Peculiarities of multiple work registration processes and death information recordings not only obfuscated accuracy, but also had serious implications and dire consequences for the management of the industry and its labor needs. Over or under supplies could not be determined until crisis situations arose. In addition, the sprawling nature of the frontier fostered discrepancies and inaccuracies in compiling employees' statistics in gold and diamond mining. At the outset, timely verification of duplicated information was absent because mechanisms were not in place to address the problem. Inconsistencies in the recruitment and registration system caused the seemingly perennial lack of enough workers for gold and diamond mining.

Consider the case of a miner registering to work for The Demerara Gold Association for a four-month period from February to May 1890. Once his stint ended, he could register with, and provide services to another company from August to November of the same year. Quite possibly, the same man could be recorded twice in one year because he worked for two different employers. Sometimes, an identical person registered for the same company at different times during the year and could also be counted twice or more. During the period June to August 1890, an estimated 712 gold-digging laborers registered at the headquarters of the Institute of Mines. The records indicated several instances of the same name recurring with similar vital statistics. Often, it could not be ascertained whether the very person registered two or more times and dishonestly worked with different companies for shorter durations than contracted for, or whether an individual was hired for the mining or forestry industry.[56] Census reports and Institute of Mines' records included other discrepancies related to the designation of workers.

Miner, laborer, gold digging laborer, and tributor were interchangeably recorded. Gold and diamond mining workers were also tabulated with employees in rock and stone quarries. When bauxite mining emerged, personnel in that sector appeared alongside diamond-seekers, prospectors, and manganese workers in various records.[57] Notwithstanding, the mining workforce in gold and diamonds, identified in Table 2.1 only serve to show the estimated number of workers registered for the labor-intensive enterprises, rather than reflecting an absolute tally. In this case, "official" registration records obfuscate the actual number of persons who independently established claims, outfitted themselves and their employees, and minimally utilized the services of quasi-government and government agencies. The labor supply for gold and diamond mining fluctuated or was insufficient and depended on regular infusions of African diaspora labor migrants because of the high attrition

Table 2.1 Gold and diamond miners registered, 1888–1910

Years	Number of Miners
1888–89	4,765
1889–90	7,224
1890–91	15,622
1891–92	22,298
1892–93	22,957
1893–94	19,471
1894–95	17,451
1895–96	17,368
1896–97	19,945
1897–98	27,147
1898–99	21,154
1899–1900	19,550
1900–01	16,588
1901–02	14,041
1902–03	11,385
1903–04	12,003
1904–05	11,088
1905–06	11,088
1906–07	8,234
1907–08	7,069
1908–09	5,282
1909–10	4,391

Source: H. J. Perkins, *Notes on British Guiana* (London: Sampson, Low, and Marston Company Ltd., 1896), 38; *British Guiana: Report of the Commissioner of Mines, 1899–1900*; and *British Guiana: Reports of the Lands and Mines Department, 1907–1908, 1909–1910*. *Diamond miners included from 1900–01. See *Reports of the Commissioner of Mines, 1900–1901 and Reports of the Commissioner of Land and Mines, 1902* and subsequent years.

rates among them. Ultimately, their presence in and their value to mining is far from miniscule. These workers made a name for themselves, set the pace and standards for mining production, and their capabilities determined the type of labor migrants to be imported. *Legislation #23, 1890,* already alluded to was enacted "to make provision for the introduction of immigrants other than from countries in Asia," was ample evidence of the determination to capitalize on and exploit African descent labor.[58]

The other sector of the mining industry, bauxite, was increasingly referred to from the period, 1913–1914, in the *Reports of the Commissioner of Lands and Mines.*[59] By 1917, bauxite production began and the next year the first shipment of the product occurred. Although the organizers of the bauxite industry exhibited some of the same controlling traits as their colleagues in gold and diamond mining, they implemented, in addition, progressive ideas and actions. The phasing out of the Institute of Mines' involvement in labor recruitment and mining activities began as bauxite emerged. Officers of the Institute of Mines who transferred to the Department of Land and Mines exhibited some of the same sentiments toward bauxite workers as they had in their previous labor and employment capacities. Nonetheless, the organizers and management of bauxite production utilized distinctly different labor recruitment policies and migration strategies from those officials in gold and diamond mining.[60]

Conclusion

At the turn of the twentieth century, colonial administrators, the plantocracy, and mining entrepreneurs were compelled to confront the issue of labor. Adequate labor was crucial to ensuring maximum profits from gold, diamonds, sugar, timber, and later bauxite. African-descended labor was preferred, not because other types of workers were unavailable, but due to unfounded and stereotypical beliefs of some colonial bureaucrats and others that only Africans had the stamina to withstand dangerous environmental situations. In gold and diamond mining, the detrimental working conditions faced by the laboring class did not receive adequate attention. Instead, the colonial hierarchy focused on recruiting labor migrants, streamlining registration, and controlling workers. This took precedence over any thought of preserving miners' lives with one proposal following the next for interior settlement and development. In succeeding decades, piecemeal infrastructure development provided inadequate services in the interior to facilitate gold and diamond mining and hinterland development. The next chapter will show that mining diversified the country's export-oriented economy as migratory movement continued.

CHAPTER 3

MINING FACTORS IN A DIVERSIFIED ECONOMY

IN THE LAST TWO DECADES OF THE NINETEENTH CENTURY, encouraging output of gold ushered in mining and diversified the mostly agricultural economy of Guyana. In 1882, two years after the enactment of *Ordinance #16 of 1880: An Ordinance to make Provision for Gold and Silver Mining*, the first gold exports were recorded. Gold output was 250 ounces in 1884, and increased to 32,332 ounces by 1889–1890. Throughout the 1890s, gold production continued. During this period, gold mining companies in operation ranged from the large Essequibo Company with capital totaling $60,000 from 600 shares at $10 per share to the mid-level El Dorado Company with capital of $3,000 from 60 shares sold at $50 each to the smaller Berbice Company with a capital of $400 from 16 shares costing $25 each.[1] At first, most of the male-dominated workforce was employed by companies, but by the mid-1890s laws permitting more access to government land were enacted and small-scale independent groups began to finance their own prospecting undertakings. In 1899, the initial law permitting women to participate in gold and diamond mining was enacted. Bauxite mining emerged in the twentieth century[2] (See Figure 3.1).

Methods of production in gold, diamonds, and bauxite mining varied. The first two commodities and their mostly labor-intensive excavation underscored the arduous and painstaking features of the work. Bauxite mining, which early incorporated the use of mechanical equipment, was no less grueling. In this analysis the importance of ongoing migration to shore up mining, workers' value to the precious metal and gems sectors, and the worth of their production, including jewels, demonstrate labor's importance to economic development and the creation of wealth.

By 1900, diamonds were added to the exports of the country. Unexpectedly, from 1904, the output of gold and diamonds began to fluctuate. From 1918, when bauxite exports began, the commodity's export was mostly stable.[3] Factors inhibiting stability and higher production levels in the first two commodities included international border disputes, limited access to a wider area of the country for prospecting,

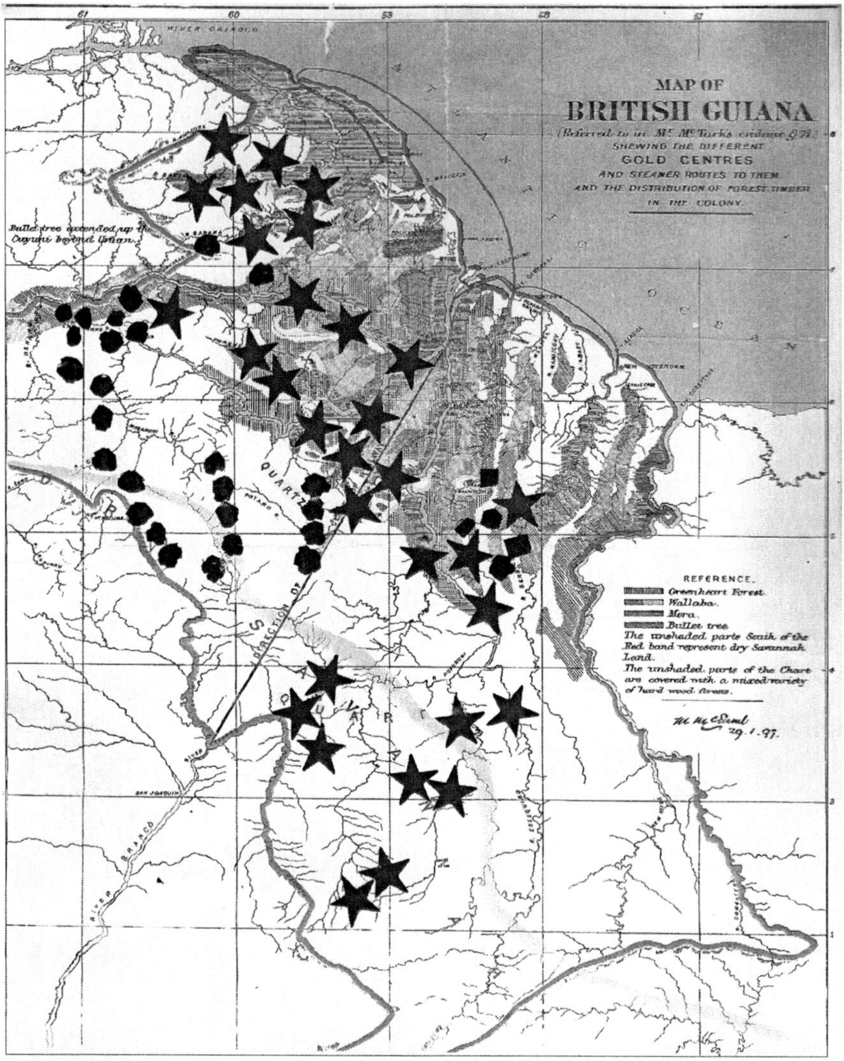

Figure 3.1 Map of Guyana with gold, diamond, and bauxite locations.

Key: ★ = Gold ⬢ = Diamonds ▮ = Bauxite.
Source: *West Indian Royal Commission, 1897*, Appendix C. Part 11.

withdrawal or lack of investments by capitalists, inept management, questionable development plans, inadequate transportation, settlement difficulties, and labor shortages.[4] Labor was vital in ensuring the viability of mining, but the astounding attrition of workers militated against increasing productivity. High mortality rates associated with unsafe environmental settings in interior mining regions caused labor shortages and shortfalls in output (See Chapter 4).

At the turn of the century, Guyana's territorial dispute with Venezuela and the loss of some land to that country contributed to the fluctuating amounts of gold and diamonds produced. By 1904, monopolizing strategies of mining magnates and financial institutions associated with Southern Africa interfered with Guyana's development of the gold and diamond industries.[5] A smallpox outbreak during August–September 1904, and the quarantine that followed prevented miners from traveling to the gold and diamond fields and temporarily halted production. Authoritarian order did not ensure progress and in 1905, riots resulting from socioeconomic tensions between the colonial hierarchy and the working class, limited the number of workers available for mining.[6] The two World Wars and labor competition from bauxite production also impacted gold and diamond mining. In addition, the problem of heavy rainfall during some years, which caused flooding and swollen rivers and creeks impeded travel to prospect for gold and diamonds. Periodic droughts also hindered production.

At the dawn of the twentieth century, the colony's African diaspora workforce was still being replenished by incoming labor migrants. Regional competition for workers in mining, railroad, and canal construction, as well as in the sugar and oil industries continued.[7] Nevertheless, to promote economic development locally, colonization schemes were contemplated to encourage labor migrants from the coastlands, the Caribbean, and afar to settle and work in the interior of the country. The *Echo* and *The Gold Mining Gazette* advertised for overseas labor, based on a Governor-in-Council Order.[8] In 1920, the Immigration Agent General reported that almost one thousand West Indians had arrived in the colony. By the 1930s, advocates from several quarters had called for the colonization of more West Indians and African Americans in the country, but the schemes also aroused controversy.[9]

Officials of a newly organized labor union held a meeting in the capital city and questioned why "money and energy" were being expended "to attract (foreign) population" when local workers were unemployed.[10] After identifying inept labor management, wage disputes, non-compromise between employers and employees, and the lack of interior infrastructure as reasons for insufficient hinterland workers, the officials proposed that funds could be better used for local development rather than recruiting labor migrants. Proponents of migration schemes invariably pointed out that constructing more railroads and roads to facilitate movement to and from the interior would benefit many parties. Meanwhile, some legislators noted that internal migration, the economy, and interior infrastructure development were undermined by planters who feared labor would move from the coast and sugar production to interior undertakings and jeopardize their viability and profits.[11]

Advocates for the curtailment of incoming labor migrants had no discernable success and until the 1940s, immigration was largely responsible for the country's population growth.[12] By that time Caribbean migrants who had been the majority of persons entering Guyana were migrating to places such as the United Kingdom. *The West Indian Census, 1946: British Guiana* identified places in Demerara, Essequibo, and Berbice Counties where the foreign-born lived and worked including gold, diamond, and bauxite mining locations. As late as the 1950s, it was suggested that Great Britain should divert "the increasing flow of Jamaican and other West Indian immigrants arriving" in that country to its South American colony.[13] Censuses confirmed that the African diaspora workforce in the country included Barbadians, Grenadians, Jamaicans, St. Lucians, and other West Indians, North, Central, and South Americans, as well as Africans. By 1946, the population of the 83,000 square miles country was estimated at approximately, 370,000 persons; with 90 per cent of them on the coast while the much larger hilly sand and clay belt and hinterland regions needed more occupants to facilitate projected economic enterprises.[14]

Aspects of Gold and Diamond Production

Michael Moohr in "The Discovery of Gold and the Development of Peasant Industries in Guyana, 1884–1914: A Study in the Political Economy of Change," considered the importance and significance of the gold industry. Moohr noted that the sector's initial high productivity level caused legislators to modify regulations governing the sale and use of government land that benefited the gold industry and other economic enterprises. In coastal and riverain areas, rice, livestock, and dairy production expanded and created occupational opportunities.[15] Moohr observed that interior mining operations hinged on the perseverance of workers who labored under precarious conditions, and were either self-employed or worked for small, mid-sized, or large companies.[16]

In 1890, gold exports from the colony were valued at $1,124,756. Table 3.1 indicates the values of selected exports and their fluctuating nature from 1891 to 1900.[17] Each miner's value to the industry was calculated from the estimated 12,003 registered gold miners for the year ending December 31, 1890. Using $1,101,652.40 from the *Report of the Commissioner of Mines, 1890–1891*, each miner's production average was valued at $91.78 for the year.[18] Table 3.2 confirms the number of miners and the value of the gold produced at $18 per ounce from 1892 to 1900. Fluctuating gold production and its value reflected not only market price variations but also the mobility of miners between gold and the ever increasing production of diamonds.[19]

In the 1880s diamonds were mostly found in Essequibo County where, by the end of the next decade, several diamond mining companies were in operation. Diamond mining companies such as the British Guiana Diamond Syndicate,

Table 3.1 Value of selected exports, 1891–1900

Years	Gold ($)	Sugar ($)	Timber ($)	Balata ($)
1891–92	1,958,254	*	96,327	75,869
1892–93	2,366,100	7,535,868	104,799	98,905
1893–94	2,454,538	7,528,898	74,343	40,105
1894–95	2,402,141	5,895,091	81,529	49,028
1895–96	2,136,954	5,021,569	63,890	42,829
1896–97	2,237,486	5,272,311	14 8,608	96,613
1897–98	2,177,869	4,912,913	93,173	147,899
1898–99	1,989,346	4,996,714	81,043	135,134
1899–1900	1,990,207	5,288,420	*	*

Source: *British Guiana: Administrative Reports, Reports of the Comptroller of Customs*, and Rodney, *A History of the Guyanese Working People*, 225–238.

Table 3.2 Value of gold and miners registered, 1892–1900

Years	# of Miners	Value of Gold ($)	Average per Miner ($)
1892–93	22,957	2,414,239	105.16
1893–94	19,471	2,493,500	122.92
1894–95	17,451	2,393,906	137.18
1895–96	17,638	2,122,492	120.34
1896–97	19,945	2,294,629	115.05
1897–98	27,147	2,186,833	80.55
1898–99	21,154	2,036,054	96.25
1899–1900	19,550	2,030,211	103.85

Source: H. J. Perkins. *Notes on British Guiana and Its Gold Industry* (London: Sampson Low, Marston, & Co., 1896), 36–39 and *British Guiana: Reports of the Commissioner of Mines, 1898–99*, 20.

Barnard's Syndicate, the Mazaruni Company, and Marshall Syndicate increased government revenue from royalties, registration of miners, leases for claims, prospecting permits, and licenses to trade in the gems.[20] Diamonds were a boon to the economy as sugar had begun to decline.[21]

In November 1894, the colony's governor, Sir Charles Cameron Lees, had appointed commissioners to devise ways to increase the production of gold and other commodities. The commissioners' report to the legislature recommended updating mining regulations, upgrading the collection of royalties and fees, improving employer/employee relations, providing more land for mining, and addressing transportation to and from the interior regions.[22] Legislation to reduce the purchase and lease- prices of government lands was also enacted to encourage mining and other production.[23] It is noteworthy that the new legislation enabled small groups of self-financed mostly African-descended persons to prospect independently for gold and diamonds. These miners broke the monopoly previously held by the larger and overwhelmingly foreign mining companies. Within three years parties of small-scale miners and their supporters successfully petitioned the government to abandon an impending increase of the royalty on gold that would have reduced their profits.[24]

In 1895, Joseph Chamberlain became the British Colonial Secretary. He decided to expand the colony's gold industry by encouraging private investors to develop the interior infrastructure. Chamberlain felt that the country's economic development depended on gold rather than sugar. Aiming to improve imperial trade, Chamberlain viewed resources from the colonies as critical to upgrading the socioeconomic situation in England. He believed that the colony's administrators were incompetent, had done little or nothing to promote its development, but had done much to ruin it. Chamberlain affirmed that "although the gold industry paid moderately high taxes, the bulk of these royalties were shipped directly abroad to the Crown."[25] Unexpectedly, however, the intentions to develop the colony's mining industry received a setback because of increasing investments in gold and diamonds in Southern Africa, Klondike, Yukon in 1896, at Nome, Alaska in 1899, and in Canada. Capitalists with an interest in investing in mining in Guyana were not forthcoming and Chamberlain's vision of large scale mining development through investment of private funds remained unfulfilled. He could not utilize public funds to implement measures to attract more investors.[26]

Venezuela, Brazil, Suriname, and French Guiana also produced precious metals and gems. Gold deposits "are located in the Guiana shield...an extensive Precambrian greenstone belt" encompassing "415,000 km^2 and extending from Venezuela through Guyana, Suriname, and French Guiana into Brazil's Amazon basin."[27] In Suriname, interior gold mining had been undertaken by Maroons (descendants of runaway enslaved Africans) for centuries. By the late nineteenth century, many Surinamese plantation workers increasingly found alternative employment in prospecting as artisanal miners.[28] In 1870, gold mining accelerated when the New York South American Suriname Goldmining Company began operations.[29] Suriname's gold output increased from an estimated 24,236 ounces in 1886 to 37,987 ounces in 1894. The estimated value of the gold exported was $413,106 and $647,516 in the former and latter years, respectively. By 1901, over 5,500 employees worked in Suriname's gold industry. Two years later a railroad constructed to link Paramaribo, the capital city, to the interior goldfields tremendously increased Suriname's gold output.[30] By 1895, French Guiana's gold exports totaled approximately, 90,263 ounces.[31] Labor migrants from the Caribbean and elsewhere were involved in prospecting in the neighboring countries, and in Colombia and Ecuador; fueling ongoing concerns regarding labor supplies locally. Guyanese were also leaving the country as part of a Caribbean-wide migratory movement.[32] Coupled with the miners' value on Table 3.2, loss from this workforce was troubling.

Estimating the effects of insufficient labor on gold production, the Commissioner of Lands and Mines calculated that "one man produced, at a minimum, one pennyweight of gold daily" (one pennyweight equals 1/20 of an ounce). The Commissioner calculated that if there was a shortage of 1,000 miners during the year, a shortfall of 15,000 ounces of gold would occur.[33] Table 3.3 indicates that gold and diamonds were among the country's main export earners. Gold's value was above the $million mark, and in combination with diamonds, the two commodities were the second most valuable export earners. Although the Commissioner viewed mechanization as

Table 3.3 Value of selected exports from Guyana, 1900–1910

Years	Gold ($)	Diamonds ($)	Sugar ($)	Timber ($)	Balata ($)
1900–01	1,886,802	12,876	5,411,792	87,374	94,007
1901–02	1,783,160	91,463	4,983,183	89,687	113,534
1902–03	1,789,587	97,710	5,001,710	111,761	195,056
1903–04	1,572,129	90,029	5,381,484	93,954	216,895
1904–05	1,690,198	78,698	6,146,860	117,287	182,607
1905–06	1,689,747	24,215	5,799,144	119,705	193,495
1906–07	1,532,266	32,627	5,085,617	74,669	240,510
1907–08	1,148,299	14,529	4,819,324	97,310	368,538
1908–09	1,302,241	40,733	6,037,574	78,881	471,016
1909–10	1,101,678	45,056	5,785,035	77,876	1,034,076

Source: *British Guiana: Administrative Reports, 1900–1910.*

a way to increase output, he conceded that insufficient labor jeopardized the mining industry. The number of registered gold and diamond miners from 1888 to 1910, is given on Table 2.1 in Chapter 2.[34]

The Gender Factor: Women in Mining

Women engineered their way among the colony's gold and diamond miners and diversified the workforce. Although for years women were discouraged from prospecting for precious metals and gems, an *Amendment to the Mining Regulation, 1899,* allowed the subletting of claims; enabling women to participate. A decade later, of the 283 licenses issued between July 1, 1909 and June 30, 1910, nine of them were granted to women. As Table 3.4 shows, none of the women was identified by race and four of them shared a similar surname to the prospector. One married female held a license jointly with a male. Emily Crossman, a single woman, had Geo. O. Patoir as the prospector. Two of the women had no male prospectors. None of the women could prospect, but six of the licenses indicated that men were authorized to prospect on their behalf; a reflection of the era in which they lived when females were viewed as unequal to men.[35] The prevention of women from legally prospecting, that is, being directly involved in the physical activity of gold and diamond recovery, was reinforced by the *Report of the Gold and Diamond Commission, 1928,* which stated that "No Mining Privilege should . . . be issued to any woman" . . . because . . . There are obvious objections to the presence of women."[36]

Two years later, efforts to restrict women in the mining region continued. A notice issued by the Warden Magistrate of the Mazaruni Diamond Fields stipulated that except for married women and holders of claims all others must reside on the premises of their employers. The stipulation caused quite a stir and *The Daily Chronicle* pointed out the farcical nature of the measure because a claim-holder could simply allow his name to be used and never employ the supposed domestic on her arrival. The new stipulation could also result with a claim-holder having dozens of (domestic) servants registered in his name.[37] At that time, it did not occur to the

Table 3.4 Licenses issued to women, 1909–1910

Date of Issue	License #	Name of Claim Holder	Authorized to Prospect
21-7-09	B1657	Matilda Ursula Cox	Charles T. Cox
01-1-10	B1556	Susan Seriant	—
07-1-10	B1557	Caroline Vieira	Robt. A. Smith
15-3-10	B1927	Alice Telman	Alf. Telman
18-3-10	B400	Elizabeth Isabella Matthews	A. A. Matthews
06-4-10	B2062	Alex Mackey and Mrs. Josephine Williams	—
18-4-10	B2068	Emily Crossman, a single woman	Geo. G. Patoir
17-5-10	B2077	Emily Hart	—
30-5-10	B2080	Rosalie Gertrude James	Edward Benjamin James

Source: *Report of the Gold, Diamond, and Forest Industries: Institute of Mines and Forests, British Guiana, 1909–1910.* Appendix viii, Licenses, July 1, 1909–June 30, 1910.

reporter that the gender of claim-holders was not restricted to males, but included females as well.

Magistrate Sydney Miller who served in the mining region explained that the new regulation "attempting at this late hour…to curb the freedom of movement…is not only unfair to the women but it was…impossible for the Warden Magistrate, even with the assistance of the police to do."[38] The *Regulations of 1930,* allowed a woman to apply for a prospecting license, which cost $5. She could then apply for a claim license costing $14.48 and would be able to remain at any Landing by herself, own claims as she desired, and move about freely. Questioning the effectiveness of the new regulations to restrict the activities of women, Miller recommended their reversal and supported married women or common-law wives living with their husbands near mining sites. He noted that all women were not dishonest or of ill repute, and many of them entered the mining regions with their male prospectors.[39]

Former gold and diamond miners from the 1920s to the 1950s such as Lionel Arthur, Joseph Watson, Darrell Grey, and Howell Wood assured that a few women had always actively mined for gold and diamonds. Any remaining legal stipulations against women in mining were waived during the global economic depression of the 1930s. Mrs. P. J. Willems' Trading & Timber Co. activities included gold mining. Willems employed a mining engineer and dozens of laborers to prospect for gold on Pigeon Island. By 1935, groups of women miners were prospecting at Honey Camp, Essequibo. Shovels, pickaxes, and other mining implements were used by them to recover gold and diamonds. Mrs. Regina Orderson, a widow of C. Vlissengen Road, Georgetown successfully litigated against G. T. Chislett of the British Guiana Section of the B. G.-Brazil Border Commission. On March 15, 1937, the defendant had prevented the plaintiff's mining party access near the left bank of the Corentyne River at Wonatobo Fall. The sum of $5,000 was granted to her.[40] In December 1939, for the first time, women were among the contestants in boxing matches promoted in the gold and diamond fields. The promoter, F. H. Burnett *alias* Mike Jacobs selected a venue at Tiboku Goldfields so that Hilda Marblestein (133 lbs.) and Matilda George (130 lbs.) could compete.[41] Women were a distinct presence, and *The West Indian Census, 1946. British Guiana* revealed that 4,139 persons (4,108 males and 31 women) were engaged in quarrying and mining.[42]

The historiographic landscape of women's involvement in gold and diamond mining in Guyana either indirectly as claim-holders or directly as miners/prospectors has received no scholarly attention until now. In Chris Curtis's, 2006 publication on the experiences of Dr. George Giglioli at the Linden bauxite complex and Odida T. Quamina's studies of two decades earlier, the majority of women in the Upper Demerara River enclave are identified as housewives. Some women were domestic servants, teachers, nurses, other school or hospital employees, or clerical staff. A small number of women were engaged in retail commercial activities especially as market vendors. The paucity of women employed directly in bauxite mining persisted until after the 1950s.[43]

Mining Women: Gender in the Development of a Global Industry, 1670–2005 (2006) edited by Jaclyn J. Grier and Laurie Mercier centered on the diverse roles of women in the industry in Africa, Asia, Europe, and the Americas, except Guyana. Characterizing mining as a mostly masculine industry, the essays explored the ways in which women challenged gender constraints in relation to their lives and work.[44] Applied to the local situation, the long-standing invisibility of women in the activity of mining is problematic because females have always worked in labor- and capital-intensive undertakings. Although gold and diamonds, unlike bauxite, were located in non–environmentally friendly areas, the paucity of women in mining points not to their inability to perform arduous tasks, but to capitalists' and investors' attitudes, and their ability to persuade legislators to enact laws to limit or prevent the active participation of women in mining. In the case of the Demerara Bauxite Company (DEMBA), its management encouraged women and families to settle nearby. Yet females remained almost invisible in the undertaking. The mining camp setting of gold and diamonds prevailed for decades. Minimum attention was paid to improving the infrastructure that discouraged the settlement of families. Legal constraints and stereotyping imposed limitations on females, but they were not unchallenged and women became active participants and contributors to mining development in Guyana and globally.[45]

Factors of Gold and Diamond Production

Throughout the colonial era, gold and diamond mining in the country depended on the African diaspora workforce and their methods of production namely, dredging, hydraulic, quartz, and placer or alluvial.[46] Unlike the other methods, placer or alluvial mining is carried out only by hand. Most of the gold and diamonds recovered in the country from the 1880s to the 1950s resulted from this method.[47] Using a 'battel' (a conical receptacle), a miner sat at the waterside and scooped up earth that was washed to separate unwanted material from gold or diamonds.[48] The 'Tom' was an open box with a sieve on the bottom that was elevated by iron hooks from a wooden stake driven into the ground. As many as five miners processed the earth with long-handled hoes to separate the stones from the gravel or soil. Gold passing through the perforation was caught by mercury or quicksilver.[49] The sluice consisted of a series of wooden boxes fitting into each other. Although ten or twelve boxes could be operated simultaneously, quicksilver was placed in the last two or three sections to adhere to the gold. At least ten miners operated the sluice similar to the "Tom"[50] (See Figure 3.2).

From 1908 to 1915, approximately 393,478 and 116,985 ounces of gold were recovered by the alluvial and mechanized methods of quartz milling, dredging, and hydraulicking combined.[51] Government officials often characterized the alluvial method of gold recovery by small-scale miners as crude and unsystematic, yet every year production by this method surpassed all the others.[52] The Commissioner of Lands and Mines noted that although hardworking miners sustained the industry, a lack of scientific and technical expertise and the withdrawal of large mining companies had caused fluctuating outputs.[53] Another factor, the intricate legal battles local small-scale prospectors waged against colonial bureaucrats and private companies' officials, possibly, caused some independent gold and diamond miners to cease operations.

Washing for Diamonds.

[*Photo by E. W. Savory.*

Placer Working, Using a Sluice.

Figure 3.2 Men washing for diamonds and gold miners.

Mining Regulations, including exclusive access to vast expanses of land and royalty concessions, were crafted to attract and to benefit foreign conglomerates or individuals who sometimes exhibited peculiar attitudes to prospecting. Land allocated to large mining companies was abandoned if gold or diamonds were not quickly discovered. Some of the abandoned lands were reallocated to and worked by small-scale miners who had toiled for years in the same location, hoping that their efforts would be rewarded by eventual discoveries.[54] At times, however, the hard work of these miners was fraught with legal difficulties as was the case of Messrs. Payne and Williams, two self-financed miners who "found a very large nugget of pure gold weighing 333 ounces in 1902, in the Five Stars District of North West."[55] The area, to which the two miners had a mining claim, was previously mined and abandoned by Sprostons Company Limited, a large conglomerate. On hearing of the discovery, company officials pursued the two miners in court to claim the nugget. Although on March 5, 1903, judges of the country's Supreme Court decided to award the gold to Payne and Williams, the company's officials appealed the decision at the Privy Council in London. The final outcome of the litigation is unknown, but the two miners spent much time and money to resolve the issue.[56] Unfortunately, a short time later, another group of miners encountered similar treatment from the colonial power structure.

Peters' Mines, operated by a group of African-descended miners—Hatton Garden Diamond Syndicate in the Puruni District—was coveted by a foreign firm, the American entity, British Guiana Consolidated Gold Concessions Company. The management coveted the sixty-four acres of land to which the local miners held claim near Majuba Hill at Strong Hope Creek on the right bank of the Puruni River. To justify the contemplated action of seizure, government officials and agents of the private company insidiously belittled the legitimate operators of the claim. Called into question was the claim-holders' inability to maximize production and profits from the undertaking. This concoction led to the acquisition of the land against the will of the claim-holders. Within a year, the takeover by the Americans was completed. The Commissioner of Lands and Mines supported the action and argued that it was much better if the enterprise was controlled by those "competent and financially able to develop it." He described "the primitive fashion" of alluvial mining done by the original claim-holders as wasteful, although most of the gold produced in the country was by this method.[57]

Denying potential profits from accruing to independent miners (and thereby perpetuating dependency) was a conniving method resorted to by colonial officials to avoid extending munificence also to small-scale entrepreneurs of this ilk. The government's failure to support the entrepreneurial efforts of the miners was contradictory and harmful. Special favors such as the providing of free and cheap land, financial subsidies, and tax concessions extended to other groups, were rarities for people of African ancestry in the colony. Instead, at certain times, the commissioner viewed mining workers as part of an undisciplined labor force and rued "the absence of any proper supervision and control" because of "a dangerous element of lawlessness" among them. That same commissioner deemed the very workforce as worthwhile when gold and diamonds were prolific and profits immense. When production and profits decreased, he conveniently characterized miners as "thriftless and lazy" and as a bad influence.[58] Although there were other contributing factors accounting for the fluctuating outputs of gold and diamonds, labor was conveniently blamed.

A significantly overlooked but dynamic factor impacting gold and diamond production centered on the resolution of a territorial dispute between Guyana and Venezuela. On June 6, 1904, the King of Italy, on behalf of an arbitration panel, announced a decision on the issue, and roughly 5,000 square miles of Guyana's territory became Venezuela's. The conceded area contained deposits of gold and diamonds.[59] This setback and the loss of valuable territory contributed to the ongoing fluctuating outputs of gold and diamonds. Ironically, it appears as though this loss caused colonial administrators to react by further exploiting the working class. Administrators decided that previous profits from gold and diamond mining had to be surpassed at any costs and *Amended Mining Regulations* were put into effect by December 30, 1905. The law reemphasized regulating and controlling mining workers because they needed added scrutiny and discipline.[60] While the cost of mining privileges (the legal authority to work) remained at 25 cents, the new law raised the costs of prospecting permits and rental licenses, required by independent individuals and groups, to $ 5.[61] Increasing fees and the prices for selective items overwhelmingly utilized by these workers were ongoing practices of the colonial power structure from the nineteenth through the twentieth centuries.

In the World War I era, unstable international situations caused European markets for diamonds to dwindle, and as sales plummeted, local banks curtailed advances for their production. Gold shipments overseas were intermittent due to increased insurance rates and shipping charges.[62] A decrease in gold miners caused a decline in production. Many of them opted to mine diamonds instead because of an existing price increase for the gemstones locally. The war aggravated an already grave situation of high prices for food and equipment required for mining operations while employment opportunities were expanding on the coast. Labor demands and attractive wages in agricultural work induced many would-be prospectors to work on the coast; except in cases of gold or diamond "shouts."[63] During the war, the bauxite industry emerged and attracted to its production thousands of current and potential gold and diamond miners.

Prior to the outbreak of World War I, representatives of multinational companies were acquiring privately owned land for bauxite production. British colonial bureaucrats allegedly conspired with Canadian and American interests to rent government land below market value.[64] The view existed that there was more to it than the mere seizing and monopolizing of land for favored clients. The contention was that political, corporate, local, and international machinations and collaborations of the economic elites ensured that the most profits accrued to them. This was true of both bauxite and diamonds. The first undertaking flourished because of the monopoly of the few bauxite companies, while the output of diamonds fluctuated due to the actions of mining magnates from South Africa.[65]

Europeans and Southern African mining magnates manipulated investments so that Guyana's diamond production would eventually stifle.[66] In the litigation, *United Diamond Fields Co. of B. G. vs. Solomon Bernato, Joel Bernato et al.*, in London, Solomon and Joel Bernato, Ernest, Otto, and Louis Oppenheimer,

Henry Claridge, and several others were part of a wealthy diamond syndicate. They were accused of restricting diamonds on the international market to their advantage that caused the plaintiffs' company to fail.[67] Similarly, a consortium including the banking house of the Rothschild, Consolidated Mines Limited, the DeBeers Company, and the Oppenheimers along with British, French, and South African bureaucrats colluded to control the supply and price of diamonds globally.[68] The commodity was not insignificant to Guyana's economy, and until 1926, the country was the second largest producer of diamonds in the British Empire, before output waned.[69]

By 1927, policy makers considered reverting to the controlling of mining by a single enterprise and not the free-for-all system of the independent gold or diamond miners. The consensus was that miners of African ancestry could provide the labor, but ownership of all claims or exclusive permissions for mining rights should be given to foreign companies and monopolies. Small-scale working-class miners saw through the veneer of colonial administrators' policies because favoring large companies at their expense also caused a reduction of the colony's revenue. They recognized the contradictory nature of the policies and viewed official disregard of their mining activities as attempts to foster their dependency.[70]

Meanwhile, the DeBeers, Oppenheimers, and their business allies created a monopoly and dissuaded other entrepreneurs from investing in mining in any country the cartel did not endorse. This prevented "serious consideration of any scheme for hinterland development" in Guyana.[71] By the 1930s, a local legislator Barrister-at-Law Joseph Eleazar tirelessly explained that the lack of attention to hinterland development emanated from the collusion of planters, businesspersons, expatriates, and colonial bureaucrats.[72] At the dawn of the twenty-first century, allegations of the manipulations of economies were validated when the Oppenheimers and DeBeers Companies were unmasked and punished by enormous monetary penalties that were levied against them for their pervasive, monopolistic, and illegal control of the diamond market.[73]

Long before this however, understanding the consequences of ignoring willful slights, miners petitioned the Commissioner of Lands and Mines to oppose "the granting of more than 200 feet on the river banks between the Tumatumari and the right bank of the Mahdia," to the Potaro Syndicate, a large conglomerate. The petitioners noted that in 1932 and 1933, small-scale producers were responsible for increasing gold production in this area. Their contentions were: the granting of exclusive permission and mining rights to one company would reduce the area available to them for prospecting; the action would result in government losing revenue, fees, and royalties from their (the small-scale miners) numerous claim licenses and production; small-scale operations would increase the amount of revenue from a range of services linked to mining such as passenger and freight services provided by government lorries and ferries.[74] The prescience of the working-class miners on government policies underscored that inept management, cronyism, and the inflexibility of the colonial hierarchy, undermined the economic development of the country. Furthermore, foreign mining entities were often accused of colluding with colonial officials to pay the least royalties and taxes on the commodities they extracted and exported.[75]

In the 1930s, colonial bureaucrats who considered awarding extensive areas of the colony's lands to large conglomerates were also challenged by other interested parties. The British Guiana Miners Association, some members of the legislative council, lawyers, medical practitioners, and others joined miners in a series of protests held at schools and Tipperary Hall, Buxton to protest the application of Bookers Bros., the sugar conglomerate and British Guiana Consolidated Goldfields Ltd. for mining concessions to the entire Puruni River basin of 460, 000 acres of land. Consequently, the number of acres requested was reduced and small-scale miners were not displaced from their holdings as they had anticipated.[76]

The attraction of gold and diamonds as a means of economic opportunity was important to small-scale miners. This probably helps to explain the attitude of prospectors such as Lampley Wade. A well-known "bush-man" from West Coast Berbice, Wade and his thirteen-man crew continuously prospected for gold and diamonds in the upper reaches of the Berbice River. At the end of April 1934, Wade and his fellow miners arrived at New Amsterdam, the main city in Berbice with diamonds weighing over 25 carats. On being questioned about the significance of their find, Wade retorted "yes, a man can make a living." Their attitudes typified the determination of gold and diamond miners to make a living despite all the odds against them.[77] Self-determination by the working class was consistent with their goals of providing for themselves. Gold and diamond miners were not adventurers involved in prospecting as some sort of lottery as an article published on mining in the country suggested.[78] They were indefatigable workers who contributed tremendously to the development of the country's major industries and they were valuable to the economy. Wherever labor was needed, this segment of the workforce displayed its value and flexibility; as was the case with bauxite mining.

BAUXITE: AN ADDED FACTOR IN THE MINING INDUSTRY

The advent of bauxite at a strategic time during World War I augured well for multinational businesses, European and North American investors, colonial officials, and potential employees. Labor was provided by mostly African-descended employees, who had worked at wood-cutting grants, in sawmills, on the railway or wharf, and in the balata industry in the vicinity. Local people and African diaspora labor migrants who had also worked in other enterprises gravitated to the new industry.[79] Bauxite became a magnet attracting many former and potential gold and diamond miners to its production. Instead of working and living or crisscrossing between coastal agriculture and employment in the two other mining sectors, they became part of bauxite's workforce.[80] Agencies recruiting gold and diamond miners provided similar services to the bauxite industry. *Ordinance #20 of 1914, July 1914, Employers and Labourers Amendment Ordinance 1914 effective 30 December 1914,* which applied to the entire mining industry, legalized the arrangement.[81] From 1917, bauxite was produced by DEMBA.[82]

The country's ore is located below the surface or topsoil and the "overburden" of trees must be removed with the use of heavy machinery. Strip mining, as the process to obtain the bauxite is called, required the use of bulldozers to clear several layers of sand and clay in order to excavate the bauxite ore, which is taken to the industrial plants, crushed, washed, and dried before export.[83] Bauxite production is also labor intensive. "Before the introduction of locomotives, wheeler-excavators" and "draglines, miners used spades, shovels and pick-axes"... "wooden cars and mule carts" all hand-operated implements and contraptions, to excavate and transport the ore to the processing plant.[84]

Amerindians first used bauxite, a clay-like ore, as part of their house building materials.[85] Between 1868 and 1873, surveyors Sawkins and Brown identified this "reddish white friable pisolitic rock in the Upper Demerara river area." Geological fact-finding missions from 1897 to 1898, discovered extensive deposits of bauxite in other areas of the colony. By 1913, particular attention was being paid to the possibility of mining bauxite in commercial quantities in Guyana.[86] Extensive bauxite deposits occur in Mackenzie-Christianburg-Wismar, (Linden), and at Akyma in Demerara, in the North West District, and along the banks of the Essequibo (at Bartica), Berbice (at Ituni, Kwakwani, and Aroaima), and Corentyne rivers[87] (See Figure 3.1).

Kwakwani, 135 miles up the Berbice River is another important area of bauxite production. American Cyanamid Company also known as The Berbice Company began operations in 1939. Everton, at the mouth of the river is the transshipment point for the bauxite overseas. Reynolds Metals Company, another American entity, bought the Berbice Company in 1953. By 1954, its predominantly African-descended workforce were persons who had been recruited originally by the Berbice Company from employees based at the US wartime-controlled, Atkinson Field, the site of the country's main airport.[88] Reynolds Metals is one of the world's leading producers of aluminum. DEMBA and Reynolds "controlled freehold lands, leases, and exclusive permission of about 3,140,000 acres in the bauxite belt," having "a virtual monopoly of Guyana's accessible good bauxite lands."[89] DEMBA controlled the majority of the land. The British and Colonial Bauxite Co. Ltd. operated Plantation Bauxite Co. Ltd at Christianburg from March 1949 until 1957, when its equity was transferred to DEMBA.[90]

Suriname like Guyana began bauxite production in the early 1900s. "By 1930 Surinam held 16.2 per cent and Guyana 7.5 per cent of world production... by 1950 Surinam and Guyana together provided 45.1 per cent of" the world's bauxite.[91] In *The American Takeover: Industrial Emergence and ALCOA's Expansion in Guyana and Suriname, 1914–1921,* Carlo Lamur discussed how the American Company, ALCOA, and its subsidiary gained control of bauxite production in the two countries while they were still under British and Dutch administrations, respectively.[92] Lamur noted that in both colonies, the collusion of Americans, colonial officials, and others led to a foreign takeover of bauxite production and exports.[93] When the *S. S. Moheigan* left Guyana in 1918, with the first shipment of 2,037 tons of dry ore bauxite bound for North American smelters, African-descended locals and Caribbean labor migrants from Barbados, St. Lucia, St. Vincent, Grenada, Jamaica, and Trinidad and Tobago predominated the workforce. DEMBA's employees numbered over 800 men, excluding supervisors.[94]

The Diversified Economy: Gold, Diamonds, Bauxite, and Jewels

During the nineteenth and twentieth centuries government commissions were convened and reported on the feasibility of establishing or expanding economic enterprises to diversify the country's economy. Commissions reported on the gold, forestry, fruits, balata, and diamond industries. A central focus of the commissions was always the question of the availability of African-descended labor.[95] As output and the value from bauxite mining indicate, those workers did not disappoint. Bauxite could not have appeared at a more opportune time than during World War I when it was especially valuable due to its high quality and the aluminum it produced for the United States, Great Britain, and the Allies. The significance of several export commodities to the economic health of a developing country cannot be overlooked in a world devastated by war. That is another reason why the value of the mining industry is so critical in demonstrating the vital role of African diaspora workers as factors in analyzing Guyana's development.[96]

Employees of the bauxite industry, as Table 3.5 indicates, contributed immensely to the country's exports, value, and earnings.[97] Their decisions to venture away from sugar plantations, to grapple with the challenges of the gold and diamond fields including death-defying journeys, and the rigors of establishing the bauxite industry were profound. By 1933, gold production revived somewhat and some foreign mining companies explored prospecting opportunities. Unfortunately, by the end of 1934, disastrous floods inundated much of the colony, temporarily curtailing gold and diamond mining. As the decade ended, government granted more land for exploration and an increased number of exclusive permits for capital enterprises to operate. The British Guiana Miners Association drew up a $20,000 plan to seek government's assistance to finance ten expeditions with twenty prospectors each to work in areas of the Cuyuni, Mazaruni, Essequibo, and Corentyne Rivers.[98]

The diversified economy of Guyana averted a national crisis during World War II. The colony, because of its bauxite production and the products' utility, made a "substantial contribution to the war effort of the United States through shipments to Canada." By 1942, employment in the bauxite industry had reached almost 4,000 persons. Production of bauxite spiraled upward during the war, especially from 1941 to 1943. By 1947, bauxite production increased to 1,290,367 tons valued at $6,729,112 and remained impressive for the rest of the decade. By 1944, the total value of all the gold produced in Guyana since 1884, amounted to $58,217,385. Diamonds produced from 1901 to 1944, valued $41,147,376. Nonetheless, the uncertainties of World War II wrecked havoc in the gold and diamond industries and "the low price for gold established by the Breton-Woods (1944) agreement" was inimical to its mining. During wartime, the cost of food, equipment, transportation, and fuel utilized by gold and diamond miners increased. Prospecting by large companies continued, but it was the small-scale miners and their hand-methods of extraction that still accounted for most of the output.[99]

By mid-century, Canadian and American interests in the colony's development led the British government to create a local Department of the Interior to develop the hinterland region.[100] Government officials believed that the colony's future

Table 3.5 Value of gold, diamonds, and bauxite, 1920–1956

Year	Gold ($)	Diamonds ($)	Bauxite ($)
1920	200,938	961,278	136,364
1921	224,990	1,643,698	57,957
1922	179,070	3,859,357	—
1923	105,369	4,956,410	530,345
1924	98,803	4,097,137	781,736
1925	131,970	4,057,285	886,343
1926	98,263	3,300,952	920,682
1927	99,422	3,614,938	806,549
1928	76,149	2,518,836	839,615
1929	100,535	2,277,566	914,035
1930	89,581	1,431,364	598,080
1931	269,313	535,603	626,308
1932	381,594	541,337	498,850
1933	633,247	551,948	263,953
1934	675,296	483,485	318,137
1935	845,494	496,313	663,911
1936	926,525	487,724	1,719,793
1945	319,477	513,678	3,660,635
1946	354,995	1,279,241	16,062,798
1947	557,850	808,360	6,729,112
1948	529,981	325,024	19,514,827
1949	622,225	1,189,184	12,009,356
1950	531,302	1,363,377	13,832,442
1951	476,324	1,698,475	16,417,492
1952	885,085	1,477,634	22,240,571
1953	730,402	1,349,579	23,539,858
1954	916,903	1,198,636	23,234,607
1955	825,517	1,345,644	24,787,365
1956	396,847	1,334,720	29,332,800

Source: *British Guiana: Report of the Comptroller of Customs, 1924–1925*, 1932–1939; *British Guiana: Colonial Annual Reports*, 1945–1957; TNA: CO941/171–241/958, *Colonial Office List, British Guiana, 1956*, 63.

mining development depended on external interests as well as local, private, and foreign capital investments, and improved communications to and from the interior.[101] An estimated 150,000 persons of employment age resided in Guyana by the end of 1950, but only approximately 4 percent of this number worked in the entire mining industry. They contributed about one quarter of the colony's revenue; with bauxite alone accounting for 75 per cent in the category.[102] Quamina's statement that "the importance of bauxite workers to Guyana is out of all proportion to their number" is appropriate.[103]

In 1952, for example, the income from mining rose to as high as 30 percent of the colony's entire revenue; rising to an all-time high of 31 percent in 1962. In 1955, bauxite accounted for 90 percent of the country's mining revenue. At the end of the next decade, bauxite accounted for 97 percent of the entire value of mining including customs fees, corporate income taxes, export taxes on bauxite, royalties and lease rents, navigation lights, and harbor dues. In *Company Towns in the Caribbean: A Preliminary Analysis of Christianburg-Wismar-Mackenzie*, Cedric Grant stated that "in over 50 years, 1917–1969 . . . of operations the bauxite company exported over 60 million tons of bauxite . . . valued at over $1 billion."[104] At that time 6,640,518 acres of government lands were allocated to mining; with gold and diamonds utilizing 103,038 acres and bauxite using 4,096,370 acres.[105]

In early 1951, an estimated 2,686 persons were employed in the bauxite industry. At the end of that year, the industry employed about 3,120 persons or roughly half of those working in mining. Wages paid to bauxite workers were high when compared with other sectors. Unskilled workers' pay averaged 27 to 29 cents per hour while skilled and semi-skilled employees were paid 31 to 35 cents hourly. In 1956, a laborer in bauxite earned an average weekly wage of $23.52 while those in sugar or government employment were paid $12.48 and $15.12, respectively. Similarly, an electrician in bauxite earned $33.60 weekly while a comparatively skilled person in sugar or government employment was paid $14.50 and $23.52, respectively.[106]

From 1884 to 1951, the value of the country's gold output amounted to more than $61 million. The amount totaling over three and one quarter million ounces was produced as follows: one-fifteenth by dredging, roughly one-thirtieth by milling, and about one-seventieth by large scale hydraulic operations. The bulk, some 85 percent of the output, resulted from the hand-methods of alluvial or small-scale miners. In 1951, B. G. Consolidated Goldfields Limited received $2,600,000 through the Colonial Development Corporation in addition to the $960,000 granted in 1948. Similarly, the Kurupung Placers Ltd., a diamond seeking concern, was provisionally granted $133, 000 from the US Marshall Plan to develop mining and other enterprises in the colony's interior.[107]

By 1955, the aquatic recovery of diamonds from the Mazaruni and Potaro rivers accounted for most of the diamonds recovered locally. Crews of men, including the Wade brothers of Berbice County, recovered more than 2,600 carats of diamonds from the Mazaruni River alone. The next year, among the forty listed companies and investors granted permission to prospect for gold, diamonds, and other valuable ores was the business enterprise of Ivan Wade, a villager from West Coast Berbice. Wade was authorized to explore 600 acres, a portion of the Berbice River bed, for gems.[108] In 1955, approximately, 6,000 persons were employed in mining, colony-wide.[109]

Significantly, in 1952, jewelry, a new category, directly related to the gold and diamond industries, was added to the country's exports. Jewels accounted for revenue totaling $27,335, $13,934, and $12,787 for 1952, 1953, and 1954, respectively. Soon, however, the jewelry industry, which initially showed promise as a valuable export earner fell in step with the fluctuating trend of the gold and diamond industries. In 1955, the export value of jewelry fell to $6,950; just over half of the 1954 value.[110] Pervasive smuggling, a characteristic of gold and diamond mining, more than likely bedeviled the jewelry industry as well. However, government, private business, mining groups, and individuals never lost sight of finding El Dorado in Guyana. As the following chapter details, African diaspora labor migrants paid dearly in their quests for gold and diamonds.

CHAPTER 4

The Perils of Labor in Mining: Migration and Mortality

WORKERS IN THE PRECIOUS METAL AND GEM SECTORS of Guyana's mining industry, their families, and associates paid an enormous price in developing the country's economy. Decreasing coastal employment and an increasing labor demand in interior mining were the "push" and "pull" factors influencing a largely African diasporan migratory movement to the gold and diamond fields. El Dorado, the reputedly fabulous and golden city somewhere in the Guyanas fatally continued to lure many men and fewer women in quests to find gold. Earlier explorers in the Americas from the days of Sir Walter Raleigh in the sixteenth century to the prospectors of the nineteenth and twentieth centuries risked death to make the legend of El Dorado a reality. Mythical or otherwise, by the 1890s, the legend though generations old, was not time worn. During the more recent period, the seasonal acceleration of workers in gold and diamond mining was accompanied by alarmingly high mortality rates. In Guyana's cultural repertoire one of the best known songs, *Itanami*, succinctly reveals the innermost anguish of miners who contemplated aborting their quests and returning home empty-handed on first glimpsing that and other deadly waterfalls on hinterland rivers on their way to the fields.[1]

Miners crossed and sailed up and down many dangerous rivers and creeks hoping to find areas encrusted with the precious metal and gems. Journeys often entailed trekking for hundreds of miles along roads, paths, trails, and through forests and jungles. Miners were cautious and endeavored to avoid poisonous plants and wild animals that could curtail their efforts to make a living. Essentially, this pendulum migration, back and forth between coast and interior took place seasonally. In hindsight, it seemed as though the miners understood that they could not all become rich, but at least they would die trying. African diaspora miners, daunting as their travel and work were, sustained the precious metal and gems industries from a spectacular start in the late 1800s, through inauspicious decades to fabulous gold finds beyond the end of the twentieth century.[2]

Defying the perils of labor's hostile environmental settings, the prospectors migrated by traveling partly on dangerous rivers in crude vessels. Thousands of them as well as other interior workers such as boat captains and their assistants, camp attendants, and companies' clerks were injured or died in the mining regions. This tragic underside to prospecting for precious metals and gems included miners who died suddenly when vessels in which they were traveling capsized and all aboard were swallowed into the enormous and fatal bowels of the country's waterways.

Follow-on migration into mining was guaranteed by local villagers, urbanites, family members, and overseas workers from identical countries who elected to work for the same companies or employers, preferring to be together in the interior. Unfortunately, in many boating fatalities deceased persons often had similar origins or were family members.[3] Former miners often recalled compatriots who died by drowning and pondered on the sorrow of families who lost multiple members when they all went to watery graves in the same mishap. In a sensational1923 river fatality, twenty-one miners drowned. Four brothers from the Primo family of Buxton Village, East Coast Demerara were among the deceased. Macabre drowning incidents resulted in sobriquets such as "Buxton" and "Plaisance" graveyards that identify interior locations where many workers from those villages met untimely deaths.[4] Still, other miners traveled periodically between the coast and the interior for their entire working lives; thus giving this migration its pendulum characteristic. A few miners struck it rich, literally finding "streets paved with gold or studded with diamonds," while some workers returned home with afflictions such as "bush yaws" which bedeviled them for the rest of their lives. Miners had perilous and varying experiences, but a resilient core of them endured the occupational dangers of the work until old age.[5]

In nineteenth- and twentieth-century Guyana, high mortality rates plagued coastal agricultural workers as well, but they died near to their families and associates and in familiar circumstances. In the case of gold and diamond miners, their deaths were often unexpected and in places distant from their homes. Most wrenching of all—even if their families and friends were notified—it was usually long after the deaths and burials. In this context, some miners' deaths are memorable because of the sheer uniqueness of the calamity. Mrs. Blackie, a villager of Nabaclis, East Coast Demerara who had just completed burying a son at home one day, received news the next morning that her other son, a miner had died and was already buried in the interior.[6] Among the perils these miners and their associates faced were dangerous travel fraught with injuries, diseases, starvation, and deaths. Added jeopardy included calamities at work and unscrupulous employers. These aspects of migration, mining, and mortality, analyzed in the chapter, are integral parts of the collective histories of the country's working class.

What the travel and work entailed to both body and mind is recounted by James Culley who, with a group of prospectors, left his village and traveled to the entrepot mining town of Bartica. Culley, a native of Buxton-Friendship Village, East Coast Demerara, who prospected from the 1920s to the 1950s, recounted the risks of the journey. The retired prospector explained that "you had to hire a boat from Bartica and travel for twenty-eight days up the Essequibo River to the gold fields." He noted that many of the boats had no engines and members of the mining party took turns alongside the boat's crew in using wooden paddles or oars to propel vessels upriver. Culley observed that vegetation in the rivers caused impediments because they could entangle vessels and delay trips. Rain fell and soaked the miners and later, the sunshine dried them. Most disconcerting of all, were the hazardous rapids, cataracts, or waterfalls the travelers had to navigate. Boats were unloaded and supplies and the craft were hauled overland bypassing impediments and reloaded for ongoing journeys on the waterway. This procedure added to the parties' fatigue, caused delays, and sometimes diminished food supplies. The lengthy journeys also entailed camping at night along the riverside.[7] After danger-plagued weeks of travel, on arrival at their mining sites, under varying conditions including bad weather, ill health, rationed food and other supplies, lack of female company, and the absence of family members, workers prospected for at least six months.[8] As absentee parents or spouses and despite daunting situations, the workers understood they had to grapple with the realities of the times and environments in order to make a living, and to take care of themselves and their families.[9]

The realities of the times and environments in which the workers lived are captured by colonial officials, private investors, business persons, and others who were part of that 1890 commission, convened at the governor's behest to foster the country's economic prosperity. Members of this commission and subsequent ones viewed labor as indispensable to prosperity. Yet the inordinate focus on profits impaired the commissions members' understanding of the horrible price injured and dead workers as well as their families and associates paid for the gains from mining. Concerns aired on the declining workforce failed to conceptualize the ricocheting effects of high mortality rates that decimated the labor supplies.[10] Seven years later, in 1897, after similar intervening commissions in 1892 and 1895, for example, questions were again raised that made the connection between insufficient labor for economic development and casualties of workers. Causes contributing to economic crises and labor shortages, and finding solutions for the occupational fatalities received a modicum of attention because of the colonial hierarchy's cavalier attitudes to the workers and their leaders.[11] Consequently, incapacitations and deaths in the gold and diamond regions imperiled lives and labor supplies for contemporary and prospective economic activities.

In a perverse way, Death Registers for the mining regions corroborate *Census Reports* and *Reports of the Institute of Mines* on the presence of African diaspora miners in the country. The registers reveal the national and international identities of thousands of persons who suffered and died while traveling or prospecting in the regions. In the midst of this carnage, the power structure focused on piecemeal infrastructure development, mostly to facilitate increased profits, but hardly to ensure workers' well-being. High mortality rates had a direct impact on the labor supply, but apparently, once policies were implemented to foster profitable returns, what befell the producers of the profits were not the concerns of the profiteers. The consensus of the colonial hierarchy was that immigration would remedy all ills associated with labor and mining.[12] Meanwhile, frequent deaths stalked workers in the precious metal and gems industries.

WATERY GRAVES ET AL.

Astounding revelations on injured, sick, and deceased workers in the gold and diamond regions of Essequibo County verified the enormous and ultimate price paid by those who developed and diversified Guyana's economy. Death Registers recorded drowning as the main cause of deaths. Malaria, Black Water Fever, Leprosy, heart and liver diseases were frequently recorded. Deaths that were also due to sheer exhaustion, malnutrition, starvation, and inadequate health care may have been preventable. However, when workers died after being buried alive by fallen trees, stones, or earth, or from the attacks of insects, mammals, and reptiles; the state of mutilated bodies indicated the excruciating pain they suffered. Sometimes, the cause of death was unknown and in many instances accidents or occurrences of diarrhea or dysentery accounted for the high mortality among miners. Untimely deaths occurred when sick miners succumbed to their illness on long journeys to riversides in efforts to obtain further transportation to medical personnel or institutions.[13] For decades, medical assistance, especially for serious illnesses or injuries, which required hospitalization was only available at far-flung interior facilities. Action to address the interminable delays, which caused unwarranted deaths of persons in the mining regions, and efforts to improve workers' health and safety were contemplated only after the airing of stupendous calamities in the media. Although periodic attempts to improve the rudimentary health care and medical services continued, facilities were only provided in disparate areas across the mining regions. Lack of timely access to appropriate services routinely or in emergencies added to fatalities among mining workers. Hinterland dispensers often recounted death-defying medical and surgical procedures they performed in remote areas of the mining regions under dangerous septic conditions.[14]

From mining's inception in the 1880s, death rates in the prospecting districts due to chronic diseases, sudden illnesses, various types of accidents, and boating tragedies had been escalating. No solution to the ongoing tragic situation seemed forthcoming and the decade of the 1890s, was physically and mentally disastrous for workers. From 1891 to 1895, a series of deadly catastrophes befell many of them in various mining districts of Essequibo County. The case of the disastrous incident at Aracaboosie (Arakabusa) Fall, ten miles up the Cuyuni River is very revealing.[15] On May 29, 1891, eleven men including Gussey Moore, an African Guyanese, and a Barbadian of similar ethnicity died by drowning in the vicinity of the waterfall when their boat capsized. According to *Death Register, 1891 Essequibo*, the men's ages ranged from eighteen to forty-nine years and they were all gold digging laborers. The previous day, mining workers of similar ancestry and occupations, two locals, Thomas George and William Primo, a Barbadian, and a French Guianese, Thomas Laurence drowned at Tabaneiro Fall, Essequibo. These workers' ages ranged from twenty-five to thirty-five years.[16]

Of 270 deaths recorded in 1891, a total of 157 African Guyanese comprised the majority of the deceased persons. Forty dead persons came from the Caribbean Islands. There were two African-born including forty-six year old Zambia; two persons each from French Guiana and Suriname, and one African American. East Indians and Amerindians also died. Miners were not the only ones who died, but they predominated, especially in the drowning category. In 1891, fifty-one persons allied with the gold industry died by drowning on thirteen occasions in four of Essequibo County's mining regions alone. Other causes of death recorded were "unknown," diarrhea, dysentery, fevers, gunshot wounds, falling objects, poisoning, and tuberculosis[17] (See Table 4.1).

As 1892 ended, the escalating accidental deaths of miners were the focus of a December 12 article in *The Gold Mining Gazette.* Aimed at drawing government officials' "attention to the dangers which travelers to the gold fields are exposed to in the rivers," the article emphasized that "the daily loss of human life and property" was "a sad thing" because "gold diggers" are "so neglected and it is to them today the colony owes a great portion of its prosperity."[18] Still, deaths in the gold fields increased the following year with approximately, 370 deaths recorded in four of Essequibo County's gold producing districts. The deceased were local and foreign-born miners from the Caribbean islands, French and Dutch Guiana, Africa, and the United States as well as some other persons who had worked in related occupations. The gravest drowning tragedy of 1893, occurred on July 17 when eighteen miners died in a single mishap. The casualties were between the ages of twenty to forty years old.[19]

In 1894, of the 173 miners who died in the abovementioned mining districts, ninety-eight of them were locals. The other dead persons included Antiguans, Vincentians, Jamaicans, and Surinamese. Deceased persons from Martinique, the US, French Guiana, and Africa were among the 1894 group.[20] The following year, approximately 282 persons who worked in gold mining and related occupations died in the districts. Similar to previous years, the majority of the deceased were African Guyanese with many dead Pan-Caribbean persons in varying numbers. April 17, 1895, was a particularly tragic day. Twenty-two persons traveling on the

Table 4.1 Selected deaths in the gold mining regions, 1891–1895

Date	Place	Cause	Occupation	#S	Origins
14/10/1891	Taminoda Fl	Drowning	Gold Digger	6	Guyana
17/11/ 1891	Goeway Fl	"	"	7	and Barbados
"	"	"	Boat Hand	1	Guyana
19 /06/1893	SmitGat Fl	"	Gold Digger	7	Various
"	"	"	"	5	Guyana
17 /07/1893	Taminoda Fl	"	Clerk	1	"
"	"	"	Gold Digger	18	Various
17/08/1893	"	"	Various	11	"
28/06/1894	Carawabo Fl	"	Gold Digger	4	Guyana
"	"	"	Boat Hand	1	"
27 /08/ 1894	Sakarara Pt.	"	Gold Digger	6	and Barbados
24 /12/ 1894	Tumatumari Fl	"	"	10	Various
"	"	"	Laborer	2	Guyana
17/04/1895	Tanamoo Fl.	"	Tributor	4	Various
"	"	"	Gold Digger	18	Various
Fall=Fl.	Point = Pt.				

Source: *Death Register*, Registration Districts, 3, 4, 16, 17 Essequibo, 1891–1895.

Cuyuni River in the vicinity of Tanamoo Fall drowned when their vessel capsized. The dead persons, ranging from seventeen to thirty-eight years old, were mostly locals including an East Indian. One Scotsman and a Barbadian also drowned.[21]

Mining workers continued to die under equally harrowing circumstances elsewhere. Some miners died in hospitals, on the way to their homes, or at home from causes directly related to occupations in the gold and diamond fields. Those deaths, it must be noted, are not recorded with those in the mining regions. Family and community members pointed out that some relatives and friends who mined for gold and diamonds, returned home, and sometimes died from work-related illnesses. Nonetheless, deaths recorded in the mining regions confirm the overwhelming presence of African diaspora labor migrants—locals, Caribbeans, and others who bore the brunt of the tragedies associated with establishing Guyana's precious metal and gems industries.

The ongoing recruitment of labor migrants from the 1890s to the 1940s, parallel to prevailing conditions that imperiled their lives can only be viewed as the pursuit of profits over people's well-being. In the 1890s, Dr. Ozzard, government medical officer at Bartica Hospital, noted that on arrival at the institution many gold miners were in a most deplorable condition and even starving. Others suffered from malaria fever and symptoms of liver and kidney diseases, dysentery, and pneumonia. Dr. Ozzard emphasized that often the food provided at the hospital to sick gold workers was unsuitable and the mining regions lacked fresh food and vegetables. Contending that the length of patients' travel added to their worsened state of arrival at the hospital, Dr. Ozzard referred to a patient with a fractured spine whose five-day journey in an open boat included the navigation of several falls.[22] By September 14, 1893, thousands of outpatients who worked in mining and other related occupations had been treated at Morawhanna Hospital, North West District and Baramani Hospital, Essequibo County. Accommodations at both hospitals were woefully inadequate to admit the large number of patients from all over the mining regions who required hospitalization. Although by October 1896, a health center and dispensary operated at Arakaka, North West District to facilitate patients from gold mining sites in proximity, Dr. Ozzard indicated one year later that the mortality rate remained high.[23]

At the hearings of the *West India Royal Commission* in 1897, Dr. John Monteith Rohlehr pleaded for government intervention in providing timely and adequate healthcare services to decrease the high mortality in the gold fields. Dr. Rohlehr produced information on the high yields of gold, and called for agricultural and other provisions to facilitate the settlement of families in the interior. He further wrote to the secretary of state for the colonies on the same issue, and identified ways in which the colony's healthcare services could be revamped. His efforts were treated with ridicule and ignored by the colonial administrators. Until his death in 1922, Dr. Rohlehr indefatigably fought for the improvement of the socioeconomic conditions of the Guyanese working class against the obdurate, inflexible, and profit-oriented capitalists hierarchy. That is why in 1906, over 6,000 persons petitioned the Colonial Office protesting the neglect of the recommendations of the 1897 *Royal Commission*. Meanwhile, workers had to contend with the insensitive attitudes of managers of mining companies. For instance, numerous incidences occurred of sick workers, for whom no arrangements had been made to house or to treat them. Many cases arose of ill and malnourished mining workers stranded at various locations trying to procure transportation to receive medical attention. Sometimes, they were forced back to work—during convalescence—in order to complete their contracts.[24]

As late as 1937, accommodation at the Public Hospital Bartica was woefully inadequate. With an increasing population due to timber and mining activities, the institution could not cater to the large number of persons seeking treatment. Everyday sick persons were left unattended, and others not quite better were discharged from the facility to make room for new admissions. Dire and dangerous situations worsened due to outbreaks of epidemics. Improvised beds were often jammed against each other in the wards while open verandahs were utilized to provide additional accommodations. Patients were exposed to rain, and at night some of them slept on benches or on the floor.[25]

The lives of miners were also imperiled by irresponsible employers on whom they relied and had contracted for regular supplies of food and transportation to and from work- sites. Sometimes, the contractual terms were violated by employers and essential commodities were not forthcoming. The only solution for affected workers in the remote areas of the country was to use their wits and rely on the mercy of nature to survive. Paradoxically, the perils of labor and the environment are intertwined with mortality as the calamities of Henry Belot, the foreman of Samuel Parris, James Lynch, Samuel Robinson, Peter Norton, and James and Ernest Williams illustrated. On June 30, 1890, when Belot first appeared in *The Gold Mining Gazette*, he was registering as a gold digging laborer at the Institute of Mines to work for Cartey, Jacelan, and Childebert Company. Seven years later, in July 1897, he was in court as a plaintiff in a case against Theophilus Hatile and A. J. D. Amil.[26]

Allegedly, in late January 1897, Belot and the above named men contracted to work in the mining operations of the defendants for four months. The terms of their contract stipulated that they would be provided with food supplies and conveyance to and from the mining site. On April 15, their food supplies ran out. The miners had no transportation to convey them from the site where, if they remained, they were likely to die from starvation. With no other recourse, Belot and the men decided to seek help. Along the way, their compatriot Peter Norton, weakened from starvation had tarried on the road. He never made the journey to the riverside. Norton was presumed dead since nothing was ever heard of him again. After wandering in the forests for days and surviving on wild plants, the other miners made a raft. While attempting to board the vessel, it capsized and Ernest Williams drowned. The survivors sailed to the mouth of the Potaro River where they boarded a boat for Tumatumari, a major township.[27]

The perils that workers faced in gold mining included accidents caused by falling trees, earth, and rocks. The mishaps were often caused when deep diggings left large boulders of rock exposed without sufficient support. The loosened missiles fell on laborers working below, crushing them to death.[28] John Sampson, an employee of Winter Gold Mines Ltd., died on the February 13, 1899, at Barima when falling earth crushed him. Soon after, James Pitt, an employee of H. Bascom died accidentally when the branch of a tree fell and killed him.[29] Alonzo F. Chase, a prospector employed in Barima died of suffocation because earth fell on him on June 19, 1899. Similarly, Samuel Rollox, another miner died at Canawarook on September 19 of the same year from being struck by falling earth. Michael Henry, a crew member of Emanuel Thomas's mining group, was killed on September 28, 1899, by a falling tree. Abraham Newland died on October 2 of the same year under circumstances similar to those of the miner Henry at Barama. Newland was employed by Chalmers and Cox Ltd (See Table 4.2). In myriad ways, death continued to stalk miners, for example, on May 27, 1939, Joseph Ferdinand, a St. Lucian East Indian stood on a bushmaster snake which bit him three times in rapid succession. As in many similar incidences, delay in obtaining medical assistance, caused Ferdinand's death.[30]

Usually, all deceased miners and their associates in the interior were buried there. In a sense, death certificates could eventually provide some closure for family members and other interested parties. However, the records remain silent on the demise of many of the workers whose deaths remain unrecorded. Surviving "old-time" miners recalled instances of seeing dead bodies floating on interior rivers. If their lives were not endangered, on encountering a corpse, the passersby would try to secure and to bury it. Otherwise, the unidentified body floated away to a watery grave and eternity.[31] *The Gold Mining Gazette*, January 21, 1897, concurred that there were instances of "deaths in the gold districts not being reported to the wardens in compliance with Sections 101 and 102 of the *Mining Regulations of 1896*."[32] By 1900, when the output of diamonds was officially recorded, a total of 104 miners had died in the four mining regions. Among the dead, locals numbered seventy-seven, while some of the others originated from Dominica, Barbados, Nevis, Suriname, French Guiana, and Venezuela. Several deceased persons including a Krooman were African-born. Multiple drowning deaths remained prevalent in the succeeding decades. In 1922, a vessel carrying twenty-four prospective workers to their mining site up the Mazaruni River was caught in a whirlpool and sank. Five of the men drowned. As Christmas 1935 approached, thirteen miners from Potaro and Mazaruni goldfields on their way home aboard the *Tacouba Express* drowned when it capsized off Parika, Essequibo River. No bodies were found and it was

Table 4.2 Deaths in the goldfields, 1894–1901

Year	Natural Cause	Drowned	Unknown	Accident	Lost
1894–95	87	15	37	3	1
1895–96	93	17	56	10	4
1896–97	69	10	38	1	—
1897–98	71	20	47	9	1
1898–99	77	6	25	2	1
1899–1900	98	9	28	4	—
1900–1901	65	27	11	6	—

Source: *British Guiana: Report of the Commissioner of Mines*, 1895–1901.

presumed they had been washed into the Atlantic Ocean by an ebb tide. Inevitably, before the decade ended, and as the presence of women increased in the gold and diamond fields, they were numbered among drowning casualties. Among the five deaths in a 1937 Mazaruni boat disaster at Mora Rapids, was a twenty-seven year old woman Iris Jones[33] On May 14, 1956, as mining compatriots, Sinclair Semple of Litchfield Village, West Coast Berbice, James Thomas, *alias* Sailor, and George Bovell of Campbellville, Greater Georgetown were on a mercy mission taking their injured colleague, Ivan Fowler, to hospital by boat, the vessel unexpectedly sank and Semple and Thomas drowned.[34]

SEQUELS TO TRAGEDIES

Computing mortality statistics in a vacuum is hardly meaningful if the results are not placed in the contexts of social and economic realities. In this case, the pervasive casualties, characteristics of mining life, and associated occupations adversely impacted survivors. The deaths affected families who had lost breadwinners, household-heads, spouses, fathers, brothers, and sons. The demise of mining workers added social and economic costs to families who experienced dire circumstances and serious consequences related to tragedies connected with interior occupations. Quests to wrest precious metal and gems from the jaws of Guyana's lengthy rivers and the bowels of its vast land expanse had deadly and unfortunate socioeconomic results. Workers and their advocates recognized the fundamental causes of their socioeconomic plights and offered solutions to the seemingly endemic problems.

By 1897, economic conditions in the British Caribbean colonies had deteriorated to such an extent that the British government convened the said *West India Royal Commission*. During the hearings in the then British Guiana, 183 working class persons had their signed petition presented. It outlined the steady decline in employment opportunities and the effects of rising taxes and cost of living expenses due to falling sugar prices. The petitioners and those they represented expressed a willingness to work in the gold fields because of employment opportunities, but noted that the environmental conditions caused too much suffering and deaths, especially from drowning. They stated that untimely departures had left "many homes destitute, many women were widows, children became fatherless, and thereby pauperism increased in the land."[35] This graphic information did little to alleviate the miseries of labor because as the twentieth century evolved mortality rates in the gold and diamond mining regions remained at high levels. Economic losses and social costs resulting from diminished and insufficient financial resources for family members occurred when miners and household-heads died in the gold and diamond fields. The tragedies severely imperiled the lives of spouses and offspring and had long lasting effects on dependents. Like tentacles, interior mishaps stretched deeply into coastal villages, urban areas, and in incomprehensible ways, the perils of labor in mining were intertwined with the dynamics and intricacies of African diaspora migration.

Such was the case of Barbadian-born, Duncan Fitzgerald Alfred King who migrated with his parents and only sibling to Guyana during 1872. In 1896, King entered the United States as a student and graduated from Howard University

Medical School in 1900. Two years later, his younger brother, a gold and diamond miner who had supported their widowed mother drowned in Guyana's interior. Unfortunately, the realities of the times in which he lived circumscribed Dr. King from adequately fulfilling his filial responsibilities. In the United States, African-descent professionals including healthcare personnel were discriminated against and often could not obtain employment for which they were qualified. As more professionally trained African-descended persons emerged, regulations were altered in the British colony to restrict or to prevent their employment. Dr. King who desired to return home and take care of his mother applied for a position at a local government medical institution, but he was denied employment. The terse reply from the Colonial Office to his application stated: "The laws of British Guiana require registration in the UK as a preliminary to registration to practice medicine or surgery in the colony. We have declined to accept the degrees of Howard University on the advice of the British Medical Association."[36] The widow and mother remained destitute.

Evelyn Angelina Wright-Toussaint was born October 2, 1910, at Paradise Village, East Coast Demerara, but never knew her father Alexander Wright, a Barbadian diamond-seeker. Wright had contracted to work for Myers and Mayers, a mining company operating in the Omai area, but died by drowning on March 29, 1913. Evelyn Toussaint knew that her father had died at Konawarook, Upper Essequibo River. He was interred far away from his family who resided in Demerara County and never had the opportunity to see him and bid him farewell. In 1996, in Mocha Village, East Bank Demerara, this writer presented the eighty-six year old widow, Evelyn Toussaint with the evidence of the cause of her father's absence from his family's lives; she cried and recalled the bitter experience.[37] Growing up fatherless and in a female-headed household was not a pleasant experience for Evelyn Wright and her three older siblings because of their mother's inability to fend adequately for all of them on many occasions. In the challenging times of the World War I era, they often had to depend on the mercy and goodwill of relatives for their survival, which was better than being on the streets or in the Alms House.[38]

Sequels of boating tragedies haunted family members in unexpected ways especially through haphazard and incomplete recordings of miner's deaths. On numerous occasions, in the absence of autopsies the causes of death remained inconclusive and the words "Not Medically Certified" and "No Medical Attendant" were often included in the records. As a result, the inheritors of miners' estates who died under tragic circumstances in the interior could encounter unexpected difficulties in collecting intended death benefits on the coast. Robert Conzius, a forty-two year old diamond-seeker drowned at Mary's Fall, Mazaruni River on February 21, 1933, along with six other men en route from Bartica to the gold and diamond fields on board the vessel, *Phantom Flyer.* The record of Conzius' death included the annotation "Not Medically Certified."[39] Unfortunately, for Cecelia Nurse, the reputed wife of Conzius, his body was never recovered. The Demerara Life Insurance Company had insured Conzius' life for $1,000, and opportunely used the phase "Not Medically Certified" to deny the payment to Nurse, the benefactress. She, therefore, then claimed the amount in court.[40] Although the outcome of the case remained unclear, it is evident that miners' dependents faced myriad uncertainties when their providers died unexpectedly. A peril of labor Conzius never envisioned, stalking his survivor, was that his partner would have to wage a costly and time consuming court battle to claim the benefits he had bequeathed to her. Indeed, at no point were miners' destinations bejeweled that is "paved with gold or studded with diamonds." The sequels to migration and mortality have antecedents and legacies.

The antecedents to the migratory movement and mortality are wedded to the travails of the working class in nineteenth-century Guyana. Survival for African Guyanese and their diaspora counterparts meant constantly struggling to ensure they were adequately employed or had the economic resources to take care of themselves and families. Therefore, in an insidious way, the high mortality rates among miners caused by the precariousness associated with their work were intricately linked to the workers' efforts to improve the quality of their lives and those of family members. Norman E. Cameron's, *The Evolution of the Negro* (1929) and Kean Gibson's, *The Cycle of Racial Oppression in Guyana* (2002) capture succinctly the lot of these long suffering people in the country. But while Gibson focused on their victimization, travails, and tribulations, Cameron investigated the multifaceted ways in which community members grappled with an array of challenges; not always unsuccessfully. He identified a number of social, political, and economic activities they undertook to better their situations. The two studies, published almost three quarters of a century apart, enable an understanding of why gold and diamond miners literally "walked between the raindrops" so long and so far to make a living.[41]

By the late nineteenth century, most people of African ancestry in the colony understood that interior gold and diamond mining and coastal agricultural work, the two major employment avenues, were both arduous undertakings. Motivations to mine included the possibility of more lucrative earnings compared with wages paid in the sugar industry and relatively fewer direct micromanagers. More opportunities existed for miners to prospect independently in their self-financed groups or to branch out as entrepreneurs. Some miners established economic linkages in varied enterprises and thereby became employers instead of remaining as employees. Miners, many of whom owned farmlands at the rear of villages, had opportunities to engage in agricultural pursuits at home when it was too rainy to prospect.[42]

Money accrued from profitable mining ventures enabled recipients to acquire financial resources, purchase real estate, and engage in other economic activities. The quality of some miners' lives improved, as shown in chapters 9 and 10, and the quest for gold and diamonds took on a life of its own in the socially and economically depressed times of the late nineteenth century. The majority of workers operated in situations where Europeans controlled the country's economic resources and activities. Contemporary literature indicates that similar to their counterparts in South Africa, gold and diamond miners labored under abject conditions and died in alarming numbers.[43] The perils of labor, which gold and diamond miners faced, extended into the twentieth century before better knowledge of the locations of falls, rocks, and other maritime obstacles in the waterways. Sturdier crafts, air, and improved land transportation minimized the dangers of river travel. Still, the legacies of migration and mining included workers whose lives were imperiled through incurable diseases contracted on the job.

Economic and Environmental Realities of a Mining Life

In 1939, twenty-seven year old Octavius Barnwell of Belle Vue Village, West Coast Berbice County joined his older brother's mining party to prospect for gold and diamonds. Numerous trips later and after twenty-two years of regularly commuting to and from the interior, Barnwell had to give up his mining life in 1961, because he was afflicted with "bush-yaws" and was permanently disfigured.[44] To the best of knowledge, "bush-yaws" is "a dreaded infection known to Guyanese for decades." The disease is, reputedly, transmitted "by a fly which inhabits the jungle." When a person is bitten by the insect a virus infects "the blood stream and spreads." Soon, a "tiny worm from the fly multiplies…and begins to feed on the red corpuscles," lesions then appear on the skin. No prophylactic precaution exists for the infection and all known methods of therapeutic treatment are palliative. Medical practitioners remain baffled about the condition.[45]

Barnwell was attracted to mining because of his dissatisfaction with agricultural work at nearby Bath Estate. He often slept overnight in cane or rice fields so that he could start working very early the next morning to boost his wages.[46] Plantation Bath in Berbice County is fifty-eight miles from the capital city, Georgetown.[47] By 1881, "Bath comprised of about six abandoned cotton estates and a former sugar estate" Later, although rice was grown at Bath, the holdings were amalgamated into a mostly sugar growing entity with a large number of indentured workers and coastal residents.[48] This partially explains why Barnwell's work on the plantation was so arduous, exploitative, and poorly paid. Colonial officials and planters had previously deliberately recruited excess workers, which created a "reserve army of labor" to keep wages low. This aspect of working class life fostered situations of grinding poverty and perpetual deprivations. As late as 1939, the time when Barnwell first headed to the gold and diamond fields, many aspects of work in this colonial society retained the vestiges of enslavement; an institution that had ended one century earlier. As a result, it did not take much urging to prompt Barnwell to leave the coast and agriculture to work in interior mining at a prospecting site near to Issano Landing where some frontier conditions persisted.[49]

Mining groups of the 1940s and 1950s, typified small-scale operations that initially began in the mid-1890s. By 1895, restrictive land policies were amended and requests for mining claims and prospecting licenses by small-scale mining enterprises increased. Until that year, the majority of African-descended gold and diamond mining workers were employees of mostly European, American, and Chinese claim-holders. The employers were usually business persons, investors, or owners of private companies. Prior to their participation at the higher level, those of African ancestry were mostly constrained from participating independently in mining activities because their financial circumstances precluded them from becoming investors or claim-holders. Enactment of the new legislation not only eased certain restrictions but also served to facilitate inroads by entrepreneurial-minded employees who become employers.[50] The opportunity to move from grueling labor status as employees to a level where initiatives and expertise meant more benefits was a major incentive for their economic mobility.

Enduring unpredictable and potentially hazardous environmental situations probably caused some employees to self-evaluate their worth. Mining workers possibly questioned the value of making sacrifices to accrue profits mostly for others. To the self-financed operators, it seemed more sensible to mobilize relatives, friends, and acquaintances to pool resources, and form their own mining parties. The emergence of these small-scale self-financed operators is an important indication of community members' ongoing efforts and strivings for improvements despite the daunting circumstances in which they functioned.[51] In a sense, both the pains and gains from gold and diamond mining accrued to them as employers-owners; rather than being only employees providing most of the benefits for others and enduring all the sufferings.

Working in gold and diamonds engendered a long-standing practice—the reliance of interior workers on the services of the post office to remit money to their families on the coast. For the most part, this was a safe exercises, but danger lurked. *The Sunday Chronicle* of March 23, 1930, reported the arrest of a village letter carrier who was alleged to have broached a package containing a Cash Order. The postal worker who had been attached to the Belfield Post Office, East Coast Demerara was arrested and charged following a police investigation into a series of reported "thefts of cash orders enclosed in letters addressed to villagers from their relatives in the interior diamond fields."[52]

Conflict among the miners themselves, on account of the destructive behavior of some individuals, added to the perils of work in the hinterland. The case of two longtime partners, Gustavus Moore of Den Amstel Village, West Coast Demerara and Randolph Griffith demonstrated that even close friends could have bloody disagreements while seeking for gold and diamonds. During December 1930, Moore was seriously wounded in the Oranapai diamond-mining district, Mazaruni and ended up in the Public Hospital. The assailant, Griffith, used a prospecting knife on Moore who was armed with a cutlass. Griffith inflicted severe wounds on Moore's left hand, nearly severing it from his shoulder. The two had quarreled and fought over the sharing of diamonds they had found. Although the outcome of the case is unknown, Griffith was arrested, charged with wounding, and placed on "$500 bail. A preliminary hearing of the case at the Kumakusa Court was set for January 6, 1931.[53] Unfortunately, the vicissitudes of a mining life could be absolutely overwhelming as the case of a diamond-seeker stranded at entrepot Bartica exemplified. He was robbed of all he possessed and hanged himself in desperation.[54]

CONCLUSION

African diaspora labor migrants loom large in the development of colonial Guyana's mining industry. Many of the miners are only statistics collected by various agencies or included in the annals of the companies for which they worked. Deceased miners live forever in the memories of families and friends who received little or no information of their deaths and were left in a quandary for the rest of their lives. Although the number of deaths recorded in the gold and diamond mining regions by affiliated entities hardly synchronized, without seriously considering mortality statistics, the colonial hierarchy was unable to resolve the perennial labor need; and costly labor migration continued. In part, this state of affairs account for their inadequacy in addressing infrastructure and healthcare issues that caused the deaths of so many workers. Miners and their leaders consistently pointed out ongoing dangers associated with the work and tragic events ensuing. The high death rates in the interior were in plain view, but for too long the problem was halfheartedly addressed.

Spectacular gold discoveries and impressive finds of gems failed to persuade colonial officials to act with alacrity to preserve the life and limbs of a critical factor of production—labor. Predictions indicated sustained recoveries of the commodities alongside commensurate development works, but colonial officials and multinational mining investors, alarmed at fluctuating profits from gold and diamonds, attributed the scarcity to inadequate labor supplies while conditions causing casualties were continuously overlooked. Workers literally fended for themselves eschewing uncertain coastal employment, conflict-driven relations, and arduous plantation labor to travel on the murky, dank, and deep waters of Guyana's rivers to mine for gold and diamonds.

The tradition of grappling with the challenges of institutional tyranny, exemplified by workers' struggles against the sugar hierarchy, and African Guyanese efforts to establish independent gold and diamond mining enterprises were ongoing. Confronted with precarious situations, the independent groups had to formulate strategies to obtain maximum rewards from their labor. At the organizing and developmental end of the industry itself, maximum profits rested on colonial officials and the mineocracy formulating an adequate infrastructure plan and program to expand mining. This was imperative at the onset of the mid-1880s migratory movement to the interior, but action in that direction was incremental, as the following chapter details. The perils of labor miners encountered tarnished the luster of El Dorado in Guyana. When bauxite emerged, many workers abandoned these two sectors in favor of the new one. The shift in the direction of migratory movement by these workers provided the panacea for a partial solution to their untimely demise. Before then, as the following chapter indicates, colonial policy makers viewed incremental infrastructure works as key to develop mining and other enterprises.

CHAPTER 5

Aspects of Infrastructure Development: Gold and Diamonds

In 1880, the inaugural enactment of Mining Regulations officially signaled gold production as a new phase of economic activity in Guyana. At this early stage, there was no indication of the uneven development of the mining industry that would follow. The emerging mineocracy failed to anticipate that the plantocracy, who were influential in the local legislature, would view mining as a competitor for labor and limit funds for infrastructure development of the mineral-rich interior regions. Investments by local and foreign capitalists as well as the British government failed to materialize in sufficient amounts to expand mining and interior development.[1] Cost-minimizing policies of bureaucrats and investors limited funds to address systematically the dangers, inherent in traveling to and from the interior and substandard conditions, which caused the untimely deaths of workers. This chapter points out that despite the shortcomings in infrastructure work in the gold and diamond producing regions, the persistence of African diaspora labor migrants—individually, as independent prospecting groups, or as employees of larger companies—contributed to the longevity of the mining industry.

From 1886, Mining Regulations and Ordinances pertaining to the production of gold, diamonds, and other ores were amended periodically. As the decade of the 1880s ended, local administrators, those in Great Britain, and other economically interested parties slowly implemented measures to expand mining and to influence profitable outcomes. *The Mining Regulations of 1892* facilitated more access to government lands and increased gold output. Within two years, a government commission was convened to find ways to develop gold mining further. Among the recommendations of that body was the upgrading of infrastructure work, especially in transportation to make access and egress to and from the goldfields easier.[2]

Since 1886, the royalty paid on gold had been earmarked for interior development. Yet the findings of the said commission again identified transportation, especially the high cost of travel, as a major problem. Mining Regulations enacted in 1896 confirmed that the long-delayed action to address transportation had contributed to the less-than-expected gold yields and hindered the industry's expansion. In the twentieth century, interior population settlement schemes and commensurate infrastructure development to expand agriculture, forestry, and mining were reemphasized. However, infrastructure developments were piecemeal, selective, and often in disparate and distant locations from workers' living or work sites to be of much use, especially in cases of emergencies.[3]

Fifty-three years later, in January 1939, on the eve of the visit of another Royal Commission investigating an economic crisis in the British Caribbean, eminent Buxtonian and mayor of New Amsterdam, Joseph Eleazar, a Member of the Legislative Council, aired an interesting expose. The contents, slated for presentation to the *West India Royal Commission, 1938–39*, succinctly characterized the colonial government's policy as one that had always supported capitalists who wanted to make a fortune and leave the colony whilst the inhabitants bore the brunt of increasing taxes. Honorable Eleazar emphasized that since emancipation over a century ago, the working class had shown an interest in order, progress, and development, but government policies had remained unchanged. He identified several proposals to develop the country which had encountered government's overt and covert resistance. One significant example Eleazar noted was the 1897 proposal to construct an interior railroad to link the colony's regions.[4]

The proposal was blocked because government sided with capitalist planters who anticipated that labor would be attracted to the hinterland, create prosperity there, and influence agitation for increased wages on the coast. Hon. Eleazar felt justified in claiming that government geared its policies "to assist the capitalists as in the days of slavery and not to develop the country for the colonists."[5] The legislator disclosed that he had uncovered evidence that proved that certain local bureaucrats including the Director of Agriculture had misrepresented information to officials at the colonial office. Eleazar noted that his proposal to reintroduce the construction of an interior railroad was undermined by the collusion of legislative bureaucrats and private interests, although they had previously supported it.[6] Hon. Eleazar was the latest voice traditionally raised from post-emancipation times through the advent of the African Association of British Guiana, the People's Association, and Dr. J. M. Rohlehr questioning colonial administrators' management and policies.[7] Ann Spackman's "The Role of Private Companies in the Politics of Empire: A Case Study of Bauxite and Diamond Companies in Guyana in the Early 1920s" and George Lanning and Marti Mueller in *Africa Undermined: A History of Mining Companies and the Underdevelopment of Africa* pinpoint the activities of local business personnel, their legislative supporters, and overseas based operatives who controlled the economic direction of the colony. They engineered bureaucratic delays that led to the slow development of interior land and infrastructure works.[8]

Aspects of the consequences of the pace of infrastructure development of the interior regions including land, marine, air transportation, and other communication systems are focuses of this chapter. Buildings—commercial, public, health care, housing, religious, entertainment, and security—in the mineral-rich regions slowly emerged during the late nineteenth and first half of the twentieth centuries.

The long drawn-out developments gave credence to the belief that an accelerated plan and program focusing on interior infrastructure work would have increased production and profits in a wide array of economic enterprises.[9] With the cost-minimizing policies of private investors and capitalist colonial governments as backdrops, the chapter discusses the local environmental context with which mining workers wrestled.[10]

As the nineteenth century ended, it was shown in Chapter 3 that some British and local bureaucrats advocated utilizing private and public funds and offering incentives to attract large capitalist companies to expand interior mining and economic development.[11] It turned out that most of the capitalists were uninterested in long-term investment in interior infrastructure. Mining ventures were systematically speculative in orientation and focused narrowly only on exploration and exploitation to gain maximum and quick profits on limited investments. Local legislators insisted on allocating a modicum of financial resources in selective interior works, and only if benefits from proposed undertakings were discernible. Funds were allocated mostly to construct specific roads or buildings and to clear selective sections of a few rivers to afford accessibility to particular areas of the interior. The endeavors, often undertaken in virgin territory, failed to address adequately health, housing, and safety concerns; vital to the well-being of a workforce.

Mining Workers and the Frontier

Increasing gold production in interior Guyana fueled largely by the migratory movement and work of African diaspora miners raised the possibility of a countrywide labor shortage that would hinder the expansion of other economic activities.[12] Renewed focus on recruiting persons of African ancestry was directly related to increasing profits from gold mining and the expectancy that incoming workers would labor as arduously as their counterparts already in the colony. As colonial administrators, agriculturalists, mining investors, and business persons contemplated the country's economic direction, they failed to recognize that the notion of a labor shortage had much to do with their managerial inefficiencies and lack of initiatives. Allegedly, their own incompetence, racism, hostilities, and insensitivity to workers had created the labor shortage.[13]

In the nineteenth century, economic forces increasingly nudged some British officials to explore ways in which to diversify the colony's sugar-based economy. The descendants of the formerly enslaved people and their contemporaries showed the way by their work in gold mining at first, and diamond seeking later. The resulting value and growth of the gold industry, which influenced the prosperity of the country, was impressive. It seemed as though El Dorado in Guyana had become a reality during the period spanning the late 1880s to mid 1890s. Recognizing this trend as a breakthrough to diversify the colony's economy and to sustain its development, some colonial bureaucrats redoubled their efforts to expand business undertakings. Coffee and cocoa production was stagnant, while sugar output was again decreasing. Conversely, rice, timber, and balata production thrived. Significantly, increasing profits from gold underscored the need to develop mining. Furthermore, gold was the commodity influencing the direction of the economy and caused the enactment of land liberalizing legislation. The 1901–1902 merging of two government departments to form a single Department of Lands and Mines synchronized their functions for expeditious exploration and exploitation of minerals and ores to promote mining development.[14] Yet infrastructure projects targeted selective road construction to give business interests access to raw materials. The bureaucracy paid considerably less attention to water transportation and the waterways that were most frequently used by workers.

Roads and trails, mostly linked to government-operated harbors and river landings, were priorities. Specific rivers were cleared only to make them navigable for a certain number of miles. When funds were allocated for public expenditure in the interior, they were earmarked to install specific means of communications and to erect a few government buildings. Waterborne and land-based government-operated transportation facilities such as steamers and lorries were meant to generate profits quickly. The welfare of workers was secondary, and most projects were geared to exploit the colony's natural resources to attain more profits for the colonial hierarchy and the economic elites.[15] Imperial policies were moving beyond only exploitation and exportation of one commodity to the diversification of colonies' resources.[16] In the local context, lack of finance from the colonial office militated against this development.[17]

The types of long-term infrastructure development occurring in California during and after the gold rush period of the 1840s–1850s, captured in H. W. Brands' *The Age of Gold: The California Gold Rush and the New American Dream,* were hardly replicated in Guyana.[18] Placed in the context of earlier and contemporary massive transcontinental projects—the transition that included steamers, railroads, highways, and the Panama Canal's construction—diversified California's mining and agricultural economy and promoted increased trade and profits. The local economy fluctuated at a time when the model of systematic infrastructure development enhancing economic output was already established.[19] For decades, in the interior regions where the majority of inhabitants and workers were non-Europeans, protracted and selective infrastructure developments, which were factors in the high mortality rates, remained pervasive. Ironically, the longevity of interior mining had much to do with the workers' efforts and persistence, especially in the pre-cyanide age of the late nineteenth and early twentieth centuries. The prolific use of cyanide and capital-intensive machinery by the late twentieth century, rather than the mercury (quicksilver) and manual labor of earlier decades caused a resurgence of spectacular and consistent discoveries of gold and, therefore, revitalized the industry.[20]

From the inception, mining workers' attitudes were different from that of most managers of large business firms. The managers were impatient, had get-rich-quick mentalities, and abandoned operations if immediate profits did not appear to be forthcoming. At a time when contraptions for gold and diamond mining were technologically challenged and techniques were mostly unscientific, immediate bountiful returns were not always possible. Although the lack of comprehensive development plans including infrastructure works in the interior limited profits accruing to government and private businesses, African diaspora workers, who had always contributed their labor to the development of all the colony's resources, were now, as miners, instrumental in breaking the almost monopolistic hold sugar had held on the colony's economy.

They worked in timber and forestry enterprises and in producing sugar, rum, molasses, coffee, cocoa, copra, coconut oil, lime-juice and oil, beef and hides, and rice. Other segments of the population participated in diversifying the country's economy, and gold and diamonds along with rice emerged as profitable enterprises. In the 1890s, rice did not suddenly appear as a viable export commodity. Walter Rodney's, *A History of the Guyanese Working People, 1881–1905*, credits "Africans from the rice growing areas of the Upper Guinea coast" with growing "rice in Guiana since slavery days."[21] Initially, rice production by Africans was opposed by planters who viewed it as competing with their labor needs for sugar. Although Africans continued to grow rice in small quantities, the modern rice industry was established mainly by East Indians.[22] By the late nineteenth century, gold mining had influenced several aspects of the colony's economic development. As government allowed more access to its lands, interior workers and small-scale independent prospecting groups had to cope with the hostile environments in which they operated.[23]

Roads in the Gold and Diamond Mining Regions

Even before the country's spectacular gold shouts of the late nineteenth and early twentieth centuries, leaders of various exploratory teams had suggested that the improvement of the interior infrastructure was critical to successful mining. Government officials, perhaps due to competing interests, failed to act with alacrity to implement a plan and program in that direction. Chapter 4 indicated that as the migratory movement from the coast to the interior got under way in the mid 1880s, waterways were the only means to take workers directly to the gold and diamond fields. Dangers inherent to this type of travel resulted in continuous calamities for workers. Funds to improve the infrastructure were often limited, hinged on immediate profitable returns, and road construction projects were selective instead of comprehensive. One can assume that because the workers were overwhelmingly of African ancestry, Amerindians, or immigrants they were viewed through the prevailing lens of scientific racism, and therefore, their lives were not worthy of being safeguarded.[24]

Every segment of the population was well-acquainted with the precarious environments with which interior workers contended, but the lackluster effort of colonial officials in responsibly organizing the industry persisted. As custodians of the populations' vital statistics, the input of the registrar general, surgeon general, and chief medical officer were critically important in any discussions relating to organizing the mining industry. Yet time and again when relevant commissions were convened, and recommendations made, their implementation was long-delayed. Adding to the delays in improving interior conditions were the *Reports of the Commissioner of Lands and Mines* and the *Institute of Mines and Forests* that contained contradictory or incomplete information on the number of active, maimed, sick, and deceased workers. The conflicting information lessened the urgency for requisite interior improvements. Hardly synchronizing with the colony's death registers, which were also incomplete, the said reports underscored the self-interests of the profit-oriented colonial hierarchy of legislators, planters, agents, business persons, steamship liaison personnel, mining investors, and government and quasi-government employees who pursued cost-minimizing policies.[25]

Selective construction of roads, railroads, bridges, riverside landings, and portages as well as the clearances of sections of waterways, the erection of very few hospitals, several dispensaries, government quarters, and post and telegraph offices signaled attempts at interior development. In 1889, the construction of paths or roads to circumvent deadly waterfalls in certain portions of the rivers Essequibo, Cuyuni, Potaro, and a few others were barely effective.[26] As the *Death Registers* for the gold and diamond regions confirmed, for a protracted period spanning the nineteenth and twentieth centuries, a large number of workers still relied on and died while traveling on more extensive stretches of the interior waterways. W. W. Kenrick argued in "Opening Up the Country," that despite the colony's reliance on sugar its economy will be greater and more valuable "when it develops...and utilizes to its fullest extent, the precious gems and valuable minerals it contains." Kenrick's vision of consistent and comprehensive infrastructure development was widely circulated and debated. More than half a century later, colonial officials and other economically interested parties were still contemplating Kenrick's advice.[27]

Suggestions that water transportation should be the primary means of interior travel were largely ignored. Instead, as gold-mining operations expanded, some roads were constructed including the Potaro-Konawarook and the Bartica-Caburi Road, which together with the Barima-Barama-Cuyuni and Cartabo-Puruni Roads, were the major ones built from 1892 to 1899.[28] The nineteenth-century pattern of selective infrastructure works continued for decades. With the completion of the Kartuba-Puruni Road by 1908, plans for the proposed Morawhanna Venezuela Road remained active because it was projected to link valuable gold, diamonds, and other ore-producing areas. Construction of a road from Morawhanna to Wanaina River progressed.[29] In the 1920s, limited road building and maintenance continued. Meanwhile, workers suffered numerous hardships associated with inadequate transportation to convey them to and from mining sites expeditiously. Inconveniences and travel delays decreased their financial resources causing some of them to abandon their mining ventures.[30]

The 1930s witnessed the construction of a road between Bartica and Potaro, a vital link to coastal and interior areas. The construction of the Denham Suspension Bridge across the Potaro River and roads to other gold and diamond bearing regions in Tumatumari and Tiboku were completed. A branch road to the Mazaruni mining district was built to eliminate the need for travel on nearby rivers and avoidance of the often deadly Itanami Fall, which had taken so many lives. The Bartica-Potaro road network was intended to enable easier access to interior prospecting regions, and to encourage population settlement as well as agriculture and forestry undertakings.[31]

While one cannot overlook the fact that road construction opened up the frontier and expanded access to new prospecting areas, it must be noted that these developments occurred over an inordinately long period.[32] Furthermore, for all the points raised to laud road construction in the 1930s,the work remained selective even when funds were available from colonial development provisions. As 1934 ended, inadequate transportation impeded interior travel and a large number of workers were stranded in the Potaro district. Many of the miners were forced to resort to circuitous and expensive travel via Wismar on the Demerara River to reach their destinations on the coast. By the early 1940s a government-run lorry service operated on the Bartica-Potaro-Issano Road, providing a nexus to several interior mining sites in the Essequibo region.[33]

It is no surprise then that it was not until 1956 that work began on the construction of a road linking Parika, East Bank Essequibo River, to Bartica on the same waterway at the junction of the rivers Essequibo, Mazaruni, and Cuyuni. The route connected the West Demerara and Bartica-Potaro-Issano Road, at opposite ends and effectively provided unbroken land transportation from the coast to the mining region. It must be noted, however, that roadways led only to particular points and connected certain waterways, without leading directly to mining sites. Workers had to engage in mixed travel, often with time-consuming itineraries. Decisions to construct roads depended on projected government revenue and profits for private investors.[34] After World War II, a systematic road development plan emerged. The Colonial Development and Welfare Fund and the US Marshall Plan provided large outlays of financial assistance to multinational enterprises to foster the colony's economic development. By the 1950s, *The Economic Development of British Guiana. Report of a Mission Organized by the International Bank for Reconstruction and Development* noted "a greatly improved road transport system is imperative for further economic development of British Guiana."[35] Economic development plans included the perennial proposal to construct an interior railroad.

Railroads As Non-Catalysis

The routes prospectors took to the gold and diamond fields using the coastal railway system have been already outlined. By the late nineteenth century, members of the mining-interested community and colonial officials focused on internal railway construction. Proposals to grant private companies mining concessions for ninety-nine years to spearhead the development of the entire hinterland required them to provide government with funds for the construction of railroads as well.[36] *Ordinances #7 and #8 of 1895* and *Ordinance #11 of 1896* authorized the construction and maintenance of railroad services in areas incorporating the Barima River, Arakaka, and the Kaituma River. By 1897 overland access to some gold and diamond mining centers in Potaro were provided by the Demerara-Essequibo Railway, which linked Wismar on the Demerara River to Rockstone on the Essequibo River, and facilitated connections to Georgetown, the capital city.[37] Surprisingly, by the end of the 1900s, plans to construct a more extensive railroad system were abandoned. As the twentieth century ensued, especially during periods of economic adversity, mining along with capital for railroad construction and labor from abroad were viewed as the means of promoting development. Often cited was the ensuing development in North America after the Canadian and Pacific Railway Company constructed nearly "4,000 miles of railway from the Atlantic to the Pacific."[38] Railroad construction and expansion had been successfully utilized with the onset of modernization and globalization. The colony lagged behind in this sphere.

In 1924, a government-convened commission discussed the feasibility of revamping the existing railway operations under the private management of Sprostons Ltd. The plan was abandoned due to a perceived lack of cooperation by the company's officials and the intransigence of some bureaucrats who did not appreciate fully the utility of railroads. Cost minimizing had affected the outcome because Sprostons Ltd. had seemingly requested exorbitant government subsidies for the service. It was instructive that "no member of the committee... ever had any... experience in the operation of a steamboat, launch, or railway service" and could not "appreciate fully... the increased costs of operation... due to wages, increased cost of raw materials and manufactured articles necessary for the operation of such a transportation service."[39] As late as 1946, in the midst of interior development activities, the *Colonial Annual Report* noted that further plans for interior railroad construction were apt. A decade later plans had stalled and only the vision remained.[40] Unavailability of alternative modes of transportation and lack of affordability to use existing ones by the workers forced many of them to travel with much greater frequency on waterways.

Water Transportation: A Network of Rivers and Creeks

The network of rivers and creeks providing access and egress to and from the mineral-rich areas of Essequibo County included the river of the same name and its tributaries. Although large numbers of persons traveled on the waterways, funds expended to make this mode of transportation more accessible and even safer were far less than the expenditure for roadways; inadequate as that was. Even with road construction and limited railroad provisions, water transportation remained the most popular way for labor to enter and exit the interior. Yet work on hinterland water transportation was unsystematic with no immediacy or urgency placed on improving the facilities to safeguard workers lives. Both the government and private companies wanted profits from gold and diamond mining but long-remained in a speculative mode or were unable or unwilling to invest consistently in wide-scale transportation improvements.

The economic elites viewed the mining regions as merely places for temporary extractive undertakings and infrequently left the environmentally safer coastal areas to endure travel under unsafe conditions to the gold and diamond fields. Many of them resided overseas, and the effects of inadequate interior infrastructure caused them no discomfort. Thus, their decisions on infrastructure development were not based on the environmental dangers of work-related travel or occupations. To this end, the piecemeal way of attending to waterways for limited navigation evolved with the selective clearing of *tacoubas* or fallen trees, weeds, and other impediments in certain rivers and creeks. However, interior workers utilized a much wider network of rivers and creeks than specific allocations covered. The practice of clearing the middle of selected interior rivers was hardly useful because multiple deaths by drowning of persons traveling in the mining regions continued[41] (See Figure 5.1).

Legislators, miners, and other interior workers knew that sometimes, because of impenetrable conditions including blocked channels, instead of a journey being a few miles up or down a river, travel was circuitous and much longer. Often, this meant going twenty miles down one river and an equally long distance up another

Figure 5.1 River scenes.

one to arrive at a particular destination. Systematic clearing of more rivers would have considerably shortened journeys and saved lives and time. A government-operated steamer existed between Georgetown and the North West District, but services were only provided to specific points, not usually in proximity to gold and diamond mining sites. Travelers still had to navigate neglected sections of a network of waterways at a time when government agencies were established in the North West District to promote the expansion of "agriculture, forestry, and mining."[42] Efforts to create access to the interior mining areas still focused overwhelmingly on land transport with limited work on specific rivers and creeks to connect them to the few roads.[43] By 1896, the Georgetown-based Sprostons Dock and Foundry Company provided steamer services for passengers and freight to Bartica. But, former prospectors indicated that no steamer service was available on Sundays and they had to make their own way from Bartica to their mining sites by land or water.[44]

The practice of allocating government funds only to make navigable specific portions of certain interior rivers and creeks, and to facilitate access to selected mining areas continued during the first half of the twentieth century. In 1902–1903, funds were earmarked to clear another section of Arimu Creek, a branch of the Cuyuni River, because it led to an important gold mining center.[45] The case of Arimu Creek demonstrates that cost minimizing and speculation facilitated public expenditure only in situations where profits were immediately forthcoming. Water travelers who wanted to use encumbered rivers and creeks did so for decades at their own peril. Alternatively, they could utilize existing steamer services taking them to certain points and then use roads and trails, or paths leading to mining sites (Figure 5.1).

By 1914 government took control of the privately run steamer services of Sprostons Ltd. and amalgamated it into the Colonial Steamer Services. Within five years, government operated steamers, launches, and boats, but again only to specific interior locations, and as revenue generating undertakings.[46] Eventually, both steamer and launch services operated fortnightly from Georgetown to Morawhanna and Kumaka. A launch service also operated from Bartica to Lower Camaria where a lorry/truck service transported passengers and freight to Upper Camaria on the Cuyuni River. Still, more connections to the vast expanse of the gold and diamond fields were required, and many miners were left to their own devices after arrival at designated terminals.[47]

In 1922, the Colonial Steamer Service became part of the Colonial Transport Department. When Sprostons Ltd. ended its launch service between Rockstone and Potaro Landing, the Colonial Transport Department inaugurated a boat service between Bartica and Tumatumari. The Transport and Harbours Department that operated from January 1, 1931, resulted from the amalgamation of the steamer, boat, launch, railway, road, and harbor services. In the 1940s, the department operated ferry services across the mouths of the three main rivers, Essequibo, Demerara, and Berbice and a steamer from Georgetown to the North West District. A privately operated auxiliary launch and boat service operated from Kurupung, Mazaruni River providing limited transportation to the vicinity of the gold and diamond fields.[48] Despite all of the plans to develop mining and the country's interior, transportation had only nominally improved by the end of World War II. Suggestions to improve riverine travel in the 1890s were still being considered in the 1940s because rivers and creeks offered a network of natural waterways. As late as the 1950s there were still extended sections of interior rivers on which travel was extremely dangerous, but the improvement of inland water transport was not contemplated because the cost to construct canals to bypass rapids was prohibitive.[49]

In 1894, discussants at the *Gold Industry and Crown Lands Commission* suggested liberalizing and altering boat regulations applicable to the Essequibo River and tributaries to encourage more travel by small prospecting parties. They recommended that certified captains and bowmen, whose services were expensive, should not be required on small boats with less than fifteen passengers—the majority of river traffic. To limit the services of certified personnel on dangerous waterways solely to encourage more travel and to increase mining production was not in the best interest of safety. Although the feeling existed that navigable rivers and creeks needed little maintenance following initial clearance, improving the most popular, albeit deadly, route taken by workers to the interior remained in an embryonic stage while road building took precedence.[50]

Boat captains, steersmen, and bowmen working on vessels plying the waterways required certification to operate certain passenger-carrying vessels. Boating and River Regulations were enacted, but inconsistent enforcement of the laws hardly lessened the dangers inherent in water transportation. Nonetheless, during the early twentieth century, crafts and their accoutrements were certified for services under the *Upper River Boat Regulation, 1912,* amended September 5, 1920; and the *River Navigation Regulations, 1924*. Still, the lack of expenditure to improve travel on the network of interior waterways constrained mining and other interior activities. Cost-minimizing policies led to fatal boating accidents for decades.[51] According to a US government report, by the mid-twentieth century the interior had no tarred roads. Of the approximate 400 miles of trails only about 100 miles were gravel-surfaced and passable year-round. The railroad system was a narrow-gauge one that ran from Georgetown to Rosignol (61 miles); Vreed-en-Hoop to Parika (19 miles); Linden to Ituni (80 miles) and one to Rockstone (19 miles). The report indicated that without a modern road or railway system linking the coast and the interior, transportation was mainly by river and air.[52]

AIR TRANSPORTATION

Air transportation provided another means of ingress and exit from the gold and diamond fields. In December 1921, a feasibility air-flight to the interior of the country was undertaken. The maiden flight to Kurupukari Island, Upper Essequibo River, which took three hours, would have taken the popularly used river transportation weeks to arrive at the same destination. Ironically, most miners failed to utilize air service to the interior because it was expensive. They continued mostly to utilize rivers, roads, and railroads while most government and business officials and members of the mining hierarchy traveled to the interior by air. By 1925 when the first seaplane flew in the country, the *Air Navigation Ordinance 1925—Air Navigation (Consolidation) Order 1923* became law.[53] In the 1920s government bureaucrats and officials of private companies focused their energies on establishing interior air mail, passenger, and freight services. They had little success, although provision of this service was a step in the right direction to establish permanent communities in the gold and diamond regions. The post master general had already suggested that improving mail services in interior locations would better serve the miners and their families. Air mail deliveries provided speedier service, but after the initial air delivery of mail to the interior in August 1927, the British Guiana Air Transport Company failed because of insufficient funding.[54]

In 1930, a government-appointed advisory board under the auspices of *The Air Navigation (Colonies, Protectorates, and Mandated Territories) Order, 1927,* met to reorganize the aircraft industry. Increasingly, from 1931 landing strips were constructed at specific areas of the interior such as Kumakusa and Enachu to facilitate access to mining sites connected to certain roads. Before long, the inadequacy of relying on selective infrastructure work was dramatically brought to the fore when outbreaks of malaria and beriberi occurred at the Brazilian border. The sick deputy boundary commissioner, Augustus J. Cheong of the British section of the British Guiana-Brazil Boundary Commission, died October 4, 1934. An ambulance had been dispatched to collect him the previous day, but overland delays made the effort futile. The Chamber of Commerce urged government to establish a passenger air service to the Brazilian border. That same year, the Department of Lands and Mines, Geological Survey Section, contracted the services of the American pilot, Arthur James "Art" Williams, for its work in the mining areas. By the late 1930s there were regular flights to certain mining areas, and some air mail services were restarted.[55]

The privately owned, but government-subsidized British Guiana Airways began services in May 1938. The carrier transported mail and freight, including heavy mining equipment, and passengers to and from the interior. The company maintained a ramp with hangar accommodations and workshop facilities at Ruimveldt, on the outskirts of Georgetown for amphibian aircraft to provide internal air services. The entity also operated a hangar and a workshop up the Demerara River, in the vicinity of the country's main airport, for land aircraft. Colonial Development Funds from the 1930s and specifically *The Colonial Development and Welfare Act of 1940* provided resources from the Imperial Treasury for economic development in British Caribbean territories and Guyana. A similar Act of 1945 extended the objectives of the earlier one. Out of a total provision of 140 million pounds sterling, budgeted for a ten-year period, Guyana's allocation was 2 million, five hundred thousand pounds. Grants approved by the Acts of the 1940s for the country's development included funds for constructing aircraft landing strips in selected locations in the interior. It was emphasized that since railroads and road services were limited at best and often nonexistent; and water transportation was hindered by rapids in rivers, the development of air transportation would be vital in opening up the interior. Though laudable, the premise overlooked the fact that the cost of air transport was prohibitive for the majority of the mining and interior workforce.[56]

Between 1934 and 1938, a number of water-alighting areas such as strips and pools, to connect airstrips and trails were constructed. By 1948, weekly, biweekly, fortnightly, and monthly flights connected the coast to Potaro, Bartica, North West District, and the Essequibo River District, respectively. The next year, as many as twenty-six airstrips existed and connected to 300 miles of trails leading to gold and diamond mining sites. The Konawaruk Airstrip was completed by 1953 and a successful trial run was made by a Dakota Aircraft of British Guiana Airways. The following year, there were fifty-eight water-alighting areas; and in 1955 there were over 400 miles of trails connected to airstrips. After mid century, *Ordinance #6 of 1956* provided "for management and control of interior airstrips, under the authority of the Controller of Civil Aviation."[57] Business executives, government officials, and mining personnel who utilized air transport to travel to the interior usually embarked at Ruimveldt, Demerara River.[58]

Postal, Telegraph, and Other Communications

By 1897, the postmaster general, noting the increasing circulation of mail in the gold and diamond regions, recommended the provision of more postal facilities to serve miners and their families both on the coast and in the hinterland. The postmaster general observed that men who formerly never left the confines of the villages in which they lived were traveling to and from mining sites and he advocated parallel development to enhance communications and settlement.[59] A trend in providing postal services based on population growth followed, but in many interior gold and diamond producing areas settlement was sporadic. Consequently, post offices were selectively constructed to serve only areas of increasing population growth. From the

1930s, Postal Agencies, Traveling post offices, and interior shops that sold postage stamps and provided mail services to specific locations were established. By the mid 1950s, there were 48 post offices, 65 postal agencies, and 17 Traveling post offices. Inland surface mail was delivered by the Transport and Harbours Department and private companies, while Bartica, Rupununi, and Potaro had air mail services.[60]

In 1893, telegraph services were provided at a government-established facility in Bartica. Intermittent upgrades to this service continued during the ensuing decades. Electricity and residential radio services were in place by the 1930s and eventually, post offices were equipped to provide radio telephone services. By 1946 such operations extended to the Venezuelan border. In 1947, the Rupununi Gold Mining Company, Porthos Mining Company, Tikwah Mining Corp., Cuyuni Goldfields Ltd., and B. G. Goldfields Ltd. operated transmitting sets that linked various parts of the interior with Georgetown by radio telephone. Government increased the number of transmitting stations it operated in the interior as well.[61]

Upgrading telecommunications were commendable, but the pervasive lack of investments to improve river transportation, time consuming and costly travel along with expensive airfares, limited the utility and impact of the communication system. Often, it appeared as though the plans and programs of private management and government administrators were not synchronized to encourage development of the vast hinterland region, which offered a multiplicity of economic opportunities. Instead of organizing the potentially profitable mining industry to attract and maintain a permanent population and implementing a family oriented housing policy, business and government officials provided mostly substandard accommodation. Inadequate residential facilities were inimical to overall interior development.

Housing in the Gold and Diamond Areas

Circumscribed by terrain and climatic conditions, the migratory movement to the gold and diamond mining regions had followed pendulum, circular, and seasonal trends for decades. Cost-minimizing decisions related to the industries' profitability and the mostly speculative nature of ventures limited the construction of the type of accommodations to enhance population settlements; giving these two sectors their transient characteristic. Private companies did not prioritize the construction of permanent housing and related facilities for workers who received no financial assistance to encourage them to settle with their families in the interior.[62] Mining companies often relied on temporary labor migrants, but as already noted, within a decade of the advent of mining, some workers demonstrated adaptability to interior environs and became self-employed by organizing their own prospecting undertakings. Progress of this nature emphasized a commitment to hinterland work. Yet half-hearted attempts by both government and private companies to construct facilities conducive to permanence prevailed and contributed to the slow pace of interior development and fluctuating profits from gold and diamond mining.

From the 1880s to the 1950s only a modicum of housing and other building construction such as rudimentary open-walled housing and cheap, flimsy, and camping-style makeshift accommodations were provided for employees by the firms for which they worked. Companies conveniently claimed that they had limited capital.[63]

It seemed that in calculating their operational expenditure for mining, company officials did not include probability costs for ancillary expenses. In 1895, the government agent for the North West District, an extensive mining region, opined that the gold industry would progress with an increased population.[64] Two years later, Dr. John M. Rohlehr, who was well-acquainted with conditions in the interior, confirmed that government had done little to develop the promising gold industry. Dr. Rohlehr affirmed that appropriate housing would lessen dangerous and frequent river travel, lead to permanent settlement, and that the gold and diamond districts would become self-supporting. He advised that most mining workers were multi-skilled in agriculture and building construction, assets necessary to expand frontiers.[65] By 1903, government officials undertook to rectify somewhat, the paucity of a settled workforce and farming in the interior by providing Homestead Grants. By 1907, residents along the Potaro-Konawarook Road increased by 77 from its previous 277 persons.[66] Without such grants, in most of the other mining areas, population growth was slower and inconsistent due to lack of suitable housing. Yet colonial bureaucrats and entrepreneurs failed to provide appreciable funds to assist in developing the interior infrastructure until long after World War I.

On the eve of World War II, the dire economic conditions associated with the Great Depression caused tremendous social unrest in the British Caribbean and presaged the coming of the *Royal Commission, 1938–1939*. The resulting recommendations included the establishment of a *West Indian Welfare Fund* for development of the colonies; provided for by an annual grant of 1 million pounds sterling from the Imperial Exchequer for twenty-one years.[67] In 1948, pointing to the recommendations for development programs identified by the aforementioned *Royal Commission*, the *Report of the British Guiana and British Honduras Settlement Commission* strongly supported accelerated development of the interior, suggested large-scale migration, and the introduction of new export crops. These important programs required reinvestments of profits gained from prior economic activities in the colonies. Effective development also meant addressing the welfare of the human factor of production.[68]

By mid-century, officials of capitalist companies were abandoning their earlier speculative and cost-minimizing approaches to gold and diamond mining. The Cuyuni Goldfields Ltd, for example, established a company town, and amenities included rental housing with electricity for miners and their families. A company-run nonprofit store was opened and an eight-bed hospital, a library, and facilities for recreational activities were provided by the company. As the population increased, the post office erected at 110 Miles Potaro added banking services for its nearby residents. When the Tikwah Mining Corporation relocated its operations near the Puruni River, its officials constructed a road from the waterside to its mining site and erected houses for their workers.[69] Employees of B. G. Consolidated Goldfields Ltd. were provided with rent-free or cheap housing and discount or credit on food supplies. Workers were granted sick and casual leave and medical attention. By the 1950s, at the behest of trade union activists and in the face of workers' protests, bureaucrats and employers of gold and diamond workers, conceded that improved housing and health care were essential for population settlement and growth to increase profits from interior enterprises.[70]

HEALTH CARE: PRACTITIONERS VS. BUREAUCRATS' DALLIANCES

In the midst of the "booms and busts" of gold and diamond mining from the late nineteenth century, medical practitioners such as Drs. Ozzard and Rohlehr had tried with little success to encourage colonial bureaucrats to focus more attention on the causes of the alarmingly high death rates among interior workers. Pointing out the need for several types of facilities, Dr. Rohlehr noted that if adequate health care services were provided in the interior workers would be encouraged to settle their families.[71] Instead of heeding the medical advice, government officials committed themselves to providing only certain kinds of hinterland infrastructure to attract investments in mining. They neglected the welfare of the workers and provided a modicum of health care services in far-flung areas, mostly distant from gold and diamond mining activities. The actions of mining companies' management were hardly better. I. R. Phimister explained in "African Labour Conditions and Health in the Southern Rhodesia Mining Industry, 1898–1953," that the cost-minimizing of expenditures on African workers and employment conditions to achieve profitable mining outcomes, precipitated an increase of disease among them. Phimister identified "minimal expenditure on" the "workers housing, diet, hospital facilities, and other amenities," as causes of the illnesses.[72] Similarly, in colonial Guyana for an extended period, gold and diamond mining workers toiled and existed under conditions where waterborne diseases, inadequate housing, lack of a nutritious diet, and insufficient health care caused them tremendous sufferings and deaths.[73]

When gold and diamond mining emerged, the tone was set to neglect "enforcing the regulations and protecting life and property in these outlying districts" because legislators, largely influenced by the plantocracy were "reluctant to spend money on the" gold "industry until it saw what revenues it could first extract."[74] Much of the early medical services in the country had been based on the needs of sugar plantations with planters providing medical care for the enslaved and indentures through the services of mostly uncertified persons. By 1886, "the Colonial Medical Service was founded; registration of medical practitioners, nurses and chemists became compulsory and District Government Medical Officers were appointed."[75] Meanwhile life for the miners was chaotic, but only sensational tragedies and news of seriously maimed or dead miners captured headlines and evoked responses. Recurring incidences of starving, sick, injured, and dying prospectors who were abandoned at riverside landings made the news. Government officials were forced occasionally to send relief boats and dispensers with food and medicine to render assistance. Advocacy by health care practitioners for the construction of more government hospitals at interior locations followed catastrophes. With little effect, miners themselves continuously pointed out the need for more appropriate facilities and adequate medical and related personnel in interior locations.[76]

In 1897 the colony had five public hospitals; three of them on the coast with the other two on the periphery of the sprawling gold and diamond regions. One was located at Bartica and the other at Morawhanna. Both hospitals served gold and diamond workers as well as all other residents, in varying radii. In 1897, Dr. P. M. Earle, the medical officer at Morawhanna Hospital, served an extensive interior area encompassing the mining region. The Medical District at Bartica Grove and

North Western had no appointee in 1897.[77] Dr. A. S. Ozzard was medical officer for the Public Hospital Bartica and also served H. M. Penal Settlement, Mazaruni, and the Town of Bartica. From the 1880s until his death in the 1920s, Ozzard strongly advocated improved public health services for mining and other interior workers, to little avail. Dr. Ozzard, a pioneering researcher in malaria, had specifically identified the disease as a main cause of miners' deaths. He continuously urged the government to provide a wider array of treatment and facilities to address the shortcomings in interior health care provisions.[78] Efforts conducive to improved healthcare services were slow in coming. Applied to the local situation, race is apt as a dichotomy in offering a perspective for the long-drawn out infrastructure works and insufficient healthcare services.

By the mid-1880s, colonial administrators made clear that without being able to extract immediate profits, they would not focus on protecting life in the gold-mining regions. When the governor-convened commission met to organize labor in 1890, focus centered not on preserving workers' lives, but on recruiting more migrants of African descent as a replacement workforce. The majority of mining and associated workers who suffered and died in the interior were of that specific racial profile.[79] Possibly, their race explains why they were expendable and the reason for the lack of urgency in attending to the interior infrastructure. Most alarmingly, at a time when a dire need existed for qualified and experienced health care practitioners, especially in outlying areas, the racial factor is particularly obvious in the inappropriate way colonial bureaucrats treated even Western-educated personnel.

In 1859, Dr. John Monteith Rohlehr was born in Berbice County to parents of African and European ancestry. Although Dr. Rohlehr obtained his medical credentials in Canada at twenty-eight years of age and returned to the country in 1887, colonial officials refused to appoint him as a government medical officer. Ostensibly, Dr. Rohlehr was denied registration by *Medical Ordinance #5, 1886,* enacted while he was abroad studying. The new ordinance required all medical personnel to be UK-trained and registered by the British Medical Association. On February 17, 1888, Rohlehr petitioned the Court of Policy to reconsider his case. An 1889 Petition and Special Act of 1890 enabled him to register as a medical practitioner. In the ensuing years, Dr. Rohlehr often clashed with the colonial hierarchy over the dire need for health and safety measures to safeguard lives in the interior mining regions. Healthcare practitioners were needed at the two hinterland hospitals where most of the sick, seriously injured, and dying were Amerindians and African-Guyanese. The white colonial authorities had refused to appoint Dr. Rohlehr to a position for which he was qualified, and to serve his compatriots. Rohlehr unswervingly served the laboring class in a number of capacities and when he died in February 1922, he was lauded as "a skilful surgeon and keen politician."[80] Still, the tremendous loss of life among gold and diamond miners increased as racial discrimination faced by local healthcare practitioners continued into the twentieth century.

In 1916, Dr. Loris R. Sharples applied for a position in the colonial medical service and also encountered difficulties. Dr. Sharples who was born in 1889, graduated from Edinburgh University, registered with the General Medical Association, UK, passed a course in Tropical Medicine and Bacteriology, and did postgraduate work in Tropical Medicine and Hygiene. He had acted as a Government Medical Officer at the Public Hospital Georgetown, since January 3, 1914, where his performance was satisfactory, efficient, and reliable. The governor of the colony refused to recommend Dr. Sharples's appointment not because of his lack of qualifications, competence, or experience but based on race. The governor claimed that Dr Sharples could not be assigned to areas where, as "the only medical practitioner, white patients would object to be treated by a medical officer who was not of their own race." After a long drawn out dalliance by colonial bureaucrats, Dr. Sharples was appointed because he was "better than no doctor at all."[81]

For the most part, "the embargo on the appointment of Negro and coloured doctors...had the effect of creating employment for Europeans recruited from other white parts of the British Empire."[82] In 1939, the Hon. Joseph Eleazar and Dr. J. B. Singh, an African and East Indian Guyanese, respectively, exhibited before the local legislature an advertisement for a medical officer placed in a British medical journal. It stipulated that European applicants were preferred for the local position. Disregarding previous qualification requirements, an unqualified European applicant was appointed to the position. Dr. Singh regarded the action as one of racial discrimination and injustice because a local well-qualified Guyanese was overlooked for an appointment in preference to a lesser qualified overseas candidate.[83] The egregious reluctance to employ qualified personnel based on their race and the unwillingness to heed their advice on ways to ameliorate the sufferings of the gold, diamond, and other interior workers was a colonial bureaucratic hallmark. Colonial bureaucrats instigated and implemented discriminatory and prejudicial polices at a time when two maladies which afflicted both the general population and miners—malaria and sexually transmitted diseases—were rampant.

The Ravages and Treatment of Malaria

In the case of malaria, it must be recalled that statistics compiled from Guyana's death registers of the 1890s identified the disease as a major reason for the demise of the gold and diamond mining population. By 1896, no less a person than Dr. Ozzard, the eminent researcher and government medical officer at Bartica Hospital, publicized the fact that many mining workers from distant areas arrived at the institution suffering "mostly from malignant forms of malaria fever."[84] Malaria is an infectious ailment transmitted by the disease-carrying *Anopheles* mosquito species. Symptoms of the infection in humans include chills, fever, and sweating. Topical application of crab oil and citronella oil and doses of quinine were among the earliest remedies utilized to prevent or to treat the malady. Frequent cutting and burning of trees and trash near residential communities were undertaken to destroy mosquito habitats. Fumigation got rid of the insects. Personnel at mining sites had to ensure that containers that could provide breeding places for mosquitoes were emptied and discarded.[85]

In 1928, Surgeon General Dr. P. J. Kelly and retired medical officer Dr. P. M. Earle were among thirty-three witnesses who provided oral and written evidence to a government commission seeking to improve the diamond and gold industries. But in the report that followed, under the section entitled "Medical and Sanitary

Services," not once did the word "malaria" appear, although the disease was creating havoc throughout the country. A transient interior population in the extensive mining region was blamed for lax enforcement of the *Sanitary Regulations of the Mining Laws* at camping sites. The health officers identified unsanitary conditions as fostering breeding grounds for mosquitoes, but disagreed with the recommendation that government should establish permanent river camps. They felt that "the capital cost of such camps would be out of proportion to the benefits to be derived" and recommended the rigid enforcing of the existing laws as the only course of action.[86] While insufficient staff, inadequate funds to construct buildings, and lack of first-aid stations and ambulance services were identified as plaguing the few interior hospitals, the doctors blamed cost constraints and the nature of the population for the limitations. They almost simultaneously affirmed that funds would be expended on projects only in "proportion to the benefits to be derived."[87]

There were only six government hospitals in the colony in 1931. The largest, Public Hospital Georgetown, had 564 beds. The Public Hospital New Amsterdam and the Public Hospital Suddie had 161 and 92 beds, respectively. These three institutions were all located on the coast. The three smallest hospitals, Bartica, Mabaruma, and Kamakusa with 14, 30, and 8 beds, respectively, served the entire interior region in Essequibo County. By this time, the government facilities at Morawhanna including the hospital had been relocated across the river at Mabaruma. Public Hospital Kamakusa was the only one serving the gold and diamond mining regions. These three hospitals concentrated largely on therapeutic treatment.[88]

Until the 1920s, due to the ravages of malaria fever and other waterborne diseases, the colony's population increase was due solely to immigration. From 1932, colonial officials and some management of privately owned companies redoubled their efforts to improve the health and lives of the working class, albeit with trade unionists involved.[89] After 1937 and especially by February 1945, a colony-wide prophylactically inspired campaign to deal with malaria eradication was under way. As a consequence of the campaign, the colony's population increased naturally. Utilizing the earlier attempts of Dr. Ozzard, under the guidance of Dr. Giglioli, and with the use of local and overseas funding, the malaria eradication campaign got under way. By 1946 the estimated population of the colony was 375,701 persons compared with 318,312 for 1931; an increase not attributed to immigration as before. The increase of the colony's population by 57,389 between 1921 and 1946 is usually credited to the curb in the malaria-carrying mosquito. By 1946, the evidence of chronic malaria among school children had decreased from 60 percent in the previous decade to 18 percent. Statistics from public and estate hospitals and medical districts confirmed the decreasing number of malaria cases. In 1946, 23,000 such cases were treated and 274 malaria deaths recorded. The previous year, 27,700 malaria cases and 342 deaths had occurred.[90]

The mosquito eradication campaign dramatically decreased the malaria-carrying insect and increased the population growth. In 1945, the growth rate was 1.55 percent per annum; by 1954 it was 3 percent. From 1940 to 1951, deaths and infant mortality rates decreased, the birth rate increased, and the trend continued for the rest of the decade.[91] With the establishment of the Mosquito Control Service by January 1947, the government aimed to control not only malaria but also yellow-fever and filariasis, caused by other species of the insect. DDT was regularly sprayed colony-wide. In their annual reports, the surgeon general, registrar general, and the commissioner of lands and mines often noted the value of quinine, other anti-malarial, and prophylactic treatments in decreasing the number of deaths in the mining regions.[92] By 1952, the Public Health Scheme included programs for an expanded antimosquito campaign, assisted with funds from the American-based Rockefeller Foundation. Following a 1954 government-approved commission to inquire into the colony's medical needs, the administration released a comprehensive plan to reorganize health care one year later. Plans were afoot to provide a network of community clinics and hospitals.[93] By 1956 there were seven government general hospitals and twenty cottage hospitals with a total of 2,532 beds; 37 dispensaries also operated colony-wide.[94]

The Mining Frontier and Sexually Transmitted Diseases

Frontier conditions—lack of adequate, affordable, and timely transportation to and from the gold and diamond fields, and inadequate residential and other facilities amenable to the settlement of families—were directly related to prostitution and sexually transmitted diseases in the interior. Although both the activity and the diseases may not have been confined to miners or remote locations from centers of population, there is an overarching tendency to characterize them as such. Ironically, too often in Guyana, effort was made erroneously and unfairly to link only specific places on the coast such as Lombard Street, Georgetown, and areas of Charlestown and LaPenitence, frequented by gold and diamond miners to incidences of prostitution and sexually transmitted diseases in the interior.[95] Lombard Street was one of the main thoroughfares in the capital city and it was also frequented by prostitutes. On Lombard Street a vast array of business places including goldsmiths' shops, clothing and grocery stores, restaurants, rum-shops, and boarding houses/hotels abutted. Hardware stores catering to the needs of the gold and diamond mining community proliferated on Lombard Street. Prospectors frequented the area to procure the tools of their trade. Many Goldsmiths and Jewelry Shops where miners could sell their gold and diamonds were located on that street. In the 1895 *Report of the Gold Industry and Crown Lands Commission*, goldsmiths located in La Penitence, Charlestown, and Lombard Street were identified as buying raw gold in small quantities from miners who recently returned from interior work.[96]

Some miners, like other members of the general population who were interested in prostitution and supported the trade, frequented not only Lombard Street but also High Street, Water Street, and the Tiger Bay area. Furthermore, the headquarters and offices of many firms that conducted interior businesses such as Sprostons Company Ltd. were located in Lombard Street and its environs. The ferry termini serving Linden (Mackenzie) and linked to the Wismar-Rockstone Railway, and the one connecting the West Coast Demerara, all leading to mining areas, were located in Lombard Street. In addition to functioning as a main commercial hub, nearby Stabroek Market was a central transportation point for vehicular and animal-drawn contraptions. Similar to other persons, miners frequented the area for a multiplicity of purposes, including prostitution. Colonial bureaucrats were cognizant of the ills associated with prostitution and venereal diseases and attempted remedies.

By 1928, a government report had indicated that in order to register for interior mining work, a medical certificate provided by a prospective employee was acceptable. While the information required on the certificate is unknown, it was no secret that prostitution and venereal diseases posed serious problems for the growth of the interior population and the gold and diamond industries. The report suggested "an intensive anti-venereal propaganda campaign... with a variety of reading materials, including magazines, newspapers, and literature... on the causes and effects of the disease" as well as the provision of treatment centers and recreational facilities... "to reduce the disease among the mostly male mining population."[97] Interestingly, interior healthcare personnel had periodically reported that venereal diseases were invariably transmitted by incoming miners. Although until a few years after 1928 it was illegal to grant mining privileges to women, it was suggested that gold and diamond miners should be allowed to take female partners in the interior. Instances of conjugal relationships between men and women in the interior had led to some population settlement, but food supplies were often costly and deficient; helping to create an unhealthy and inefficient workforce. Lack of nutritious food contributed to an insufficient number of both sexes settling in the mining regions and also caused a variety of diseases.[98]

Some members of the 1928 commission believed that the price of food was reasonable, especially rice, sugar, and flour, although the distance from Georgetown and the difficulty of transportation between the mining areas and the sources of supplies caused concern. Other members contended that too much of the food available consisted of salted and canned goods and recommended a varied diet including fresh vegetables and fruits.[99] Urging the cultivation of greens and ground provisions, fruits, and plantains by resident miners, the officials noted that it would decrease the cost of food in the districts. Officials of the Department of Land and Mines, in particular, suggested that allocating funds to provide for the distribution of plants and seeds would improve the food supply and the health of miners. They envisioned that an increasing population, including family units, would lessen incidences of prostitution and venereal diseases and increase profits from gold and diamond mining.[100]

The Colonial Development and Welfare Act, 1940 and *1945* provided financial assistance to operate and improve healthcare facilities including Sexually Transmitted Diseases Clinics. By 1946, attendees at the government-run Social Disease Clinics for the treatment of both syphilitic and gonorrheal infections, numbered 39,824; an increase of 925 over the previous year. Specific treatment for syphilis included "NAB" Bismuth injections, and potassium iodide. suphathiazole and penicillin were used to treat gonorrhea. During 1947, although no regional breakdown is available from the Social Diseases Clinic, Georgetown, as many as 43,004 persons attended for the treatment of both of the infectious diseases; an increase of 3,180 over the previous year's. Although the program continued indefinitely, follow-up treatment was sometimes difficult for patients from outlying districts who could not afford the time or money for weekly bus or train fares to Georgetown.[101]

The *Manchester Guardian* of May 30, 1952 explained, "We are getting out of the colonies in... trade more than we put into them in... investment. Our sterling balances... increased in the year from eight hundred and fifty million pounds to one thousand and eighty-five million pounds"... "the expenditure from the Colonial Development and Welfare Fund is fourteen million" (*sic*) pounds sterling.[102] This statement indicated that expenditure to promote the colonial development meant different things to various people. The mining hierarchy believed hydrographic, cadastral, and geological surveys and the provision of hydroelectric power were necessary for the industry's development. Colonial administrators felt that expending resources to improve some communications to, from, and in the interior would attract private and foreign investors.[103] Opinions in newspapers such as *The Labour Advocate, Sentinel, The Chronicle,* and *The Argosy* conveyed that locals were skeptical and did not all favor development plans of the 1940s and 1950s. Objections were made to specific plans envisioned to provide mostly gains for foreigners. Many working-class persons felt that the planned projects including those for the gold and diamond sectors were designed to continue labor exploitation and oppressive taxation, and development plans were skewed in favor of fortune-hunting capitalists.[104]

Conclusion

At best, infrastructure works carried out in the colony endeavored to facilitate access to the gold and diamond fields and ensure more profits, which mostly benefited external economies. Although workers' efforts to recover the two commodities depended on their good health, managers of private companies and the colonial administration pursued cost-minimizing policies that were inimical to labor. One can hardly blame the inhospitable terrestrial and aquatic features of the mining regions for the lack of consistent development. The interior region had long been considered as potentially beneficial for a variety of economic enterprises, but for almost a century speculative investments and haphazard development militated against this. An examination of the organizing of bauxite production that follows indicated that in some ways, its development was an antithesis to that of gold and diamonds, but there were similarities in the three sectors of the colony's mining industry.

CHAPTER 6

Another Approach: Organizing Bauxite Production

IN THE TWENTIETH CENTURY, bauxite was the only major export-product added to Guyana's economy. With the coming of bauxite, the search for El Dorado began to lose luster because the precarious environmental conditions associated with the precious metal and gems industries pushed workers into the newer and less hazardous undertaking. From the inception of bauxite production, its organizers who had the inadequacies of the two other mining sectors as reference points followed a different trajectory. Unlike the fiscally stingy colonial bureaucrats and cost-minimizing private investors in the gold and diamond sectors, bauxite's organizers invested in providing amenities which were conducive to population settlement. The management ploughed back enough financial resources in the infrastructure to increase profits and to enhance the industry's viability. Better living and working conditions in the stable and safer environs of bauxite producing areas, compared with the gold and diamond regions, served as magnets pulling African diaspora labor migrants to work in the third sector of the mining industry.[1]

In the realm of labor, the emerging situation was problematic for the precious metal and gems industries which were unable to maintain a consistent workforce. Bauxite's management could rely on an adequate workforce, and the industry's advent was beneficial for private officials and investors, local and foreign governments, and the workers themselves. A major aspect of the profitable evolution of the new industry was that its organizers were the beneficiaries of the consequences of the ills which befell gold and diamond mining. This chapter focuses on the ways in which the organizers of bauxite mining utilized another approach and created a worker-friendly industry. Unlike their solely profit-oriented counterparts who organized gold and diamond mining, the organizers of bauxite production took immediate action to address infrastructure issues in ways advantageous to the company and its employees. Bauxite's management provided essentials which were conducive to a reasonable standard of living such as housing, healthcare, and transportation as well as educational, commercial, religious, and entertainment facilities in a timely manner. Utilizing the continuous conflict characterizing socioeconomic relations in the other sectors of the economy as a backdrop, the bauxite hierarchy was more considerate and amenable to its workers' welfare.[2] Bauxite's organizers understood the nexus between reasonable living and working conditions and a workforce's ability to produce optimum profits on investments. Consequently, for decades the initiatives of the organizers of bauxite mining ensured stable production and exports of the commodity.

Bauxite's strategic and timely emergence during World War I, its utility, high quality, and sustained and expanding production for prolonged periods beyond the World War II era, justified its place as one of the country's major export earners by the 1950s, as Table 3.5 in Chapter 3 indicates. Enlightened "order and progress" and modernization were applied to the bauxite industry from the inception. The capital-intensive factor largely involved in bauxite production precluded individuals or small groups from pursuing independent production as in the case of gold and diamonds. The few bauxite companies in operation ensured their mining monopolies through the early legal rights they acquired. This assured relatively strife-free relationships for decades and minimized the number of litigations, between employees and management, which were characteristic of the other two mining sectors.

Colonial bureaucrats, personnel from Great Britain, Canada, and the United States, as well as officials of multinational companies were among the organizers of bauxite mining in the colony. Staff of the Department of Lands and Mines and the Institute of Mines and Forests and other personnel associated with gold and diamond mining became part of the bauxite hierarchy. In combination, they steered the bauxite producers away from the trials and errors of the two other sectors of the mining industry. Adding favorably to increased bauxite production, were wartime needs, innovations of the era, the emerging aeronautical age, the qualities of the commodity itself, and its value.

ORGANIZING A MONOPOLY

British bureaucrats had long contemplated extending mining in the colony, but North American and local businesspersons as well as colonial officials were involved in all facets of the establishment of the bauxite industry. Ultimately, American and Canadian interests controlled much of bauxite production, exports, uses, and profits. Representatives of Republic Mining and Manufacturing Company of America, who initially masked for the Aluminum Company of America (ALCOA) and its subsidiary, the Aluminum Company of Canada (ALCAN), mined bauxite through the Demerara Bauxite Company (DEMBA) at Linden (formerly Mackenzie) Ituni, Akyma, and Two Friends on the Demerara River from the inception until its nationalization by the government in the 1970s.[3]

The acquisition of bauxite lands in Demerara County by American and Canadian concerns followed the coming of prospectors, speculators, investors, and protracted legal disputes to obtain mining rights. British officials were gravely concerned about foreign entities' control of the commodity in its empire. Still, the Americans, Canadians, and their allies outmaneuvered not only the British, but also local entrepreneurs interested in acquiring bauxite producing lands. The foreigners also swindled local residents of their land. Odida T. Quamina in *The Mineworkers of Guyana: The Making of a Working Class* explained that "the legal battles... the compromises and the false representations of the 'real' value of the land, were all part of the most sinister land grab."[4] According to Quamina, local "applications for leases of 500 acres to prospect for bauxite were never given the same efficient treatment afforded the foreigners." And "in every attempt to obtain some mining concessions; their applications were rejected for one reason or the other." Unfortunately, "in every case, these local speculators lost their investments and life savings."[5] This development in bauxite had an uncanny resemblance to and was in fact part of the ongoing strategy of British colonial administrators aligning themselves with American and other foreigners to deny local gold and diamond miners access to mining rights as discussed in Chapter 3. Ultimately, "ALCOA was successful in its bid and controlled 95 percent of bauxite mining in the Demerara region."[6]

Mohammed Shahabudeen in *Nationalisation o of Guyana's Bauxite: The Case of Alcan* and two studies by Quamina contended that when the bauxite industry was organized, the mostly British, American, and Canadian expatriate officials implemented policies that had the same features as those in what later became known as Apartheid South Africa. The claim is that these policies negatively impacted on race and social and economic relations in the bauxite industry.[7] In fact, bauxite workers were paid better wages and enjoyed more benefits than employees in most other sectors of Guyana's economy. Organizers of the new industry mostly avoided the types of ongoing conflicts that pitted the colonial hierarchy against the working class. Unlike the organizers, owners, and managers of gold and diamond entities, the bauxite hierarchy, early constructed housing for their employees, and provided educational, health care, business, and entertainment facilities.[8] While the environmental factors that bedeviled gold and diamond mining were largely absent in bauxite producing areas, its organization and production did not encourage small-scale operators as was the case in sugar, gold, and diamonds.

Claims have been made that the colonial government never fully grasped the extent of the bauxite companies' profits because of disingenuous auditing by its officials. Royalties, export duties, and income taxes paid to the government by the bauxite producers were regarded as extremely low in comparison to other such producing countries. Still, at face value, by the mid-twentieth century, bauxite was the colony's second greatest export-earner.[9] Managers of the bauxite conglomerates were also accused of ploughing back a minimum of their profits into the local economy and refraining from industrialization because they controlled and monopolized output, marketing, and the selling price of the commodity.[10] This was in keeping with the tenets of European and North American industrialization in the pre-Cold War era.[11] While it is true that the intervention of the Soviet Union in Latin America and the Caribbean after World War 11 unleashed forces for changes—they were overwhelmingly in the personnel themselves. All other structures—economic; educational; social; political; administrative; and security—remained almost intact. From the Cold War era socialists, Marxists, and Communists also fostered racism, authoritarianism, mayhem, underdevelopment, and dependency.[12]

Appearing as it did in the World War I era when its utility was invaluable to the aircraft, munitions, and manufacturing industries—investors and the bauxite organizers understood the potential benefits of reformed managerial techniques. The high quality of Guyana's calcined bauxite made it valuable for lining steel furnaces and other mechanisms requiring high temperature applications. It is rich in alumina, low in impurities, and indispensable in the refractory and abrasive industries. Aluminum, an end product of bauxite, is essential for the manufacturing of automobiles and aircrafts. Aluminum is lightweight and noncorrosive making it extremely valuable to the nautical and aerospace sections of the military. In addition to its utility to conduct electricity, aluminum is used in manufacturing household utensils and equipment such as pots, pans, bottle caps, cans, and foil.[13]

Plans and programs for bauxite production left little room for cost-minimizing and its inadequacies because of long-term market demands and the multiuse of the ore. Accessibility to bauxite deposits was never a problem and that enabled consistent production. The speculative nature of gold and diamond mining, the dangers associated with travel and work in the environs of those undertakings, and the predisposition of government and private entities officials to limit funding for infrastructure development militated against the permanent settlement of workers and the expansion of economic activities in the two sectors. This chapter analyzes how environmental conditions, created by nature and humans, changed historical antecedents and the trajectory of Guyana's mining development.

Organizing Bauxite Mining: The Demerara Bauxite Company Ltd., (DEMBA)

From the earliest stages of its mining operations in the country, DEMBA organized production to assure workers attractive wages, housing, health care, and other benefits. Bauxite employees were recruited locally and overseas. On the verge of World War I, this instance of better-than-usual employment and reasonable living conditions in the colony served as pull-factors for thousands of workers from gold, diamonds, sugar, and other economic activities. More than two decades after bauxite' advent, the *West India Royal Commission, 1938–1939* issued findings on the colony's mining industry. Bauxite was considered the most prosperous enterprise of the three sectors. Housing, health care, and other amenities were regarded as modern, and working conditions and wages gave little leeway for criticism of the way bauxite production was organized. In the realm of gold and diamonds, the findings indicated—undesirable conditions pervaded the small and remote mining communities, which were plagued by transportation difficulties. Labor contracts, which were a requirement for employment in the two enterprises, were deemed as draconian, obsolete, and obnoxious. Reforms were proposed for gold and diamond operations because employers overwhelmingly benefited from arrangements that were disadvantageous to employees.[14] Migration to bauxite mining was influenced by the frequent calamities, unpredictable circumstances, inhospitable physical environment, adverse labor relations, and the long-drawn out frontier conditions with which gold and diamond workers grappled.

Census Reports and those of the Department of Lands and Mines, the Institute of Mines, and the bauxite mining companies indicated the new trend in migration flow. Fueled mostly by company-sponsored, quasi-government influenced, and follow-on, or "grape-vine" migration, the inflow of workers, many of them with families, ended up living permanently at Mackenzie, Wismar, Christianburg, Akyma, Two Friends, Ituni, and environs of the Upper Demerara River District. Other workers migrated and were employed in bauxite mining operations at Kwakwani on the Berbice River[15] (see Figure 3.1). Granted that geological surveys established the ore's preponderant presence and technological innovations readily applied to its excavation in the form of capital intensive procedures as well as the needs of World War I and industrial complexes were important factors; a guaranteed resident workforce was critical for bauxite mining.

Housing Facilitating Production of Bauxite

Residential, healthcare, and other amenities provided by employers for bauxite workers contrasted in several ways with those which were available to their counterparts in gold and diamonds. Bauxite employees were able to utilize vital services such as the registration of births, marriages, and deaths in a timely manner in proximity to where they lived and worked. This was a marked departure from the significant delays endured by gold and diamond workers for similar services. Construction of schools, recreation, entertainment, business, and religious facilities and roads in the bauxite producing areas indicate that the organizers had moved beyond migration only to replace or supplement an existing labor force to ensuring permanent settlement.[16] Although this working class segment of the population was similar to those associated with gold and diamond mining, a different set of techniques and strategies were applied to all of the factors of production in bauxite. Cost-minimizing strategies were not particular features in housing or health care in the bauxite industry. Therefore, the pervasive high mortality rates, which bedeviled gold and diamond miners because of inadequate infrastructure works, were addressed during the fledgling years of bauxite production.

The insensitivity to the housing needs of workers in the precious metal and gems sectors is worthy of comparison with the situation in bauxite. Bauxite mining was established decades after that of gold and diamonds. Unlike the situation in the other two commodities, as soon as bauxite production got under way in the early 1900s, workers were provided or assisted with housing and other necessities through the establishment of the company town at Mackenzie (Linden) by DEMBA. However, the advent of bauxite mining was a disjuncture to and contrasted with gold and diamond operations in interesting ways. Bauxite, fostered its own unique characteristics. For instance, topographic and geological conditions made bauxite production a year-round activity, unlike the weather-dictated gold and diamond industries. In bauxite mining, frontier conditions were short-lived and production decisions included the permanent settlement of a large resident workforce that was family oriented.The expansive gold and diamond mining areas long remained sparsely populated due in part to inadequate housing. DEMBA established Mackenzie with bauxite operations at the center and residential facilities located at two sides. On the northern side was Cockatara, better known as "the village" and on the south side was Watooka. While Cockatara housed a largely African-descended workforce and their families, Watooka accommodated the households of the managerial staff who were mostly white expatriates.[17]

Accommodations at Cockatara were affordable and of a good standard compared with other similar dwellings colony-wide. The one drawback was that the company-furnished housing at Mackenzie was insufficient to meet the needs of an expanding workforce. Workers and their families who could not be accommodated at Cockatara lived mostly across the river at Wismar or Christianburg. A number of DEMBA's employees lived at various not-too-distant locations along the banks of the Demerara River. Some from among the male-dominated workforce chose to commute weekly by the *R. H. Carr*, the steamer plying the Demerara River between Linden and the capital city, Georgetown.[18] The steamer's journey lasted only eight hours, whereas, it took gold and diamond miners weeks to travel to and from interior prospecting sites. Mining workers attached to DEMBA never had to walk interminable miles overland or to navigate deadly waterways as their counterparts in precious metal and gems. Overall, conditions in gold and diamond localities discouraged large resident communities such as Wismar, Christianburg, and Mackenzie in Linden's bauxite producing regions on the Demerara River.

Traditionally, officials of DEMBA have received mixed reviews for their attitudes to the infrastructure across the river from Mackenzie because the company concentrated narrowly on developing the company town and barely, if ever, expended financial or technical resources to improve the standard of living at Wismar or Christianburg. Although since 1929, the two areas were under the jurisdiction of the Local Authority, a colonial bureaucracy, which infused minimal developmental resources. Still, Wismar and Christianburg, which had been settled long before bauxite mining began, provided additional living space for the industry's work force, their families, and others. By the 1950s, government- and company-assisted housing units were constructed at Wismar Hill and later, additional units were erected in adjoining areas on both banks of the Demerara River.[19]

In the body of literature, Odida T. Quamina and Mohammed Shahabudeen among others were not enamored with certain arrangements at the residential facilities DEMBA provided at Cockatara and Watooka; nor the regulations controlling access to and egress from them. In particular, nonresidents of Watooka, the more affluent residential complex, required permission to enter the area. The facilities were often characterized as segregated and bearing striking resemblances to Apartheid-era South Africa.[20] In reality, the housing DEMBA provided for its employees at Mackenzie was not substandard, but bore resemblances to the sugar estate environment where occupational stratification determined company-housing allocations. In the twentieth century, similar to bauxite employees at Wismar and Christianburg, many sugar workers lived in adjoining villages or in estate-provided dwellings while overseers, dispensers, chemists, and primary school teachers were housed in the company's compounds or nearby.[21]

The opinion of the company's newspaper, the *Mackenzie Miner*, is insightful on relative situations of residential facilities and environs. Viewing housing at Mackenzie including Cockatara as attractive, the report noted the existence of "orderliness and cleanliness." The newspaper pointed out that Mackenzie had a comparatively healthy population living in sanitary surroundings.[22] In fact, Mackenzie's housing, medical services, and communal facilities were superior to those in most other parts of the colony where the working class resided. Nearby residential and other facilities at Wismar and Christiansburg where bauxite workers also lived were comparable to various areas colony-wide, and far better and a distinct improvement over those available in the gold and diamond regions. It is telling that unlike those two sectors and its nonresidential officials, bauxite's managerial staff lived in the producing area.

Watooka, with Richmond Hill as the most exclusive and affluent section, were initially created as enclaves for the white and mostly foreign segment of DEMBA's employees. Dr. George Giglioli, a long-term medical practitioner attached to DEMBA, offered a perspective on the different approach by the organizers of bauxite mining. Dr. Giglioli observed that "the layout of Mackenzie"...which caused a "wide separation of the staff quarters from the main native village, was rational," because "it aimed to protect the nonimmune expatriates from contracting diseases" that "were heavily endemic in the local population, such as malaria, filariasis, enteric fever, and hookworm." Dr. Giglioli said "the only Guyanese quartered in Watooka were the staff domestics, all of whom were medically examined before employment."[23]

Bauxite management utilized and applied the knowledge of the value of a resident workforce in proximity to production by providing modernized residential facilities. The long established sugar industry had benefited from a large workforce resident on the estates and in surrounding villages. As a counterpoint, an inconsistent labor supply and lack of residential and other facilities contributed to an unpredictable output of gold and diamonds. Workers' commitment to gold and diamond mining seemed to be tempered by the nonchalant attitudes of the power structure and the lack of interests in their well-being. By utilizing the experiences of the plantocracy and the mineocracy to reform their socioeconomic plans and program, the organizers of bauxite mining acted judiciously.

From the start, the company controlled the construction, allocation, and maintenance of their real estate at Mackenzie. Considering that the general population growth in the country had never been significant, it also crept up slowly in the Upper Demerara River District. Thus, the number of company-controlled houses increased almost in tandem with the population's need. In 1921, DEMBA provided forty-eight dwellings and the number increased to seventy units a decade later. By 1938, the company was expanding its operations to satisfy global market demands for the commodity. To cope with their production targets, additional workers were hired, and housing facilities were extended. Specifically, additional accommodations were provided at Cockatara for newly recruited workers. Offsetting the known situation of scarcity of food and starvation in the gold and diamond fields, land was also offered to bauxite employees for their own agricultural pursuits.[24] By 1943, DEMBA had erected another thirty-five houses in Mackenzie and 103 additional dwellings were constructed within four years. From the late 1940s, with the upswing in production, the number of workers in the bauxite belt increased and more housing was in demand for employees and their families. In the ensuing decades, especially in the 1960s, the company assisted in the construction of additional housing units at Kara-Kara and Amelia's Ward within the bauxite mining environs.[25]

At first, three categories—race, income, and occupation—mostly determined the allocation of residences in Mackenzie. The higher wages paid to the mostly white expatriate employees were linked to their managerial/administrative responsibilities and technical expertise. These criteria determined the allocation of residential quarters to those employees in Watooka. The predominantly African-descended workforce, mostly occupying the mid- and lower-level positions at DEMBA and who earned less wages than the employees residing at Watooka, lived at Cockatara. At Cockatara, the company's accommodations were comprised of individual " houses for the ordinary staff and families; range houses for ordinary employees and their families and separated by a thin partition; and Bachelors' Quarters." Cockatara housed the Constabulary Compound and residential, barracks, and office facilities. In Mackenzie, nurses had a hostel, and domestic servants, secretaries, teachers, and other workers also lived there. DEMBA collected rent for its housing and retained responsibility for the maintenance of the properties. Electrical work, carpentry, plumbing, and the painting of buildings as well as spraying with DDT, other pesticides, and chemicals to eradicate disease-carrying mosquitoes, cockroaches, and rodents fell under the control of the company.[26]

Housing in the Upper Demerara River District and especially at Mackenzie, could be characterized as segregated because of race, occupation, and income variables. Houses may be deemed as suitable, insufficient, or substandard, but unpredictable working situations elsewhere served as "push-factors" for migration to bauxite mining. The critical "pull-factors" of work in a stable environment, and wages better than in other comparable economic undertakings countrywide attracted increasing numbers of workers to bauxite mining. Although, suitable housing in the Upper Demerara River bauxite mining region outstripped its supply, lack of accommodations could not staunch the flow of incoming labor migrants and their families from local areas and overseas who often displayed sheer ingenuity in creating housing for themselves. An African diaspora complexity was captured in the circuitous routes and diverse roots of the bauxite workforce that found 'ways to make a way.'

For example, 'bottom-house' (ground-level) accommodations were added to existing structures to expand living space for newly arrived family members and friends. In general, "internal immigrants from the county of Berbice," for instance, "tended to live in close proximity with each other and to help newcomers to the community find jobs and accommodation." West Indian labor migrants further exhibited the same kind of "solidarity by acquiring a building at Wismar, which up to the 1960s, was being used as a sort of reception centre for newly arrived West Indians."[27] So many West Indians from the smaller Caribbean islands arrived to live in Linden and worked at DEMBA that to this day at Wismar, there is a residential area called the "Pala-Wala" section. "Pala-Wala" is the name locals call migrants from certain West Indian islands. Still, for most of their sojourn in the bauxite belt, Caribbean and overseas workers, unlike local internal migrants were unable to return home at weekends or even infrequently.[28]

Darrell Gray, a native of Queenstown Village, Essequibo County, worked in all three sectors of the mining industry, gold, diamonds, and bauxite. As an employee of DEMBA, he infrequently returned to his primary residence in Queenstown, Essequibo. Grays' mining experience began in 1939. At seventeen years old, he had followed his father and older brother into the gold and diamond fields. The elder Gray, born in 1882, had died at thirty-eight years of age while prospecting in the interior, leaving the older son Samuel Trenton Gray to be the breadwinner of the family which included the mother Agnes Cecelia Patterson, born in 1883, and three daughters. Unfortunately, in October 1941, Samuel died from complications of black-water fever in the interior. Viewing gold and diamond mining as an arduous occupation filled with unpleasant memories; by 1943, Darrell Gray decided to seek employment at Mackenzie. At DEMBA, Gray trained as a mechanic and worked in that occupation until he retired in 1981. Gray supported a wife and two offspring in Essequibo; he rented accommodation in Linden and, infrequently commuted to his family.[29] George E. R. Adams of Victoria Village, East Coast Demerara crisscrossed the country to work in all three sectors of the mining industry. From the 1940s to 1950s, Adams was a gold and diamond miner, but he switched to work at the Kwakwani Berbice River bauxite site.[30]

Some women who were born and resided in other areas of the country ended up living in the bauxite mining area. Eunice Williams, born Murray at Zorg-en-Vlygt Village, Essequibo County married Albert "Mike" Williams on December 2, 1949, in Georgetown and had seven children. "Mike" Williams, grew up with his grandmother in nearby Danielstown Village, but left his home at the age of fourteen to live with a relative at Linden. By 1940, he completed his primary education and trained as a mechanic at DEMBA. Williams rose to the rank of foreman by the time he retired in 1975. The couple met when Eunice Murray visited an older sister living at Mackenzie. At first, the Williams family rented accommodation at Wismar, but through the DEMBA's Saving Scheme and other company-financed projects, they built their own home. As a vendor in the local market, Eunice Williams added to the family's income. The Williams were part of the trend of family members and friends from the same coastal localities internally migrating to live and work in the riverain bauxite mining area.[31]

Such was the case of the Proctor family. Electrician, J. A. M. Proctor, joined DEMBA in 1919. In 1921, when the company temporarily ceased operations, Proctor obtained a position with the Demerara Electric Company in Georgetown. On the resumption of DEMBA's operations in 1922, Proctor returned. In 1926, he was appointed as a foreman. Ten years later, he was elected as an associate member of the American Institute of Electrical Engineers.[32] Almost a centenarian by June 2010, the electrician's cousin "Frank" Francis Hector Proctor who was born in Princess Street, Georgetown, was interviewed by the author in London. Frank Proctor was raised by his aunts and attended Brickdam Roman Catholic Primary School. In 1930, he left the city for the bauxite producing area and lived at Wismar. At first, he was a tailor, but later trained as a mechanic with DEMBA, and was employed in the mines. In 1949, after his unsuccessful bid to obtain a position in the Electrical Department of the company, Proctor uniquely reversed the trend of migratory movement. Leaving his wife, the former Urline Gobind and five children in the bauxite belt, he became a machine-shop fitter for B. G. Consolidated Goldfields Ltd., at Potaro, Essequibo County. After the company ceased operations, in 1955, Proctor viewed mining as coming to an apogee for him and migrated to the United Kingdom. Reminiscent of the memorable times he spent in the bauxite belt, in his twilight years Frank Proctor has a sewing machine and enjoys playing his clarinet, saxophone, and piano.[33]

Harold "Paris" Thompson hailed from Almond Street, Queenstown, Georgetown. After a primary education at the city's St. Georges Anglican School, he became an Apprentice mechanic with the Cuyuni Goldfields Ltd. Between 1946 and 1952; Thompson worked for the company at Potaro, Barima, Barama, and Wariri in Essequibo's gold and diamond fields. In 1952, he started a family and worked in Georgetown at the Demerara Foundry for one year. Eventually, he moved his entire family comprising his spouse Elaine Matthews (my mother) and their five offspring to live at Wismar Hill, Retrieve, and Mckenzie, respectively. Thompson worked as a carpenter for both DEMBA and Sprostons Ltd. When his spouse died in 1959, he could not endure living in the bauxite belt without her. Thompson migrated to the UK in 1961, where, at eighty-five years of age in 2010, he still lives. Three of his offspring joined him later.[34] Proctor, Thompson, and many other incoming labor migrants to the Linden environs viewed their sojourn in the bauxite belt in relation to the quality of life generally available colony-wide and considered overall conditions in Linden as highly satisfactory. Although housing was periodically insufficient for the expanding workforce, DEMBA ensured that its employees had access to one of the most modern health care facilities in the colony.

Bauxite Mining and Path-Breaking Health Care Services

In May 1925, nine years after DEMBA was established, one of its priorities, the modern Mackenzie Hospital was constructed, equipped, and staffed by the company. Not only were services provided for DEMBA's employees and their families, but other residents in the district and elsewhere were treated at the facility.[35] Prior to the construction of the new facility, DEMBA had procured the services of Dr. Giglioli who arrived at the mining town in November 1922. On his arrival, Dr. Giglioli had considered the existing hospital antiquated. He observed that "it was a long, rough, wooden building and comprised a ten-bed ward, a dispensary, a treatment room and a tiny office."[36] The staff at the existing facility had included a dispenser, a male nurse, an attendant, and a cook. The apparatus was obsolete and there were no sterilizing facilities for dressings, syringes, and other medical and surgical equipment. The hospital overwhelmingly catered to the needs of male patients.[37] As early as 1921, plans had been in the offing to construct a multilevel hospital; and work had actually started at a site on the right bank of the Watooka Creek, at its confluence with the Demerara River. After Dr. Giglioli's arrival, he collaborated with DEMBA's officials and submitted the blueprint for the erection and equipping of a new $480,000, sixty-bed hospital.[38]

The hospital, 290 feet long and modern in its features, was made of reinforced concrete. Initially, it was envisioned that the two public wards and six private two-bed ones would accommodate fifty-six patients. However, provision was made for a maximum of seventy-five beds. Among the features of the facility was an up-to-date orthopedic table, complete with instruments for attending to fractures including steel plates and screws. Provision was made to perform many other surgical procedures at what was essentially a healthcare complex. Other amenities at the institution included a Bacteriological Laboratory, an Operating Theater, and an Electrical Steam Steriliser Room. For maternity cases, the complex had a delivery room. There was an operating room for septic cases and minor operations as well as an X-Ray room with equipment for diagnosis and therapeutics.[39]

The healthcare complex supported a dispensary and an outpatients' department. The central location of the dispensary with its large storage area facilitated the issuing of supplies to the Mackenzie Clinic and apothecaries at Akyma, Wismar, and Butukari, Upper Demerara River which were partially owned by the private concern, Messrs. Sprostons Ltd. In addition, the complex's modernized dispensary supplemented similar services provided by a government-run establishment at Christianburg. A mobile dispensary plied the Demerara River on alternate days for fifteen miles on either side of the hospital's location and transported gravely sick persons to the facility especially during epidemics. Among the hospital's staff were a chief medical officer, surgeon, matron, nurses and trainees, orderlies, and wardmaids—some of whom had accommodations.[40]

The multi-complex healthcare facility underscored DEMBA's intention of long term bauxite operations in the area. It demonstrated that bauxite officials had observed the adverse effects of inadequate health care nationally and were committed to providing improved services to their employees. The officials avoided the frequent shortfalls of labor due to preventable and epidemic diseases, characteristic of gold and diamond production. They understood the importance of timely and adequate healthcare services in close proximity to employees' residences and work sites. Providing enhanced services that were beneficial to their employees' welfare assured workers' commitment to meet production deadlines at short notice. Furthermore, staffing Mackenzie Hospital with trained personnel and installing modern equipment in the facility, provided services that were far better and a direct contrast to what had previously been obtainable anywhere in the colony. DEMBA's modern hospital ensured services that supported permanent settlement of its workers in proximity to their employment and residences.[41] The officials understood that a healthy resident workforce augured well for increased profits.

Healthcare services, both prophylactic and therapeutic, provided by the company's facilities were extended to employees, their families, and other residents of the Upper Demerara River District. Many West Indians were among the mostly male recruits for employment with the company. They were overwhelmingly between the ages of eighteen and forty years, and were medically examined before employment. In 1925, many of them were among 1,285 patients who received treatment at the hospital. Patients were hospitalized for a total of 10,803 days. Of this number, 5,568 of the days were utilized by employees of DEMBA and their families; and the remainder by nonemployees. Dispensers attended to 25,027 persons who needed minor medical, surgical, and dental procedures and the filling of prescriptions. Approximately, 20,200 employees of DEMBA and their families attended the dispensaries; the other 4,827 were nonemployees of the company.[42] DEMBA's Sanitation Department had twenty employees including a foreman, sub-foreman, and a laboring crew who carried out preventative work around residential and other facilities. Sanitation crews disinfected living quarters and routinely carried out mosquito control spraying throughout the bauxite plant, hospital complex, and the entire mining area. In addition, the company provided food inspection services.[43] By 1939, DEMBA was spending thousands of dollars yearly to ensure adequate pure water supply for drinking and domestic purposes.[44]

A Focus on Malaria et al.

As the nineteenth century ended, international research revealed the role of mosquitoes in the transmission of diseases such as malaria, filaria, and other debilitating fevers. In the colony, "malaria reduced the birth rate, and increased the rate of abortions, miscarriages, premature deliveries" caused "infant deaths from prematurity and congenital debility" and "accounted for 40% of all infant mortality."[45] At the time, two local medical practitioners, Drs. A. S. Ozzard and J. F. Ferguson were conducting research in the causes of hookworm, typhoid, malaria, blackwater, and intermittent fevers. Most of what Ozzard wrote in 1893 remained valid up to the eradication of malaria in the colony's coastlands, more than half a century later. In Guyana, by the late nineteenth century the widespread use of quinine as a cure for malaria began, and a mosquito ordinance went into effect in 1909.[46]

By the 1920s, the colonial administration had already begun sporadic and limited mosquito eradication services in some areas of the country. In the case of the gold and diamond mining regions, government officials were unwilling to provide expansive services without assurances of forthcoming financial benefits for the expenditure. In 1925, public health care improved with the acquisition of the services of Dr. Giglioli and the establishment of the medical facility by DEMBA. Dr. Giglioli became known as a malaria specialist for his initial services at the mining company and in the environs of Wismar, Malali, Rockstone, Butukari, and Tumatumari. He built on the work of his predecessors in the colony especially that of Dr. Ozzard's, and introduced *Atebrin, Paludrine*, and *Chloroquine* to treat cases of malaria. After 1925, at the behest of DEMBA, Dr. Giglioli was prophylactically and therapeutically treating malaria in the Upper Demerara District with much success.[47] Consequently, the general health of the company's workers improved and few absences were due to sickness or injury.[48]

From its earliest years of operations, DEMBA was on a path-breaking course and in the forefront of providing unparallel, appropriate, and up-to-date healthcare services in the bauxite belt of this colonial society. Unlike earlier government bureaucrats who paid scant attention to professional advice, bauxite's officials' initiatives in procuring the services of Dr. Giglioli influenced drastic changes in the provision of malaria eradication in the entire country. From the World War I era, US organizations such as the Rockefeller Foundation had begun to provide funds and expertise to improve public health in the colony. By the time of Dr. Giglioli's arrival, overseas assistance had increased. From 1958, Dr. Charles C. Nicholson, the country's chief medical officer of health, built on Dr. Giglioli's work and more wide-ranging projects were carried out. Later, British and American medical experts and grants from the Colonial Development and Welfare Fund extended important public health work including malaria eradication nationwide.[49]

Another health care priority of DEMBA was the control of hookworm and its infectious disease. Between 1923 and 1925, all new employees of the company, approximately 5,500 persons, were satisfactorily treated with Carbon Tetrachloride. Prior to the initiation of the mass treatment campaign, the infection index among the workforce was 68 percent, but the most recent hookworm survey revealed a low infection rate of 3 percent. Apparently, the diagnosis, treatment, and cure of hookworm infestation led to increased bauxite production at reduced costs. Treatment was also available for typhoid fever. Isolation, disinfection, and vaccination of infected persons were promptly undertaken to prevent the spreading of the disease[50]

During 1925, the company had constructed two large community latrines. Each contraption contained a septic and chlorination tank and carried forty gallon automatic flush capacities which were suitable for 300 users. By that time, company-controlled residences were upgraded. Improved facilities "leading to a system of soaking-pits, for the disposal of kitchen waste" were installed. This innovation eliminated the accumulation of stagnant fly breeding water from open surface drains. Aptly, the commissioner of Lands and Mines stated thus, "the establishment of this highly organized, well-equipped and progressive industry marks an entirely new and important chapter in the mining development of the colony."[51]

Within the first decade of bauxite operations, the mostly Canadian managers undertook several projects to provide a reasonable standard of living for employees, their families, and other area residents. In contrast, British colonial officials and the plantation hierarchy took an inordinately long time to address adequately health care, housing, and other quality of life issues in the sugar industry.[52] Similarly, government officials and the mineocracy pursued policies that were detrimental to workers associated with the sugar, gold, and diamond industries. In the case of the latter two, haphazard organizing inhibited large-scale settlement in proximity to production. Infrastructural development of the interior mining regions proceeded at a slow pace compared with that in bauxite's environs. From the 1920s, a few years after its operations began, and until it was nationalized by the government in 1971, and renamed the Guyana Bauxite Company (GUYBAU), DEMBA provided and maintained effective healthcare as well as educational facilities including academic institutions, trade and nursing schools, and libraries.

The Bauxite Mining Community: Educational and Recreational Factors

The construction of schools, places for recreation, entertainment, business, religious worship, and roads encouraged permanent settlement of workers for DEMBA's bauxite operations. On October 5, 1940, a crowning achievement of the company, an impressive educational facility was erected in the area. The new and modern educational facility, the Mackenzie Primary School, rivaled the best among such institutions colony-wide. The school had domestic-science and handicraft centers and a well-equipped gymnasium. Many extracurricular activities in Mackenzie and its environs took place at the school. Across the river, St. Aidan's Anglican and Christianburg Church of Scotland (schools which eventually came under government control) traditionally provided primary education. On the banks of the Demerara River, north and south of Mackenzie, government-operated schools existed at Dalgin and Coomacka, respectively. DEMBA established Echols High School in 1946 (later, renamed The

Mackenzie High School). A Mackenzie Nursery School existed. The DEMBA Trade School was founded in 1958, and the company established a training school for nurses.[53] Chapter 8 provides a detailed discussion on education and professional training in relation to gold, diamond, and bauxite mining.

Wismar and Christianburg, the original residential areas, prior to the coming of the bauxite complex, had long-established places of worship. Building on this base, the entire mining area which included Mackenzie continued to support various religious denominations. Church bodies included Episcopal/Anglican, Church of Scotland, Roman Catholic, Ethiopian Orthodox/Coptic, Pilgrim Holiness, Seventh Day Adventist, and Methodist. Religious institutions were viewed in some quarters as "part of the Company's policy to cater to the fullest extent to the general welfare of its employees."[54] Sections of the churches' property such as the halls and grounds served as additional recreation centers for the entire community.

DEMBA constructed a recreation hall and two swimming pools; one located at Watooka and the other at Cockatara to serve its employees, residents, and others. The Mackenzie Inn, the Mackenzie Hotel, and Crescent Cinema provided venues for entertainment and recreation. Similar to housing arrangements, the Mackenzie Sports Club and a Supervisors' Club served employees according to their status. Later, the Salvation Army operated a guest house. The residents of Watooka, in addition to having their own swimming pool, had a guest house, cinema, staff club, a golf course, and a tennis court. Brownies, Girls' Guides, Cubs, Boys' Scouts as well as the St. John's Ambulance Brigade and the Red Cross functioned in the mining district.[55]

For several months in the early 1920s, due to the ending of World War I and lack of demand for bauxite, production by DEMBA ceased. In 1922, soon after the resumption of operations, steamer excursions were organized to Mackenzie City. One such trip, slated for Sunday July 30, 1922, promised trippers three hours of merriment in an enchanting city. A jazz band would be in attendance and refreshments were to be sold at modest prices. To cool the tempo, ice would be supplied free of cost. Tickets for the excursion cost $1 each. A trend began, and by 1939, under the auspices of Cyril Lord, secretary of the Transport and Harbours Dept. Sports Club, the *S. S. Tarpon* was chartered for a day trip to Wismar. Excursionists were to be treated with a visit to DEMBA's operations at Mackenzie. Significantly, the festivities to be held on the first Monday of August coincided with a national holiday commemorating 100 years of Emancipation from enslavement of the African component of the population. During 1937, a stage was erected at the Cockatara Recreation Club Hall as a venue for boxing-matches. A resident orchestra, under the direction of Herman Paris held monthly concerts at the facility.[56] Parallel to recreational and entertainment amenities, commercial outlets featured throughout the Linden area.

Commerce as a Factor

In the initial years of the organization of the bauxite industry, the company store emerged and eventually gave way to a general store. At first, employees of DEMBA and their families relied on a ration store, owned and operated by the privately managed, Messrs. Sprostons Ltd., for their supplies. By the early 1920s, employees who were entitled to subsistence were issued tickets to collect their rations of rice, pork, salt-fish, sugar, and other items. Subsequently, DEMBA organized a general store to replace the one which only issued rations. By 1924, the general store supplied not only food items but also footwear, clothing, hardware, utensils, haberdashery, and other commodities. Cash and charge transactions were utilized. Dollar-value coupons were issued to employees and goods taken were charged to their accounts. After 1948, DEMBA relinquished the general store's virtually monopolistic operations and allowed private business enterprises to operate. The commercial operations included William Fogarty's Ltd., Bookers Company Ltd., and A. Choo Kang and Co. Other businesses, which eventually served the bauxite mining environs were Singer Sewing Machine Company, Edwards Garage, and stores operated by Sandbach Parker and Company Ltd., Chapman, Hamilton, and Bata Shoes.[57]

The market constructed at Mackenzie was another one of DEMBA's undertakings. Vendors who rented from the company purchased a variety of wholesale supplies from the general store, other enterprises, or procured them from Georgetown. Along the banks of the River Demerara, at Wismar and Christianburg, various retail outlets including shops and stores existed and supplemented those at Mackenzie. There was also a market at Wismar. DEMBA operated a dairy farm and supplied products to the mining community. The Royal Bank of Canada and British Guiana and Trinidad Mutual Assurance Company provided financial, actuarial, and associated services for residents of the bauxite belt. In short, DEMBA provided homes and healthy surroundings, medical, surgical, dental, and other health care services, social, economic, educational, recreational, and commercial facilities as well as pension benefits and a Life Assurance Scheme.[58]

Aspects of Infrastructure in a Company Town Environment

Connecting all activities in the Upper Demerara River District was the lengthy waterway bearing the same name, Demerara River. For the most part, transportation on the river between the mining complex and the capital city was predictable. The few rapids on the river were hardly a problem for navigation by vessels not exceeding specific tonnages. The steamer, *R. H. Carr*, owned and operated by Sprostons Ltd., provided transportation for passengers and goods along the sixty-five mile route between Georgetown and Mackenzie with a stop at Wismar. Decades later, the river was replaced as the main artery of transportation by the construction of the Linden-Soesdyke Highway.[59] Myriad numbers of small motorized crafts provided transportation services for passengers and goods between both banks of the river. A bridge, spanning the waterway that catered to vehicular and pedestrian traffic, was constructed in the vicinity of Linden between Akyma and Two Friends. Travel took place beyond Linden to the upper reaches of the Demerara River. Sprostons Ltd., operated "a weekly launch service between Wismar and Malali, and a thrice-weekly launch service from Rockstone to Tumatumari, Potaro River."[60]

A long winding thoroughfare between Wismar and Christianburg connected the two areas with many alleys, valleys, and hills in the vicinity. They include D'anjou Alley, Valley of Tears, and Blueberry Hill. A network of roads such as Mora and Greenheart streets and Arvida road existed at Mackenzie. Initially, some road surfaces comprised of reddish-pink clay material—(bauxite-residue), but during rainy weather the resulting slush made movement problematic. Later, longer and wider roads were constructed with more durable and better maintained asphalt surfaces.[61] During the first three years of its operations, DEMBA constructed a railway that would extend for ninety miles to facilitate transportation of ore from the mines to the factory.[62] Organizing the bauxite mining community in the Upper Demerara River District included the provision of post offices and police stations to serve and protect the increasing population. The Constabulary Police Force was established by DEMBA solely to protect its properties and interests.[63] The law-enforcing body caused conflicts in labor and social relations at Linden as the next chapter explains. It presaged the coming of trade unions at the insistence of the workers.

Conclusion

In the bauxite industry strategies of modernization and "order and progress" were applied somewhat differently to other economic undertakings in Guyana. While elements of labor control existed, bauxite management took action to ensure workers' welfare and increased profits. In some ways, DEMBA's officials successfully utilized the experiences and tactics of colonial officials, sugar barons, and the mining hierarchy, but crafted different techniques for the long-standing profitable production of bauxite. To achieve this, the organizers of bauxite mining executed their plans and programs with finance and immediacy; provided amenities and paid attractive wages. By the mid-twentieth century as Table 3.5 and the discussion in Chapter 3 indicated, the value of bauxite exports outstripped or competed favorably with the colony's other major export commodities.

The dynamics of migration and mining included permanent settlers in the bauxite regions who may have previously worked as gold and diamond prospectors; and those who initially worked in bauxite, then switched to the precious metal and gems industries, before migrating overseas. Granted that the physical environment in the bauxite areas was unlike that of the gold and diamond mining, its organizers' different approach to housing, health care, education, entertainment, and the like encouraged permanent settlement because of the better standard of living available. Therefore, bauxite's organizers utilized another approach and created a personnel-friendlier environment compared to that of gold and diamond mining. The colony's other major export commodities, sugar, rum, timber, rice, gold, and diamonds, had emerged in the nineteenth century. In the twentieth century, demand for those commodities and bauxite in a world punctuated by two global conflicts and several minor ones continued, but it was the different approach of the organizers of the latter product that enabled its prosperity and timely benefits continued. Bountiful returns from gold mining would only occur as the twenty-first century approached, but this time with far less catastrophic environmental consequences for the workforce. Nonetheless, in the ensuing years, adjusting to a more sophisticated workforce proved enigmatic for bauxite management. The following chapter indicates that similar to other enterprises in the colony, employees and their associations challenged the bauxite hierarchy to gain even more varied benefits from the proceeds of their labor.

CHAPTER 7

Evolving Relations: Mining and Trade Unionism

In colonial Guyana, evolving, but less than harmonious socioeconomic and labor relations between the owners, agents, and managers of larger mining enterprises and the African descended workforce led to the advent of trade unions in the industry. Coming as they did out of a rancorous situation where the hierarchy exploited and oppressed labor, mining workers remained vigilant in safeguarding their welfare. Discord, rebellions, and strikes had punctured the enslavement, post-abolition, and pre-mining eras when sugar production dominated the economy. Conflict was pervasive because of workers' dissatisfaction with planters' attitudes and a capitalist system that favored the plantation hierarchy. The enactment of more stringent laws to deal with the strife militated against compromises and caused festering resentments. Disputes between the colonial hierarchy and the laboring class were not confined to the workplace. Continuous conflicts erupted into public spectacles over the imposition of arbitrary taxation, restrictive legislation, unfair treatment, and a host of other matters that working class persons felt were injurious to their well-being. Throughout the mostly agricultural era which had preceded that of mining, adversarial relations between the power structure and this segment of the population worsened.[1]

Conflicts were inevitable when gold and diamond mining emerged because members of the mineocracy aligned themselves with colonial administrators and the plantocracy to maintain the status quo. As a group, the expanded power structure propagated the same attitudes toward the laboring class which, by the late nineteenth century, included gold and diamond miners.[2] Similar to their predecessors, the newly reconstituted working class was committed to making decisions about their labor and improving their socioeconomic conditions. As outlined in Chapters 4 and 5, with the coming of interior mining workers environmental struggles continued alongside an intransigent management and an indifferent colonial government. Workers and their advocates who kept stressing the need for improving the infrastructure in the mining regions perceived that the lack of timely action to address their concerns had much to do with their status. The intentions of the power structure to perpetuate their profit-oriented agenda instead of judiciously addressing quality of life and work-related problems impacting the miners led to adversarial relations in the new industry. This chapter analyzes how strategies devised by colonial administrators and managers of private companies to maximize profits from mining presaged acrimonious relations and the coming of trade unions.

Bureaucrats and business persons had erroneously believed that mining workers would fail to see through the veneer of their profit-oriented strategies, discriminatory laws, and one-sided policies which benefited employers. The chain of events that followed encouraged some workers, employers, agents, and managers to disregard contractual obligations. Short of taking the law into their own hands, litigation was an option to settle disputes in the gold and diamond sectors. Often, the law-enforcement apparatus tilted to the advantage of the mineocracy in outcomes. Viewing this state of affairs as unacceptable, workers and their leaders continued to challenge the status quo. Ultimately, disputes due to the illegal acts of offending parties in the industry required both judicial and trade union interventions.

The specter of race was bound to appear in the relationships because the economic elites were overwhelmingly white Europeans and North Americans. The miners comprised mostly of people of African ancestry. On occasions, colonial bureaucrats made certain types of stereotypical assumptions concerning the capabilities or lack thereof of African-descended persons.[3] In part, the assumptions explain why socioeconomic and labor relations were problematic. The chapter elucidates that the inability of the power structure to make decisions and to enact laws without the tinge of race evincing, subliminally and overtly, contributed to adversarial relations in the gold and diamond industries. Although, the experiences of bauxite workers were not identical to those of the two other sectors, eventually relations between its management and employees became stormy.

A glance at the European power structure's attitude to workers during the Industrialization era adds some perspectives to their local counterparts' actions. In socioeconomic and labor relations in the European setting, class played an important role in the way the economic elites or the power structure, that is, aristocrats, the bourgeoisie, and government officials viewed and treated the working class. The economic elites paid little attention to the inhumane conditions which caused those workers to engage in wide-scale protests. When Karl Marx and Friedrich Engels took up the workers' causes and offered socialists/communists ideologies to reform the societies, the European power structure budged. Eventually, to minimize the radical changes Marx and Engels had proposed, and at the urgings of advocates of workers' organizations, the hierarchy made some efforts to improve the employees' socioeconomic conditions and instituted some reforms.[4] In the European milieu the majority of persons were of the same race, but optimum profits for themselves and class differences had motivated the upper echelons to exploit and oppress the workers.

As the struggles of the working class evolved in the Guyanese setting, it was evident that trepidations about socialism/communism that had been encroaching in the Latin American and Caribbean region by the late 1800s, had gained a foothold by the Cold War era and influenced a reversal in some imperial policies. Although the trade union movement was an integral catalyst for that change, African diaspora institutions and organizations played critical roles in preparing the Guyanese people for impending changes. The forces unleashed earlier by both the colonial hierarchy and the working class and their leaders are germane to an understanding of how the struggles of the African component of the country's population contributed to changes. In this context, while the class factor is not absent, the issue of race had to be reckoned with in a forceful way.[5]

In the Guyana of the nineteenth and twentieth centuries, race differences elucidated in the very writings of the power structures fostered adversarial relations between the mineocracy and the working class. Coming on the cusp of European labor developments and reforms, in the local situation, elements of authoritarianism, racism, class, paternalism, and profiteering impinged on relations between the colonial hierarchy and the workers.[6] Therefore, perspectives on the race/class dichotomy and "profits-over-people" are inevitable. In this analysis, a major premise is that while self-representation was important to ensure mechanisms promoting equity and justice for all, both traditional workers' organizations and European models were utilized in the evolution of trade unionism in the country.[7]

After the initial years of gold mining, a series of government commissions were convened to address varying aspects of the country's economic direction. Invariably, members of the commissions were mostly expatriate colonial and mining officials, legislators, planters, and businesspersons. Healthcare personnel and clergymen gave evidence at the hearings, but the clout of the plantocracy and their legislative supporters often influenced outcomes and decisions beneficial to themselves, but inimical to workers' welfare. Often, implementation of the decisions perpetuated adversarial relations. Most critically, financial expenditure for interior infrastructure work to facilitate mining was usually insufficient to construct much needed facilities. Inadequate facilities compromised the safety of the workers.[8]

Gold and diamond mining workers did not passively accept their conditions. After all, many of their former enslaved ancestors had been active participants in the emancipation process—by no means a smooth transition from their previous conditions. Following the enslavement era, socioeconomic, and labor relations had remained acrimonious throughout the British Caribbean for the remainder of the nineteenth century. Long standing organizing mechanisms of African Guyanese visible throughout the 1800s helped them to withstand the power structure's ongoing onslaughts which jeopardized their welfare. While many working class persons focused on economic pursuits, some of them acquired formal education and were professionally competent. Others owned newspapers and were leaders in political and community organizations. A sense of creating "opportunities from misery" existed even as conflict continued on the plantations and elsewhere.[9]

In the twentieth century, their traditional organizations functioned parallel with trade unions (some with international affiliations) to address issues vital to working class life.[10] Labor exploitation, adverse socioeconomic conditions, and conflict were unavoidable, but this chapter analyzes some activities of a segment of the population who endeavored to empower themselves and bring about reforms. It offers additional conceptualizations of the working class other than as victims of authoritarian regimes whose intentions were to keep them impoverished, dependent, and powerless. Scholarly studies, particularly those of Ashton Chase, Vere T. Daly, Odida T. Quamina, and Maurice St. Pierre address the emergence of trade unions in the country. Not lost in those scholars' analyses are the long standing struggles of workers against management who acted as though they had monopolies to control the country's direction only to their advantage. In devising measures to constrain labor, the plantocracy and the mineocracy enlisted colonial administrators and legislators to ensure their privileged and beneficial positions in perpetuity; while continuing to remain in control. Many workers combined their resources, and, as always, confronted the exploitation and oppression bedeviling them, often, in the face of hostile law enforcers. Resistance, coupled with violence and the organizing abilities of the enslaved, apprenticed, indentured, free, and unionized workers had much to do with their ongoing efforts to improve situations for succeeding generations.[11] In the case of mining workers, the trade union movement evolved as a valuable resource to reform socioeconomic and labor relations.

Factors Influencing Relations in the Mining Industry

In 1890, suggestions aired at a government-convened commission to address labor shortages in several enterprises, set the tone for long lasting impaired relationships in the mining industry. Strategies which had been utilized to construct the trans-isthmus Atlantic-Pacific link and in Guyana's sugar industry such as the benefits from importing more labor migrants than required, to undercut workers' wages were detailed. Importing excessive labor migrants increased competitiveness among workers which could result in more profits for owners of mining companies. This tactic would ensure large numbers of under- and unemployed persons to serve as replacements for gold and diamond workers who died because of the unsafe conditions under which they travelled, lived, and worked. Eventually, in the bauxite mining sector, excess workers were utilized by management to break the strike action of full-time employees.[12]

Family considerations, alternative employment opportunities, and lifestyles in general influenced the ways in which mining workers responded to workplace discord. For example, unlike their bauxite counterparts, most gold and diamond miners had no family ties in the interior where they worked; making it much easier for them to opt out of company-affiliated jobs. Laws enacted to allow more access to government land and the expansive precious metal and gems producing regions offered alternative opportunities to the entrepreneurial-minded. As a rule, gold and diamond miners were seasonal migratory workers and those who originated from rural areas were farmers at home during off seasons from mining. Although work-related disputes arose, some mining managers were not oblivious to worker's options, and, likely tempered their attitudes to employees.

Usually, bauxite workers established families near to or within easy travel distance to their workplaces, thus, relocating was not such a viable option for them. Some bauxite miners had already worked in the gold, diamond, sugar, or forestry industries or on the waterfront. Work in those enterprises could be precarious or wages could be less in those sectors, which offered little or no benefits compared with the bauxite industry. Although working conditions in bauxite may have had advantages over other employment enterprises, for some, the situation was far from ideal. A number of bauxite workers demonstrated that they were going to remain in the industry and improve their situations by challenging unfair management practices through activism and unionism. Strikes and other industrial actions in the bauxite industry became a reality.[13]

The body of literature focusing on unrest in the bauxite industry identified several reasons for the workers' activism. Reasons included management's eventual methods of mobilizing and using replacement workers to arbitrarily end labor disputes. Regulations which circumscribed living and working conditions of some employees were also cited. In addition, unfair treatment and limited upper mobility were given as reasons for discontent among some bauxite workers.[14] Employer/employee interactions and unresolved matters led to complex relationships among mining personnel. Eventually rancorous relations between management and some workers fostered the coming of trade unions in the bauxite belt.

Gold and diamond prospectors also operated their own enterprises, thus labor disputes involving employees of privately owned companies largely arose at mid-level and in larger mining entities. Compared with the prospectors, bauxite employees were in monopolistic situations where strike action, if taken, were in locations where workers had little recourse to alternative employment. With the backdrop of family considerations uppermost in workers' decisions on industrial action, arrogant managers believed that fear of dismissals and few job options in the area would limit bauxite employees' activism. Nonetheless, trade unions that emerged to represent workers mostly addressed employer/employee grievances occurring in privately owned entities. With the coming of trade unions, its officials would also exert some influence on the framing and application of laws impinging on the rights of diverse sections of the workforce.

Applied to the mining industry, ways to control, exploit, neglect, and manipulate workers ensured the continuation of turbulent relations—pervasive in the colonial society. Workers devised ways to cope with the contrived situations which caused them hardships. Some gold and diamond workers may have resorted to illegal means such as absconding with monetary advances and stealing to deal with their losses caused by employers who unjustly withheld wages, but others formed labor unions for the same purposes. Dishonest and unscrupulous practices by both employers and employees in the gold and diamond industries necessitated legal measures to stymie illicit activities. Those involved in and associated with the two sectors, required surveillance, apprehension, trial, and punishment for committing illegal acts.[15] Employers and employees were pitted in adversarial roles with the criminal justice system as the only arbiter.

Absconders in Gold and Diamond Mining

Absconding bedeviled gold and diamond mining and soured labor relations in the industry. The prevalence of absconders—mining workers who accepted monetary advances, contracted to work for a specific employer, and never reported to prospecting sites—required legal measures to curb the practice. Dangers caused by travel and work environments; unpredictable financial returns; deplorable living conditions; inadequate health care; uncaring and dishonest employers, and questionable labor and managerial practices led to absconding by workers.[16] Some potential workers inanely felt that it was not in their interest to be honest and used loopholes in the recruiting and registering of employees to commit acts of dishonesty. In the embryonic stage of organizing the gold and diamond mining industries, it was almost impossible for employment agents to detect recruits who registered to work with one company for a certain period, took monetary advances, never reported for work as specified, and repeated the offence with impunity. This development led to distrust of prospective employees and caused ill-feelings in the industry.

A perusal of the registration of 712 gold miners by the Institute of Mines from June to August 1890, indicated the confusing ways in which the mining industry functioned. Repeatedly, the same name recurred with suspicious similarities in terms of personal characteristics including ages; heights; places of residences; and so on.[17] Registering offices were ill-equipped to ensure timely verification or to detect duplications and discrepancies, even if the clerical staff suspected irregularities. Although workers cited mitigating circumstances for their infractions, relations in the mining industry were impaired due to recruiting and registering methods which caused employers to lose money when workers exploited loopholes and absconded.

Dishonest workers, who had no intention of honoring their obligations, could also register several times at various locations, collect monetary-advances at each place, and never report for duty as arranged with any of their prospective employers. Such practices could remain undetected for some time because prospecting was undertaken in far-flung locations which were often hundreds of miles away from administrative centers. Absconders faced dire consequences including incarceration, restitution of advances, monetary penalties, and the threat of corporal punishment, if apprehended. Laws were enacted at the behest of employers to address the illegal practices of workers.[18] Laws not only punished guilty workers who absconded, but served to deter others who entertained similar ideas. Eventually, the names of accused absconders, their physical characteristics, vital statistics, and last known addresses were published in *The Gold Mining Gazette* and local newspapers. This made it somewhat easier to detect absconders and enabled employers to prosecute them. Between April 1 and June 24, 1892, approximately 287 notices of arrest warrants were issued. Breach of contract cases were pursued at Magistrates' Courts to the full extent of the law. Penalties under the *Masters and Servants Ordinance of 1853,* were applicable to the guilty. Many of the workers prosecuted were described by a kaleidoscope of colors including Black, Brown, Yellow, or Red-skinned, and even "Coloured."[19]

Provisions embodied in *Section 1, Mining Ordinance, 1896,* which was amended in 1903, 1905, and 1909 overwhelmingly targeted infracting workers. The 1909 *Amending Ordinance,* ensured that workers guilty of absconding paid court fines, reimbursed employers, or were incarcerated. Restitution was waived if the worker agreed to fulfill his obligations. An employee who was found guilty of breaching a contract could not seek further employment in the industry because the records were updated to reflect the transgressions. Soon, advertisements by the Institute of Mines to recruit mining workers included provisions for executing and enforcing labor contracts, and prosecuting those guilty of infractions.[20] Ultimately, employees' resentments related to suspicions of their potential dishonesty. and stringent laws enacted and applied to workers contributed to ongoing tempestuous labor relations and the necessity for trade union intervention.

Many viewed with distaste mining workers who absconded with monetary advances after signing work contracts. It must be noted, however, that unscrupulous employers were culpable of contributing to the abuse of the advance system by absconders. Dishonest practices at the managerial level caused friction as well. Consequently, an expose of the dynamics of absconding revealed that sometimes criminal intent was not initiated by workers. Certain employers deliberately created the conditions and tempted prospective employees of other firms to abscond and go to work for them—acting otherwise than they originally intended. There were many instances of eventual absconders who had, in fact, reported for work as agreed with their employers and after working a few days, were enticed by a another manager, claim holder, or agent to leave and work for them instead. The periodic scarcity of labor prompted some dishonest employers or their representatives to offer bribes and to encourage workers already committed to other companies to work for them. Eventually, the identities of those employers and agents who offered larger advances to encourage workers not to honor previous commitments were circulated. The effort aimed at dissuading employees from succumbing to the cutthroat tactics initiated by unscrupulous employers.[21]

Some of the absconders had had previous criminal records or were loafers who had deliberately planned to take advantage of the advance system as a way to obtain money.[22] Between 1914 and 1920, the number of absconding gold and diamond miners increased. Bauxite mining had created additional competition for workers and aggravated the employment situation. At the same time, some dishonest gold and diamond employers continued to offer higher advances to the already employed.[23] The persistence of those employers in this practice strained labor relations even further and demonstrated the ineffectiveness of more stringent laws to punish workers. These employers' actions caused chasms in socioeconomic and labor relations which trade union activists would eventually address.

Strained labor relations in the gold and diamond mining industries continued due to employers who committed other willful, intemperate, or unjustified acts. Such acts led to premeditated instances of dishonesty by workers who had experienced or were knowledgable of incidences of employers cheating employees of their wages. Those would-be workers probably decided to preempt similar occurrences from happening to them by pretending to be *bona fide* employees and absconding with the advanced money. Maybe they felt that it was a credible way to recoup previous losses they or fellow workers had suffered at the hands of dishonest employers. In addition to teaching a reverse lesson in wrongdoing to employers, the action showed a fallibility in contracts. Labor contracts guaranteed miners their wages, but sometimes as the expiration of the contracts approached, employers would order employees off the mining sites on one pretext or another to avoid paying them. Most grievous to employees was the situation in which the employer would initiate warrants for their arrest and prosecution as though the workers had abandoned work and broken contract arrangements. The onus was now on the worker, who often could ill afford to hire legal representation and pay court costs to prove that the intentions and actions of the employer were to deny the wages. Before long, gold and diamond miners "could not prosecute their employers for non payment of wages" unless they "first obtained a certificate or permission from the Commissioner of Lands and Mines."[24]

Ashton Chase insisted that *The Amending Law of 1909, Employers and Labours Bill*, from which the above provision emanated, demonstrated the inequality of justice. Unlike the workers who had to obtain "a certificate or permission from the Commissioner of Lands and Mines" before proceeding to prosecute employers, the employers needed no such authorization to do the same.[25] While infractions of the law should be persecuted and punishment commensurate with the crime meted out, the scales of justice were tilted in favor of employers who had higher political and social status and more economic leverage and power than the workers. Employers who were guilty of failing to pay workers or those who breached contracts had to pay the overdue wages and a fine. Unlike employees 'cruel and unusual punishment' was not dispensed to employers who never had to pay wages that had been denied to employees, fined by the courts, incarcerated, and flogged, in unison, as punishment for their crimes. Significantly, the 1909 amendment, referred to above, was promulgated at the behest of influential bureaucrats such as the attorney general as well as mining elites including George Garnett, a onetime legislator who fraternized with members of the Court of Policy.[26]

Before the appearance of trade union activists, officials of the Institute of Mines, as arbiters, had been able to fill somewhat the breach in workers' representation and obtain recompense for defrauded employees. A few employers were charged criminally to enforce judgments against them, and in a rare occurrence, one of them was imprisoned. By 1914, mining laws were "further amended to provide for the registering of labourers by the Commissioner of Lands and Mines" and only contracts registered accordingly could be enforced.[27] Government's decision to limit the involvement of the Institute of Mines in workers' affairs was viewed with dismay by the agency's secretary James Winter who claimed that it precluded the agency from assisting employees. Members of the Institute of Mines felt that workers needed their advice and guidance in wage disputes with their employers. They felt that destitute employees who arrived in the city and were unable to find their defrauders would endure undue hardships. In particular, Winter argued that his entity was the one which was fully equipped to provide a wide range of services including the prosecution of employers, registering workers, arranging labor contracts, and refusing further services to transgressing parties. This claimed even-handedness of the officials of the Institute of Mines mattered little to government officials who transferred the agency's functions to the Department of Lands and Mines from 1915.[28]

By 1917, the Commissioner of Lands and Mines explained the public angst of officials of the Institute of Mines by stressing the amount of financial benefit that had accrued to that body. When the registration of workers had transferred to his department, the move had financially crippled the agency and its surplus funds were utilized to pay the salary of Winter, who did little work. The Commissioner observed that Winter was gravely dissatisfied with the government's action; therefore, he took every opportunity to characterize the transfer as a retrograde step which injuriously affected the workers' interest. Although with the transfer of duties, Winter was appointed and still was the registering officer for Georgetown, he also functioned as a private agent on behalf of the workers and pocketed payments for the service.[29] Meanwhile, members of the Institute of Mines were incorporated into the Department of Lands and Mines and the privately owned bauxite mining company, DEMBA.

This transfer of functions did not mean workers were unable to fend for themselves. The very notion by the agency's rank and file that workers were unable to take care of themselves had ignored their traditional collectivism methods such as pooling resources, self-employment, and organizing boycotts and strikes. They had their own mechanisms to redress grievances against employers. Throughout the nineteenth century, the working class had challenged unjust treatment meted out to them. Members of the group took action to empower themselves with or without others acting in their behalf.[30]

In this context, the colony-wide 1905 protests, which included mining workers and activists associated with the Peoples' Association must be introduced in anticipation of the coming of Hubert Nathaniel Critchlow and the Trade Union Movement by the 1920s. Workers had long aired their grievances and ably demonstrated that they were prepared to use appropriate measures to cope with the unjust actions of the power structure. The very next year after the protests, the workers, leaders, and their organization sent a "monster petition" through the colony's governor to the Colonial Office. In their missive on the colony's ills and the impact on them, they offered solutions to ameliorate the situations.[31] In the same vein, the high visibility given to the case of Edward Harvey in the local press in 1939 is instructive. Harvey, a well-off claim-holder in the Potaro District appeared in the Georgetown Magistrates' Court charged with breach of contract. Allegedly, he failed to pay wages due to thirty-two of his aggrieved prospecting employees. The plaintiffs hired a lawyer, L. A. Hopkinson to represent them in court against the transgressor.[32]

Many employers made contractual agreements to pay their employees' families at their urban offices, but some of them reneged and this often caused serious socio-economic consequences for the workers' family members. For many of them, the interior miners' wages intermittently remitted to city offices of the employers were their only means of livelihood. Failure of the employers to meet their obligations sometimes caused dependent family members to behave disorderly at the urban offices; risking police action against them. Meanwhile, some workers' family members were left wanting although most reported accounts of transgressions made it appear as though dishonesty was more rampant among employees than employers.

A local newspaper took umbrage with the unfairness of information circulating about the mining industry and its workers. The main issues were the constant efforts to pinpoint inappropriate methods of prospecting, the unending complaints by owners of mining operations and their managers, and, in particular, the criticizing and characterization of workers as though they were all dishonest. Furthermore, it was noted that workers' grievances were usually overlooked and the biased information in circulation was an injustice to the workers and to the public. While the defrauding of workers by companies' owners and management were too often ignored, some employees were forced repeatedly to visit the offices of their employers to collect portions of their wages—often withheld for no valid reason. Comparing the prompt attention employers received when absconders disappeared, the article called the treatment of workers a travesty of justice.[33]

In contrast, *The Police News* always identified "the many successful cases against absconders." Sometimes at steamer stellings (wharves), as many as a dozen policemen, acting on behalf of employers, were on the lookout for absconders. Often, the law-enforcement officers would apprehend and interrogate those who fitted a certain racial profile; that of African ancestry. In many instances, newspapers, official, and quasi-government publications, selectively noted all the negative aspects of workers' behavior and the triumphs of employers against them. Unpaid workers were supposedly protected by filing a petty debt form to recoup their wages, but it was difficult for those workers to collect money after obtaining judgments in their favor. After 1915, unlike employers who often had easy access to the legal system to recover their money, individual workers encountered undue difficulties in obtaining money to pay lawyers, to serve summonses on erring employers, to travel and to attend court hearings, and in the case of judgment, to pay for making levies.[34] Unjust laws, bureaucratic measures, and exploitative and oppressive labor conditions as well as aspects of capitalism vouchsafed the need for trade unions to protect workers' rights.

Interior workers who were defrauded by unscrupulous employers sometimes faced starvation. Nonpayment of wages could force workers without other choices to remain in the interior longer than planned and to seek employment elsewhere. Many times mining workers had to sign on to work with other companies or individual claim owners for periods which lasted much longer than they had anticipated. This was the only way for them to remit money to their dependents. Meanwhile, family members and others were left in a terrible quandary at home. They did not know if the breadwinner was sick, had died, or had simply abandoned them.[35] These were among the wrenching experiences of long-suffering miners and their families who endured all kinds of deprivations. Such experiences soured labor relations in the colonial society.

Wages were sometimes paid to mining workers grudgingly and this added to the controversial nature of evolving socioeconomic and labor relations. In 1896, wages of 64 cents per day were paid by owners and managers of large companies to their workers. Added inducements such as food and sleeping accommodations in the interior were sometimes available. By 1897, however, management of mining companies considered the wages they paid to their workers as being too high. They influenced the enacting of legislation to reduce the workers' wages to 40 cents daily.[36] The same year as the legislation was enacted; the British government had sent a Royal Commission to examine deteriorating socioeconomic situations in its Caribbean colonies. By 1904, the situation worsened in the country and the overwhelming advantages mining companies had over the workers were noted. The opinion existed that bereft of and in "the absence of trade union protections for mineworkers" the situation "additionally favoured large-scale operations in the colony."[37] The next year, the Protests of 1905, broke out. The outbreak of this wide-scale unrest was largely due to the ongoing socioeconomic disadvantages workers faced.[38] Yet it was sixteen years before the first trade union was officially registered in the colony.

Happenstance was a factor which made it appear as though potential mining workers had deliberately broken their contracts. For instance, sometimes on arrival at entrepot Bartica on their initial prospecting ventures the would-be miners were regaled with and confronted evidence of drowning mishaps and tragedies which had befallen other workers in the gold and diamond fields. The dead were often characterized as "drowning like rats."[39] The dilemma for some of those arriving workers was whether to proceed as arranged or to abandon their work agreements altogether. Most of the aspiring prospectors had been agricultural workers in areas where traveling on dangerous rivers and encountering deadly insects and animals were not features of their lives. Many of them were unprepared for the shock and likelihood of untimely death, and operated only under the premise of gaining employment. Suddenly, for some of them, that type of work would seem too risky. Not inclined to dishonesty, they willingly returned the advances; and never considered precious metal and gems mining again.[40] Quite a few employers verified that the majority of miners who received advances reported for work and abided by the terms of their contracts. Many times, "men from Mahaica walked all the way to Georgetown," a distance of twenty-five miles, "when they missed the train" because "they were so anxious to fulfill their contracts."[41] Still, in discussions on amending mining laws, the criminal elements among the workers were touted to justify the enactment of draconian and unfair laws.

Ironically, as late as the 1930s, the gold and diamond industries were the only ones still utilizing labor contracts in the colony. Workers committed to contract for 120 days, accepted monetary advances on their wages, and were transported to their work sites. Between the time of the signing of the contract and its fulfillment, situations could develop and place workers at disadvantages. For example, employers claiming bankruptcy and leaving workers stranded, unpaid, and starving. The 1939 *Report on Labour Conditions in the West Indies,* made recommendations which were beneficial to miners. The report identified gold and diamond mining contracts as horrific and archaic; advantageous to employers; often, containing unfair provisions and leaving workers' unprotected. In addition, the report identified working conditions, wage rates, rations and traveling provisions, and advances for reforming. Penal sanctions against workers were adjudged as too severe, especially since many errants were willing to pay for their infractions.[42] Unlike earlier decades, the reforms were recommended at a time when workers had trade union representation.

LAWS, RACE, AND ILLEGAL PRACTICES IN GOLD AND DIAMOND MINING

Within a decade of the inception of gold mining, government and business officials had felt that rampant stealing, smuggling, and illegal trading of the precious metal reduced investors' profits and revenue. Existing punishments to those identified as the offenders, mostly workers and their associates, were considered lax. Increasing surveillance staff in the Department of Mines, engaging paid informants, using the courts and law-enforcement techniques, and utilizing flogging for the guilty workers—similar to the Cape Colony and Transvaal, South Africa—were deemed as worthy deterrents to illegal practices pervasive in the industry.[43] The prevailing belief was that African-descended workers were mainly responsible for the illegal activities, but gold and diamond stealing, smuggling, and other dishonest practices were carried out or aided and abetted by a variety of persons. The overwhelming focus on workers as the only perpetrators of criminal activities presaged the coming of organizations advocating their rights and exposed other wrongdoers.

Ramifications of race seeped into discussions and beclouded the policies, intentions, and actions of colonial officials and their allies on matters concerning workers. A perusal of the literature on contemporaneous British mining development in Southern Africa and Australia confirmed the negative/positive attitudes of local officials and their associates based on the race of workers. In the former area, the majority of workers were of African ancestry and other non-European groups such as East Indians and Chinese. In the latter place, most of the early labor migrants were Europeans. Stringent mining laws were enacted and harshly applied to the Africans and non-Europeans in Southern Africa compared with the much more flexible and less punitive ones in Australia and Canada, areas dominated by persons of European descent. It was not accidental that the living and working conditions of mining workers in Southern Africa were woefully substandard compared with those in Australia and Canada.[44] Disparities in the accommodations of racial groups involved in mining and the disparate application of laws on workers in diverse areas of the British Empire based on race offer an appropriate segue into local developments in the pre trade union era.

Globally, it was known that dishonest miners stole varying quantities of gold nuggets from their employers. Locally, dishonest employees illicitly sold the items to interior shopkeepers or other claim holders who were in fact accessories to the crime. Although it was against the law, dishonest claim holders would then record and utilize the stolen items as though they were obtained from their own holdings. Shopkeepers concealed their illegal activities amongst other commercial transactions. Miners who illegally secreted and conveyed gold from their work sites also sold the commodity at goldsmiths' shops located in Lombard and Bugle Streets, adjacent sections of Charlestown, other parts of Georgetown, and coastal-wide. The situation continued although by the mid-1890s, search-stations had been established at Bethany Island, Koriabo, Bartica, and Baramanni, Essequibo to apprehend gold smugglers on their way to the coast.[45]

By 1894, the colony's detection or search centers were considered failures and in need of overhauling.[46] Resorting to racial profiling based solely on the number of offenders apprehended and estimates of likely losses of precious metal and gems, colonial authorities and mining officials considered the draconian laws of mineral-rich Southern Africa as worthy of emulation. Legislation crafted and applied in South Africa, *The Diamond Trade Act,* was designed to curb such infractions by employees. Searching centers were established by mining companies and white police observed, stripped, and searched the mostly African workers. Locally, the hierarchy expressed interest in examining and applying "mechanisms relating to Labour (*sic*) and Prevention of Theft"...in the most comparable country "South Africa because—the other great mining Dominions—Canada and Australia—rely exclusively on white labour and this alone constitutes a difference so great that they cannot be usefully compared."[47]

The local expatriate commissioner of mines proposed placing a white man in charge of searching the mostly African-descended workers and recommended the enactment of stringent laws to include three to fifteen years incarceration and fifty lashes to punish workers guilty of stealing gold or diamonds. This suggestion was strikingly reminiscent of similar sentiments expressed almost a decade earlier by another expatriate colonial mining official. In 1885, H. I. Perkins asserted "the role of whites in mining was as supervisors and blacks were to do the hard work under any conditions."[48] By 1894, the commissioner of mines observed in his report that he foresaw "no development in the gold industry unless" at least "one white miner" was available "for every five blacks" in the districts.[49] Even Colonial Office bureaucrats "were pessimistic regarding the capacity of future development of the West Indian Negro," because they "viewed with concern any decrease in the resident white and Indian immigrant population of the West Indies which might follow a decline in sugar workers."[50] In 1890, as previously explained, race and national origins were critical factors utilized by colonial administrators and their allies in recruiting and migration strategies. It is noteworthy that at the time due to the rigors of the prospective work, African-descended persons were largely viewed as superior to all others. This perception was based not only on assumptions of their physical capabilities, but also on the ways in which those workers were exploited and deemed expendable according to the whims of the colonial hierarchy.[51] A few years later, race as a factor to curb and to condone illegal behavior was again applied.

Allegedly, European businesspersons, many of whom were migrating Portuguese passengers on their way to Barbados and Madeira illegally obtained and smuggled gold out of the country. In the 1890s, unlike procedures in French Guiana and Suriname, travelers out of colonial Guyana, who were mostly Europeans, were not searched for the detection of illegal gold or diamonds. A cross section of those involved in mining including businesspersons and others were culprits. Yet laws aimed at curbing losses of those commodities through theft overwhelmingly targeted certain employees for apprehension and punishment. In the early years of mining, laws to deter stealing and smuggling of the items were hardly ever enforced against Europeans or to curb the offenses of owners, managers, businesspersons, and their associates.[52] But, personnel at various levels of the gold and diamond industries had long initiated ways to commit transgressions. For instance, "claim-raiding" the practice of groups or mining companies illegally occupying and prospecting on land for which they had no claim or license or without the consent of the lawful claim-holder, continually engaged law-enforcement personnel and the Courts. By the 1930s, miners were not the only ones found guilty of stealing and smuggling the commodities. The case of Ricardo Sanders is instructive. Sanders, a diamond buyer for Kraskosky Ltd., a German diamond-buying firm, went missing for a short period of time during which he converted 620 carats of the company's diamonds valued at $9,000 to his own use.[53]

Discriminatory policies embedded in the administrative and judicial systems to curb losses were harmful to mining operations. Failure to craft and apply laws equitably against all persons who committed illegal acts based on the race of the perpetrators was symptomatic of the unjust and unequal society in which the working class toiled. Following a trend from the enslavement period through the nineteenth and twentieth centuries, race remained a critical factor in the country's socioeconomic and labor relations. In the mining industry, racial profiling in crafting laws and viewing persons through racist lens contributed to the loss of revenue by government and profits for private investors because illegal activities were carried out by people of European, African, Asian, and other origins.

Law-enforcement personnel, police stations, and jails were vital to maintaining law and order in interior mining districts and to securing the country's borders from smugglers. Although for decades, government funds were allocated for the construction and maintenance of such services, the resources were insufficient to deal with hinterland crime. The porous borders Guyana shared with Brazil, Suriname, and Venezuela needed effective control because much of the gold and diamonds smuggled out of the colony was cross-border. Colonial bureaucrats and law enforcers were mostly unable to deal adequately with foreigners and others who were smuggling the commodities out of the country. Shortfalls in production, revenue, and profits were disproportionately blamed on workers. The justice system and law enforcers largely ignored the rights of the laboring class and indiscriminatingly targeted them for prosecution. This underscored the need for employee representation.

By World War II, it had become apparent that Brazilian nationals were illegally crossing the border and entering the country in the Rupununi area. The Brazilians unlawfully prospected for diamonds in the Ireng River, and illicitly transported the gems to their country. By 1948, efforts were made locally to legalize the operations of all Brazilians who were prospecting in the country. Yet within a few years, it had become apparent that rampant smuggling of both gem-quality stones and industrial diamonds had continued.[54] Guyana's industrial diamonds were regarded as "superior in quality to those from South Africa and the Belgian Congo."[55] Large quantities of the Guyanese diamonds were smuggled to Boa Vista, Brazil and taken to Rio de Janeiro for transmission to the Soviet Union. As a result of undetected smuggling, the official output of gold and diamonds from the country remained questionable. Ultimately, government officials attempted to enforce the border-control laws. In the 1950s, measures were enacted to locate more police stations and officers in areas bordering Brazil along the Ireng and Potaro Rivers. Immigration laws also dealt more effectively with foreign smugglers.[56]

Laws promulgated on a concept that only one racial group had the propensity for prolific stealing resulted in tensions and disagreements in the colony. Law-enforcement techniques geared mostly to entrap workers contributed to adversarial labor relations in the gold and diamond industries. But, belatedly, after decades of prospecting and turmoil in the industry, colonial administrators realized and conceded that there were other culprits in the industry besides the African-descended workers.[57] By then, the workers and their associates were represented by trade unions.

Trade Union Activism and the Gold and Diamond Industries

No trade union movement existed in the colony when the mining industry began operations in the 1880s. As the industry evolved, tumultuous labor relations engendered by the prevailing attitudes of the colonial hierarchy and the laboring class facilitated the coming of trade unions. But even without such organizations this segment of the workforce had had an ongoing tradition of activism, mobilizing, and organizing to remedy injustices against them. The eminent medical practitioner, Dr. John M. Rohlehr who was a staunch advocate for improving the conditions of the laboring class, is credited as an early supporter of trade unions. By 1905, a time of wage disputes and strikes, leading to riots, major social unrests, and economic dislocations, advocacy was made for "the formation of a trade union."[58]

Hubert N. Critchlow, legislator Patrick Dargan, and Dr. Rohlehr were at the forefront of the strike called on November 28, 1905. Critchlow was a waterfront worker, former gold and diamond miner, labor organizer, and eventually a legislator. Following the public protests, petitions were sent by the People's Association of British Guiana to the Colonial Office on the unrest which was caused by the colonial hierarchy's systematic unjust actions against the working class.[59] On December 12, 1905, in the midst of the mayhem, some miners who were employed at a privately owned company in the vicinity of the Puruni River, Essequibo engaged in strike action against their employer. The company's management acceded to some of the workers' requests. Alexander Jones, James Basil, Richard Jones, and Robert Semple were among sixty-seven miners in a workforce of 225 men who decided to quit rather than resume working.[60]

In 1908, miners employed at the Barima Mines claimed they were starved and underpaid and abandoned their work. Allegedly, food rations to which they were entitled had been denied them and the wages of those workers were arbitrarily reduced in contravention of their contractual agreements. Management blamed their financial expenditures as a major factor for the arbitrary decision, and promised to rectify the situation. In this way, the dispute was resolved.[61] In 1910, some gold and diamond mining companies reneged on paying interior workers, boat captains, and bowmen for their services. The workers formed a labor organization. Although the organization did not gain legal recognition, it ushered in the age of trade unionism.[62]

The first trade union, the British Guiana Labour Union (BGLU) was established by Critchlow in 1919.[63] When the Court of Policy enacted *Trade Union Ordinance #17, 1921*, the local organization became the first one registered in the Caribbean. Initially, union affiliation and representation for all workers including miners generally fell under the jurisdiction of the BGLU. The British Guiana Miners' Association (BGMA) was formed in 1928, by and for workers attached to midsize and large gold and diamond mining concerns. The British Guiana Miners' Union (BGMU) and the Mining Association of British Guiana formed in 1944 and 1946, respectively, were specifically the first ones to represent workers in the entire mining sector. In 1946, there were thirty-three registered trade unions.[64] One year later the number of unions increased to thirty-seven. Within three years, the union representing workers at British Guiana Consolidated Goldfields Ltd. (BGCG) negotiated increased wages and better conditions of employment including leave with pay, free medical attention, affordable housing, and discount on shop credit purchases for its members. At the end of the 1940s, there were fifty trade union organizations.[65] By 1954, the British Guiana Gold, Diamonds, and Mineral Workers' Union emerged. During the same year, the Guianese Miners' Development Association was formed to represent small-scale prospectors.[66]

Unionizing in the local context had reached a crescendo by mid-century. In the 1930s, British Caribbean workers' militancy and advocacy of their affiliates in Great Britain such as the League of Coloured Peoples (LCP), influenced the passage of some reforms by the British government.[67] The Cold War which began in earnest at the end of World War II influenced increased US activities in the Western Hemisphere to thwart Soviet intervention. This development critically advanced workers' causes. Coming out of a long tradition of activism and mounting challenges against unjust officials, the working class and their leaders utilized trade unions to further their causes during the Cold War era. In Guyana, leaders such as Critchlow, Frank and Dr. Claude Denbow, the future ambassador and diplomat Sir John P. Carter, Linden "Forbes" Sampson Burnham, late president of the country, and Winslow Carrington incorporated the *modus operandi* of foreign trade unions in their activities.[68] Although bauxite workers functioned under different environmental conditions from those of gold and diamond miners, were paid better wages, received health care, and could reside with family members, they too were subject to labor policies which presaged the coming of trade unions in that industry.

Trade Unions in the Bauxite Belt

Bauxite production was organized to ensure company control of the entire operations. The company-town, and in some ways the paternalistic arrangements which emerged, caused dissatisfaction among some workers. Bauxite management invested just enough funds in the infrastructure to create and maintain the "pull" factors which made employment attractive and to ensure a reserve army of workers. DEMBA utilized part-time and temporary employees, adjusted the full-time workforce to suit market demands, and wielded control over social and labor relations to guarantee increasing profits.[69] The situation could not remain static and in the ensuing decades trade union activists wrought changes in the bauxite belt. Historiographic accounts of bauxite mining operations in the country, mostly explored labor, social, economic, racial, and class relations. Some of the studies pointed out the unflattering aspects of housing, health care, educational, recreational, and business facilities which were viewed as substandard, inadequate, and insufficient. The tendency was to regard the workplace as bedeviled by pervasive employee dissatisfaction stemming from exploitation and oppression by management. Sometimes workers' consequent insubordination led to punishments which further damaged labor relations.[70] Comparisons between the South African situation and the local bauxite industry suggested that draconian and inhumane mining laws and working conditions in the former locality were replicated in Guyana.[71]

While deadly working environments are comparable in both countries' gold and diamond mining industries, bauxite work in Guyana was different and relatively safe. Regulations in the bauxite mining belt caused some workers to grumble, but as will be discussed, until the retrenchments during World War II, those workers' dissatifaction was relatively low-key. Housing for African workers in Apartheid-era South Africa was abysmal, compared with living quarters provided by DEMBA and the private residences of non-company persons. It is true that the model for a mining policy to regulate the behavior of gold and diamond miners in the colony was South

African rather than Australian because of the racial composition of the workers (predominantly African versus European). Bauxite management implemented policies similar to "order and progress" that were prevalent during the modernization era in Latin America.[72]

Cost-minimizing by private companies and government agencies adversely affected the well-being of mining workers in both Southern Africa and interior Guyana's gold and diamond fields. Punishment for infractions of precious metals and gems' laws was more stringent on workers than on management in colonial Guyana because legislators crafted and enacted regulations to favor the mineocracy. Thus, some comparisons with South Africa have credibility in gold and diamond mining rather than in bauxite operations.[73] International comparisons should be tempered by the contexts of varying environments, different time periods, and differences in the commodities produced. Cognizance should be taken of the value and ultimate utility or disposal of the products. In this context, managerial attitudes and propensities for draconian laws were spawned by the illegal actions of those gold and diamond workers who pilfered the commodities which resulted in losses for companies and their investors. Bauxite workers could not steal or dispose of the commodity owing to its bulk and the monopolistic operations; therefore, managerial behavior reflected their overarching interest in controlling all the factors of production.

Some workers felt that DEMBA's rules and regulations were irksome. Conversely, a number of longtime Upper Demerara River residents and bauxite workers mostly regarded favorably their experiences under Canadian-run DEMBA. In fact, many former bauxite workers, residents of Linden, and their families tended to recount their social and economic experiences by making comparisons with other areas of the economy. They were the ones who pointed out that the coming of the bauxite industry provided their grandparents, parents, and themselves with a viable alternative to the drudgery of plantation work and the dangers of gold and diamond mining. Generally, they expressed satisfaction with their basic housing, health care, educational, and employment situations and identified the ways in which they furthered their opportunities. This perspective is worthy and provides valuable insights.[74]

Contemporaneously, workers in the country's other economic undertakings had myriad causes to complain—on the waterfront; in the sugar industry; and in the two other mining sectors. However, those workers' wages were less than those of bauxite employees, and although the cost of living was higher in the bauxite belt than on the coast, it was much less than in interior locations where gold and diamonds were mined. DEMBA undertook farming and dairy activities to supply its employees with produce. Occasionally, the company provided its workers with land for their own agricultural use. In other sectors of the economy, it took an inordinately long time before housing, health care, working conditions, and wages were adequately dealt with. Suitable housing was not even considered for waterfront workers, one of the oldest occupations in the export-oriented economy, until after World War II. By then trade unionists had made suitable housing accommodation for workers a priority. The government's Central Housing and Planning Authority was not established until 1948. Six years later, the British Guiana Credit Union began operations which helped to facilitate housing construction, at a time when government launched aided self-help house building schemes.[75] DEMBA, from the inception of operations, anticipated its infrastructure needs and had a plan and program during a time of increasing anticolonial activities and union activism.[76]

Time is a critical element in investigating evolving relations and trade unions in the bauxite industry. Long before bauxite's advent socioeconomic and labor relations were rancorous and sometimes violent. Increasingly, during World War 11, and because of ongoing activism by workers and the appearance of more trade unions—economic disparities, injustices, and social and labor inequalities began to receive more attention. Local bureaucrats and their associates were advised to institute reforms to avoid escalating upheavals. More overseas-educated Guyanese were returning home and mounted further challenges to colonialism. Turbulent social and labor relations colony-wide became factors of life and work. It is within this national context that militancy among bauxite workers fomented.[77] Workers assessed their living and working conditions as discriminatory and utilized trade unions to improve their situations.[78]

Trade union activists endeavored to address workers' grievances. Some workers resented their treatment at work sites. Discontent with management included the failure to increase and upgrade housing, especially across the river where bauxite workers also lived. Workers and residents of Linden were dissatisfied with what they perceived as differences in treatment accorded to patients at Mackenzie Hospital. Whites and employees at and above staff or supervisory levels were viewed as being given preferential treatment at the company's health care facilities.[79] In 1938, on the eve of World War II, an attempt to establish a trade union for bauxite workers at Mackenzie signaled the coming of changing relations in the industrial complex. By 1942, representatives of the BGLU established a branch at the Upper Demerara River operations of DEMBA. The Man Power Citizens' Association (MPCA) soon followed, and later, Guiana Mine Workers' Union (GMWU) represented bauxite workers. By the late 1940s, the employees of Reynolds Metals Company on the Berbice River, the next largest employer of bauxite workers, gained union representation. Bauxite workers increasingly solicited assistance from the BGLU to aid them with grievances they had against DEMBA. Various employees expressed dissatisfaction with their schedules, the nature of their work, long working hours, limited job categories, and inadequate wages. Other workers were displeased with their health and medical benefits. While unsafe conditions, insufficient break-periods, and poor employee/management relations bothered the workers, before long, they were most aggrieved by incidences of large scale retrenchments.[80]

An unpublished *Report of a Committee of Inquiry* in 1942, confirmed that disputes between DEMBA and its employees were increasing. Not to be ignored was that at the outset of production a group of activists had aired dissatisfaction with emoluments and working conditions. By 1942, the hostility of the company's security forces made other employees more determined to change existing relations by obtaining union representation.[81] DEMBA's management resented what they considered as encroachment on their private domain by unauthorized persons such as BGLU's officials seeking to represent workers. The trade unionists' requests for permission to enter Mackenzie were denied by the company and the colonial police. They were forced to conduct their organizing activities across the river at Wismar and Christianburg.[82]

Bauxite was a strategic commodity in the war effort and the necessity for sustained and optimum supplies was a foregone conclusion. Yet in the midst of the war, in 1943 and until 1947, management retrenched a large number of workers, ostensibly because of decreased demand for the product. DEMBA had 3,600 employees in 1942, but by 1946, only 1,600 persons were fully employed by the company.[83] The retrenchments heightened long simmering hostilities between employees and management. At the behest of its members, BGLU officials approached bauxite officials seeking recognition but, they were rebuffed. A subsequent decision by the workers to take strike action was unsuccessful. More retrenchment of employees followed and the workers turned to the Man Power Citizens' Association (MPCA) which until then had mostly represented workers in the sugar belt.[84]

The workers and their union were determined to change the status quo, but DEMBA's officials remained obdurate to most of their demands. This led to a sixty-four day strike by employees lasting from April 13 to June 16, 1947. Striking employees received support and aid from other workers and organizations in the colony. The commissioner of labor appointed a Committee of Enquiry on May 27, 1947, and the issue was resolved the following month. The MPCA activism at the Mackenzie and Ituni operations of DEMBA preceded recognition by Reynolds Metals Company and The Berbice Bauxite Company at Kwakwani and Everton for their employees . By 1953, workers at the last named entity left the MPCA to form the Berbice Mine Workers Union. Three years later, in September 1956, this union merged with the British Guiana Mine Workers' Union (BGMWU).[85] Most notably, after trade union activism, socioeconomic problems were addressed and workers received a cost of living bonus. Proposals for job evaluations and a non Contributory Retirement Income and Life Assurance Plan were discussed.[86] Before mid-decade, the full-time bauxite workforce at Mackenzie increased to 3,104 persons.[87]

Registration of Trade Unions fell under the Registrar of Trade Unions, functioning under the Registrar of Deeds Office by provisions of the *Trade Unions Ordinance Chapter 57*. At the end of 1955, there were fifty registered trade unions; ten of them were employees' associations; thirty-seven of them represented workers; and three others were identified as miscellaneous.[88] In 1956, *The Colonial Annual Report* noted that the bauxite "industry is one of the most important from the point of view of production value, but employs a much smaller labour force than the sugar industry."[89] During the year, officials of the BGMWU negotiated a 14 percent wage increase for DEMBA's employees. Unskilled workers received hourly wages ranging from 40 to 49 cents; semi-skilled and skilled employees were paid 50 to 57 cents and 57 to 60 cents, per hour respectively. A bauxite laborer earned $25.52 weekly and an electrician was paid $33.60 per week. A government-employed laborer and electrician were paid weekly wages of $15.12 and $23.52, respectively. A sugar estate laborer was paid $12.40 and an electrician $14.50 weekly.[90]

The Annual Report of the Department of Labour, 1956 confirmed that a cost of living survey by DEMBA and BGMWU was used in negotiating wages and allowances for workers. Tenets of the agreement were implemented retroactively, from October 1955, with wage increases for unskilled, semi-skilled, and skilled workers.

A three-year agreement with provisions for revised wages and a shorter work week was mandated.[91] The colony achieved political independence from Great Britain in 1966, but workers and their organizations continued their struggles against new government regimes and the management of private companies for more equitable distribution of the country's economic resources.[92] Although conflict was unavoidable and engendered by centuries' old rancorous relationships between workers and the power structure, this was not the entire story of development in Guyana. Parallel to ongoing strife, the working class did make progressive strides.

CHAPTER 8

Internal Migration and Village Dynamics: Families and Communities Coping

THE ABILITY TO COPE AND TO IMPROVE THEIR CIRCUMSTANCES was no easy feat for persons of African ancestry in colonial Guyana. After enslavement ended in 1838, their lives were exacting, but they used their initiatives and experiences to grapple with and survive socioeconomic hardships. Although they faced a plethora of challenges, formerly enslaved persons sought and found ways to make a living and to provide for the future. Long accustomed to agricultural work, many coastal villagers undertook internal and seasonal migration to the gold and diamond fields from the 1880s. From 1916, many of them worked and settled in bauxite producing areas. In this chapter, the activities of a rural segment of the Guyanese population are analyzed through the prisms of migration and mining. Villagers' prescience in anticipating beneficial outcomes from investing portions of their mining incomes on targeted projects such as farming and furthering the education of their offspring were reflections of their long range goal of self-reliance. Women were factors in the aspirations and achievements of families as well as the nexus enabling migration and mining.

Migration to seek work opportunities provided a safety valve for thousands of persons of African descent to avoid the harshness and conflict of plantation labor and discord in alternative employment. By the late nineteenth century, not only had they migrated internally, but they had ventured abroad to Venezuela, Suriname, French Guiana, Brazil, Africa, and the United States of America, motivated no doubt by vistas they envisioned as profitable.[1] Villagers' involvement in the internal migratory movement was consistent with their tradition of doing for themselves and fostering empowerment rather than dependency. With gold and diamond mining, workers faced myriad challenges in balancing their lives and work in the interior and on the coast. Although miners missed their families and familiar environments of the coastlands, many of them understood that the work was an opportunity to provide for their economic needs. The pendulum migratory movement of villagers to mining which emerged allowed many of them to continue their seasonal involvement in home-based agricultural and other economic undertakings. Migration to bauxite mining in tandem with infrastructure development of the environs fostered permanent communities unlike the long existing frontier conditions of the gold and diamond regions.[2]

This chapter focuses on the multidimensional abilities of village miners, their spouses, and others who initiated coping mechanisms to deal with the vagaries of life on the coast and in the interior. Viewed within the concept of enduring pain to achieve gain, the refrain of former migrating miners, "we missed our families, but we had to do it," expressed their resolve in finding ways to make a living.[3] In their efforts to cope with the cycle of economic travails which periodically confronted them, village miners engendered further opportunities for their progeny by educating their children to function at different and higher socioeconomic levels. A premise of the chapter is that by the mining era, villagers had long made the connection between management of their resources and the improvement of their collective conditions. Analyzing the painstaking ways in which some workers invested their earnings concomitantly from agriculture and mining to achieve specific goals is a central feature of the chapter.

Enslaved Africans in the Americas never received financial compensation for their centuries of unrequited labor. To this day, their descendants have not been recompensed for the exploitation of their enslaved ancestors' labor. From the 1830s, when indentured Asians and other labor migrants arrived to replace or supplement the formerly enslaved in plantation work, they were paid. After complete emancipation in 1838, and in the ensuing years, the granting of free or cheap land was never extended to the African component of the population, in contrast to what was done for others. African Guyanese were mostly not the beneficiaries of land settlement schemes supported by government funds. Instead, wherever they ventured on the coastlands, challenges awaited them. Still, they endeavored to grapple with the morass in which they found themselves. As a group, villagers never allowed their surroundings to decay nor did they abandon agriculture, as they willingly grasped other economic opportunities including interior mining work.[4]

Village Agriculture and the Mining Economy Intertwined

In the decades after emancipation, independent agricultural pursuits were complex undertakings for the villagers. For one, coastal drainage and irrigation became increasingly problematic for farmers, livestock producers, and other residents. Colonial administrators hesitated to install appropriate flood-control mechanisms, and unlike the bureaucrats' acquiesce to planters' demands; they rejected reasonable requests by African Guyanese for finance to improve their environs and economic enterprises. How this segment of the population survived freedom needs investigating beyond capitulation and failure.

Alan Adamson's "Monoculture and Village Decay in British Guiana, 1854–1872," Brian Moore's, *Race, Power, and Social Segmentation in a Colonial Society* and "African Guyanese Political Disempowerment in the Nineteenth Century,"

Kean Gibson's *The Cycle of Racial Oppression in Guyana*, and aspects of Walter Rodney's and Kimani Neheusi's studies, illuminate abject conditions and responses in the form of public protests.[5] Prior to these studies, Norman E. Cameron in *The Evolution of the Negro* had observed that most of the writings spanning the nineteenth and twentieth centuries projected the colonial hierarchy's points of view on village development and overwhelmingly neglected the opinions and activities of African Guyanese. Cameron noted the group's travails as well as their triumphs.[6] Following Cameron's trajectory this chapter focuses on the ways in which this group wrestled with their circumstances to improve their conditions from the agricultural through mining eras.

After emancipation villagers continued economic ventures to take advantage of expanding opportunities created by increasing domestic demands for fruits, vegetables, ground provisions, and livestock, especially in the urban centers of Georgetown and New Amsterdam.[7] Throughout the period spanning the 1840s to the 1880s some villagers exported their products to the Caribbean islands and Nikerie, Suriname. In the 1880s, even with migration into mining, villagers continued farming. Some farmers from villages on the West Bank Demerara such as Klien, Pouderoyen, Schoon Ord, Vauxhall, Bagotville, and Canals #1 and #2 built their own boats and transported their goods for sale at Stabroek market across the river. In 1889, village exports totaled 4,482 bunches of plantains, 15,800 pounds of arrowroot, 3,039 bushels of corn, and fifty pounds of coffee.[8] In the 1890s, villagers including residents of West Coast Berbice and Mahaicony who had earned money from mining returned home and invested their earnings to improve their farming and supplement their incomes.[9] Local factories and mills were supplied with sugar cane, rice, coconuts, corn, cassava (yucca), and other products from the villagers' holdings for processing. The manager of Plantation Lusignan, East Coast Demerara asserted that cane farmers from the adjacent village of Buxton earned over $12,000 some seasons from supplying the sugar estate with produce.[10]

In 1844, concurrent with the cultivation and sale of villagers' crops, The Royal Agricultural and Commercial Society (RACS) was formed and held local, intercolonial, and international exhibitions. By 1855, residents of African-Guyanese villages were active in ensuring the success of the exhibitions that served as marketing venues for their products. Items they produced from sugar cane, bananas, plantains, cassava, corn, and coconuts were exhibited. Some of the villagers were skilled mechanics, engineers, blacksmiths, sugar boilers, coopers, and carpenters who had their manufactured items displayed. At the local exhibition held in 1866, one of the displays was a locally invented steam engine. In 1888, a RACS exhibition was held at Buxton Village and by 1897, another such event was staged at Golden Grove Village. On May 30, 1898, the Victoria Belfield Horticultural and Industrial Show was inaugurated. Villagers of Bagotville, West Bank Demerara held exhibitions, some of which were promoted under the auspices of District Agricultural Societies and Farmers' Associations; organized by the end of the nineteenth century. Established in 1901, an aim of the Board of Agriculture was to encourage peasant industries on the coast and in interior mining areas.[11] The intertwining of agriculture and mining by villagers supplemented incomes and released vital resources to them. It is within this ambiance that African Guyanese empowered themselves.

Early in the nineteenth century, farming and commercial transactions had provided the resources for villagers to acquire real estate. Later on, utilizing available opportunities for self-improvement, villagers' prolonged agricultural activities even as mining expanded. Consequently, many of them avoided relying on a single way to make a living. As the twentieth century evolved, farming by villagers sometimes attracted offers of assistance only after their hard work had demonstrated the viability of the undertakings. *The Small Farmers Committee Report of 1931,* indicated that village farmers whose cultivations were six to eight miles distant from their homes had for almost a century walked or used boats to travel to their farms. Expressing concern for the small-scale agriculturalists', government assistance and reform of the system of land tenure were recommended.[12]

At a time when some residents of Mocha, East Bank Demerara—a village established by formerly enslaved Africans—were risking their lives in gold and diamond mining, their compatriots declined unsolicited assistance from the very parties who had caused their ancestors so much pain. During a May 1934 meeting, villagers discussed the unrequested assistance offered to them by the management of the adjoining private-owned estate. With alacrity, the villagers unanimously declined the gesture of help with their cane-farming. They surmised that the offer was a ruse, and their acceptance of it "would give the estate a better chance to take away their land" because estate officials "had already asked for certain facilities which were not granted."[13] Villagers' suspicion that the gesture of help by the estate personnel was a ploy to encroach on their land emanated from the pervasive hostilities they had traditionally encountered from economic elites in their efforts to empower themselves.

The significance of these twentieth-century villagers' perspicacity in spurning the unrequested assistance had much to do with the adverse situations they and their ancestors had endured as they grappled with the realities of life in this plantation society. Memories of the pain of enslavement and its aftermath persisted and underscored the villagers' scepticism of the motives of the plantations' representatives. Mocha's 1930s villagers captured the long struggles they and their progenitors faced to maintain their integrity. At no point in their protracted involvement with the plantocracy and their allies were villagers' interests ever considered. Rather than paying them living wages after enslavement ended, plantation owners, who were financially compensated for freeing the enslaved, imported indentured workers to undercut their earnings and to replace them. Even more vexing, was the fact that taxes on products utilized by the freed people were increased to subsidize indentured immigration which made them under- and unemployed and caused them economic hardships.[14] Therefore, they were skeptical of unsolicited offers of assistance from estate officials who had traditionally and unconscionably trampled on their entitlements in order to control them.

African Guyanese had paid dearly for freedom. They understood that decisions concerning their existence and well-being centered on their self-reliance and self-help. Among them were those who would prefer to die trying for themselves rather than waiting on or allowing others to act on their behalf. Long before 1934, many villagers from Mocha had established ways and means of independent living; and had initiated ways to survive freedom and to provide for the future. Many of them journeyed to and worked in the gold and diamond fields while others toiled in the bauxite producing areas. In the face of economic uncertainties and ecological hostilities, villagers remained undaunted and made a living from agriculture and mining, and ensured the survival of their habitats.[15] To reiterate, the refrain of miners—"we missed our families, but we had to do it"—eloquently expressed their motivations.[16] This group cannot be pigeon-holed as only struggling and helpless victims of circumstances who were knocked about by social and economic changes as though they had no input in shaping their lives. Similar to other segments of the society, agricultural, mining, and other workers had an understanding of changing socioeconomic situations and made choices for their long-term benefits.

Guyana's mining industry emerged at a timely period of global development when diversifying of economies enhanced the wealth of nations. Economic advances in societies and the satisfying of human needs for food, housing, health care, education, transportation, and employment, depended on a nations' wealth and on the use of additional minerals and ores.[17] In this connection, a related development is important. The gold standard emerged by the 1870s to estimate the wealth of a nation by the value of its reserves of that commodity. Gold was used to indicate the prices of domestic currencies in relation to the amounts of the item a nation held in reserve. Gold became a "universal and stable unit of valuation."[18] As the nineteenth century ended, the worldwide need for and the value and utility of gold and diamonds served as motivating forces to attract an African diaspora workforce to local employment opportunities. In the twentieth century, the international demand for bauxite and its proliferation created additional openings for this workforce. "Push-factors" of arduous work and non-commensurate recompense in other employment sectors as well as under- or unemployment along with the "pull-factors" of possible increased financial rewards and the flexibility of mining migrants to work in other intersectoral economic undertakings enhanced this migration.

Theories of internal migratory movements usually address rural to urban migration and mostly involve agricultural laborers moving to industrial complexes. This type of migration often impacts negatively on a country's economy when migration outstrips employment opportunities.[19] In Guyana's case, the rural to rural migration to mining while farming continued created additional employment opportunities; precipitated population growth; and tepid infrastructure development in the interior. Specifically, migration to and from the gold and diamond fields, pendulum-like in its feature, enabled miners to remit or to take home earnings and to work in agriculture and livestock rearing seasonally. Such activities sustained the workers' rural communities. Still, the mobility of this male-dominated mining workforce encompasses several theories and categories of migration.

Internal, rural-to-rural, urban-to-rural, frontier, circular, serial or chain, follow-on, seasonal, and short-or long-term were all features, which contributed to what was conveniently characterized as cyclical migration. Cyclical migration, generally, hinges on periodic employment requiring a workforce willing to shift from areas of dense population to places inhabited by few people so that they could satisfy current labor needs. Cyclical also denotes shifts in an agricultural work regimen, which is seasonal and depends on preparing land, planting, nurturing, and harvesting at a specific time with workers returning home until the next cycle of production begins.[20] This definition was not wholly applicable to gold and diamond mining in Guyana because cyclical suggests the fulfilling of labor requirements for the establishing and functioning of industrial, manufacturing, or agricultural undertakings as long as the demands for the products existed. The specificity of gold and diamond mining in the country followed a different trajectory because globally, the quest for precious metals and gems and the profits thereof have been insatiable. Thus, although this migration and mining hinged on weather cycles, the ongoing demand for the commodities makes the cyclical description inadequate. Cyclical inadequately describes this migratory movement because while crops can be cultivated according to climatic and natural seasonal variations or demands, no matter how favorable the weather and fertile the soil, gold cannot be grown on demand to suit prevailing market conditions. The commodity cannot be manufactured and the local output of gold, which has been in demand for centuries, flourishes to this day.[21] Gold and diamonds, unlike agricultural products, can be stored in perpetuity.

The unique definition of pendulum migration for this movement was fostered by the inadequate interior infrastructure and hazardous environmental conditions that dissuaded workers from settling with their families. Many workers had no other recourse than to head home from the mining fields at the onset of bad weather; only to return when conditions improved. Hence, it is apt to characterize this migratory movement as pendulum. Migration to work in the gold and diamond fields by villagers was fueled by socioeconomic factors—some of which were previously outlined. Subsequent villagers' movement to bauxite production created additional socioeconomic vistas. Labor migration to bauxite work led to the permanent settlement of families in those regions more so than in the gold and diamond areas. Families who relocated to the bauxite areas were no longer directly part of village dynamics. Nevertheless, factors influencing the internal migration fell under several rubrics.

The human aptitude or propensity to seek opportunities and improve socioeconomic circumstances is omnipresent. In Guyana, escaping from conflict associated with working on plantations was one factor. Low-paying and arduous work, in general, on the coastlands that contrasted unfavorably with benefits and incomes from bauxite mining were important factors. Gold and diamond seekers could work independently and periodically, allowing them the flexibility and time not only for farming and livestock producing, but also for public works.[22] Newspaper accounts captured the essence of miners' abilities to utilize their earnings to make up for shortfalls from agriculture. It was noted that surplus laborers were no longer visible on the streets because work in the gold fields had created opportunities. Returning miners were "repairing and enlarging their old dwellings, adding outhouses . . . and buying furniture."[23] Many villagers including miners, upgraded their habitats and provided for the future by educating their offspring to function at different and higher levels of the socioeconomic scale.

Education Enhancing Expectations: The Role of Mining

Villagers understood that their own initiatives and the resources available to them could be utilized to empower themselves and their offspring. Some former miners explained that braving the perils associated with prospecting for gold and diamonds, had much to do with educating their children and enabling them to pursue less hazardous occupations, and at professional levels. Both urbanites and ruralites envisioned that investment in education would alleviate or lessen likely disadvantages of their offspring and enhance their quality of life. Thus, in the 1890s when colonial officials suggested that as potential agricultural laborers, it was ill-advised for parents to educate their children, James Corry of Den Amstel Village, the president of the Village Chairmen's Council, dissented.[24] Villagers expected education to offer added solutions to their healthcare, legal, and other predicaments that plagued them.

As outlined in chapters 4 and 5, many villagers including mining workers suffered because of inadequate healthcare services. In the interior, the lives of workers in the precious metal and gems industries were imperiled due to the unavailability of adequate and timely healthcare. From the late nineteenth century, in spite of bureaucratic roadblocks, African Guyanese were among the professionally trained healthcare practitioners who emerged and endeavored to fulfill the need for improved services in the entire colony. Education assured such ascendancy.[25] Education enabled A. B. Brown of Den Amstel Village, West Coast Demerara to become a legal practitioner and one of the first African Guyanese elected to the local legislature. Brown was an entrepreneur who co-owned cocoa and coffee producing, Plantation Middlesex, No. 2 Canal Polder, West Bank Demerara. He invested in the coconut, sugar, gold, and diamond industries. Brown financed groups of miners to prospect at sites he owned in the interior. Education enabled Brown to aspire, to achieve, and to contribute to socioeconomic, political, and mining development.[26]

Villagers' pursued education in the twentieth century by establishing private and fee-paying secondary schools. By the 1920s, Matthews High School, Den Amstel Village, West Coast Demerara was established. In the 1930s, J. P. Gray, a graduate of Howard University, was the principal of a private secondary school in Vergenoegen Village, East Bank Essequibo. Classes were held at the Liberty Hall of Marcus Garvey's, Universal Negro Improvement Association (UNIA) which served residents of nearby villages such as Parika, Farm, Tuschen, and Uitvlugt. Cumminsburg College emerged in Buxton Village, East Coast Demerara during the 1930s as a precursor of County High School established in 1956.[27] Urbanites and their rural counterparts ensured their offspring utilized educational institutions available to them through organizations such as the Negro Progress Convention (NPC) and the League of Coloured Peoples (LCP) and graduates of Howard, Shaw, Tuskegee, and other Historically Black Colleges and Universities. While the LCP established academic institutions at the primary and secondary levels, the NPC organized a local Tuskegee (Institute) University to prepare its graduates for agricultural and industrial careers. Residents of the Upper Demerara River bauxite mining complex also utilized educational institutions, including technical schools to enhance their socioeconomic opportunities.[28]

By the mid-twentieth century, the training of more technicians in geology and other mining-related occupations was promoted as a way to exploit the country's natural resources and to enhance its economic development. Some locals trained overseas as cartographers, geologists, and hydrologists, while under the auspices of the Department of Lands and Mines and The Geological Surveys Department, similar training opportunities were created locally. In April 1951, the government established the Technical Institute in Georgetown to provide training in various skills and trades. By May 1956, the East Coast Demerara villages of Golden Grove-Nabaclis had a branch of the Technical Institute operating in its community center. Anns Grove Village, a few miles distant, already had a similar facility. Occupational training was also provided by other government agencies including the Education, Medical, Public Works, and Transport and Harbours Departments. *The Annual Report of the Labour Department, 1956* identified thousands of persons who had completed or were still in training.[29]

In Linden, the bauxite mining area, there were several primary schools and the Mackenzie High School.[30] In March 1956, the Demerara Bauxite Company (DEMBA) began a four year mechanical engineering apprenticeship training scheme. Many of the trainees were recipients of primary and secondary education from the local schools, identified in Chapter 6. The training scheme was projected to include electrical engineering. By 1958, the DEMBA Trade School was established at Mackenzie. Training at the facility specialized in skills required for the operation of a proposed alumina project.[31] Increasing global demand for minerals and ores and expanding opportunities in the local mining industries warranted a better educated and trained workforce. Government and management of private companies, at the behest of trade unionists, recognized the necessity of training more local engineers and technicians. In the gold and diamond industries, the training of geologists, geological field assistants, mechanics, and boat captains were implemented to facilitate an expanding mining industry.[32]

Mining life was harrowing and many villagers viewed education as a means to experience vicariously their other aspirations through the younger generation. Still, in the evolution of secondary education, the institutions were mostly established in urban locations and somewhat interspersed in a few villages. Knowing that specialized training would be a reality with the foundations provided by primary and secondary education, miners spent portions of their earnings to ensure their offspring had access to education. Children of rural miners and others who had no such facilities nearby traveled to the cities to attend academic or technical schools. This was a complex undertaking for the family.

For the students, daily, weekly, or monthly travel was sometimes necessary, depending on the proximity of their homes to the schools. At other times travel occurred at the beginning and end of a school semester. Students who lived at home in villages had to awake early each morning to travel varying distances by foot, bicycle, car, bus, train, or ferry to attend school in Georgetown. Some students whose parents or guardians wished to avoid the rigors of frequent travel arranged for their children to board and lodge on a weekly or monthly basis, or to stay for the entire semester with relatives or friends. Some of those persons were not always kind and understanding. In addition to expenditure for travel and board and lodgings, parents had to provide money for school uniforms, other clothing and shoes, books, equipment and fees, as well as for illnesses, miscellaneous items, and emergencies.

Consider the case of the aforementioned Darrell Gray of Queenstown Village, Essequibo who sent his daughter, Yvonne to Cambridge Academy High School in Georgetown. The father provided the financial means for her to attend the institution for several years while residing in the city. As a parent, Darrell Gray never shirked the responsibility of giving his daughter all the accoutrements necessary for success. Ultimately, Ms. Gray became an educational administrator. In turn, she raised several children who became professionals. Yvonne's father credited his earnings in gold, diamond, and bauxite mining for providing the wherewithal to ensure his daughter's academic achievements and those of succeeding family members.[33] The elder Gray's decision illustrates the dedication and contributions of miners to the achievements of their offspring. Although Prince Albert Daniels, a gold and diamond miner from Perseverance Village, Essequibo County had six other children, he sent the eldest, Paulette to Tutorial High School in Georgetown. From the late 1950s, Paulette Daniels resided with her father's relatives in DaSilva Street, Kitty during the school year and returned to her village for vacations. Eventually, she became a healthcare professional in Canada.[34]

Linden, the bauxite mining area had only one major secondary school, the Mackenzie High School. It could not meet the educational needs of all the residents. Similar to mining workers in other parts of the colony who lived in places with insufficient educational facilities, Ulric Griffith, a bauxite employee, sent his nine children to secondary schools in Georgetown. Some of the Griffith siblings attended Cambridge Academy. The father rented a house at Campbell and Sussex Streets, Albouystown in the city to accommodate his children where he or his wife visited them periodically. To supplement the family's income, the grandmother worked as an independent stallholder at Wismar Market. Ulric Griffith was born at Anna Regina Village, Essequibo and lived and worked at nearby Suddie Village before migrating to the bauxite belt. Similar to many residents of this diversely populated mining community, Mr. Griffith's wife's ancestors were Caribbean migrants from Barbados.[35]

Members of the African component of the Guyanese population had mixed experiences, contemplated their environments, and made decisions according to their realities. However, their actions in ensuring their children had access to educational opportunities were not unique. Parents from all walks of life utilized education as a way of ensuring that their offspring could function in different and better circumstances. Understanding the changing nature of their environment and the wider society and equipping succeeding generations to function effectively was a hallmark of the Guyanese people. As a workforce, the offspring of miners who emerged after the mid-twentieth century did not as a rule become mining or associated workers. They are included in the entrepreneurial, managerial, technical, industrial, and academic sectors as facilitators of the ongoing development of the country. Alongside their male counterparts women had ensured their offspring were equipped educationally to function in all sections of the society. Another important facet of the evolution of Africans in Guyana parallel to the mining industry's development is the intertwining roles of men and women.

Women, the Dynamics of Village Life, and Aspects of Mining

The contemporaneous roles of women who were less visible in mining, but were largely responsible for keeping families together when the men migrated for their prospecting ventures,, cannot be overlooked. Mining men from the villages were sons, spouses, fathers, and agricultural and other workers who often viewed their earnings from labor migration as a means of fulfilling economic responsibilities as heads of households. Birth and marriage recordings for village residents reveal the existence of a preponderance of male-headed households and miners who took responsibility for their families.[36] Village women performed multiple roles as daughters, partners, wives, mothers, nurses, and caregivers for the young and aged. As breadwinners, women were farmers and vendors whose diverse roles enabled them to cope with the internal mining migration of their male counterparts. Women sustained their families and ensured the longevity of the villages. They played equally important roles in the dynamics of village life whether the men were away for a short time, as in bauxite on a weekly or monthly basis, or in the gold and diamond fields for much longer periods, spanning months or years.[37]

Villagers were engaged in teaching, nursing, security, construction, and clerical occupations, but agricultural and related work and the contributions of women who had connections to the mining industry is a focus of this section. The activities these rural dwellers undertook, whether a majority of men were present or absent for short or long periods, are important. Women's input in their families and villages' progress were part of the dynamics of life because those who remained at home as well as the ones who migrated to work were intricate components of their communities. To understand how these inhabitants survived and enabled succeeding generations to ascend socioeconomic ladders requires placing women in the fabric of village life. After all, they were multitaskers who took care of families, relatives, and friends' food, clothing, medical, and educational needs. Women cultivated crops, tended poultry and dairy animals, went fishing, and were vendors of farm, homemade, and other products.[38] The activities of various segments of the villages' population ensured much more than a monolithic or stereotypical community. Individuals and groups, both males and females, exhibited diverse abilities and undertook varied economic ventures in addition to agricultural and mining pursuits to grapple with the vicissitudes of their lives.

Traditional perspectives on out-migration by males, even though temporary at times, generally pinpoint the consequences and implications for rural communities, especially the impact on family organization. A key concern is that this migration forces women and children to assume added responsibilities in a mostly matrifocal environment. Another concern is that the absence of a skilled component of the workforce—adult males—decreases road and building construction and their maintenance as well as tax payments.[39] In rural Guyana, even when funds were available for public works, assignments were never strictly divided along gender lines. Furthermore, mining men would delay their departure to the hinterland to shore up sea defenses and to engage in other public works. Farms and villages were never entirely devoid of male family members because they did not all depart simultaneously. Males sometimes rotated their departures to the interior in order to assist women in agricultural and other activities. Revenue collecting by local government authorities showed a decline when males were away in the gold and diamond fields and an increase on their return home, but payments on village properties never ceased entirely through men's absences. Women, many of whom worked outside of the home, continued taking care of all family related activities and had long been active participants in the upkeep of their families and habitats.[40]

From the inception, women inserted themselves into the fabric of their villages. When the first African-Guyanese village, Victoria, was bought on November 7, 1839, five women were among the sixty-four purchasers. Similarly, when the mortgage was fully paid and transport obtained by the purchasers of Buxton Village in 1840, Whiliminner, Mary Diamond, and Corobal (Corabell) M'Kenzie were proprietresses.[41] Countrywide, the purchasers of villages and owners of real estate always included women who asserted themselves and maintained their families. Although some evidence suggests that migration can decrease marriage possibilities and births; resulting in imbalances in gender and age ratios in the communities from which the migrants originated, the localities examined in this study did not undergo feminization or senilization.[42] In Guyana, the exception occurred because of the nature of what can be appropriately called pendulum migration, that is, the ongoing forward and backward movements of workers between the mining fields and their residences. The dual movement revealed here, usually occurred between peak periods of agricultural production—and the dry or rainy seasons—when climatic conditions alternatively favored mining. During dry weather labor migrants traveled to the gold and diamond fields to prospect. With the rainy seasons, when swollen waterways made travel too dangerous and the earth too soggy for the excavation of gold or diamonds, labor migrants remained at home to pursue other economic activities.

As mining emerged, some villagers continued to fish in nearby canals and the Atlantic Ocean with locally made boats, fishing-nets, and seines. Other villagers remained in artisan, construction, and business enterprises, or pursued diverse forms of self-employment. The strivings of the residents predated and complemented the efforts of significant numbers of villagers who chose to work in the gold and diamond fields. Village women were always involved in farming, fishing, and livestock and poultry rearing for both family consumption and commercial purposes. Women, as part of the workforce, promoted and sustained cottage industries. Similar to men, women utilized *Manihot Esculenta* or *Euphorbiaceae* (cassava or yucca) and manufactured cassareep sauce (a meat seasoning and preservative) or made bread and starch from the root of the plant. They produced coconut oil for domestic, commercial, and industrial use including the manufacture of soap. Women and men made and sold pepper sauce, pickles, jams and jellies, straw hats and baskets as well as confectionary and baked goods. Both males and females were active participants in the improvement of their villages and taking care of themselves, their families, and others.[43] Economic and varied activities by women who functioned as head of households when males migrated were no phenomenon.

In the Caribbean, Elsa M. Chaney affirmed that when men migrate women "must continue to manage the household and care for the children ... cultivate the family food and produce the cash crop" and utilize "off-farm employment to supplement what they can earn ... from the land."[44] In developing societies with enormous quantities of hinterland mineral resources such as Guyana or Suriname where employment opportunities fluctuated, women were involved in activities that were essential for their families and communities' survival. Local women coped with the men's absences, although migrating males caused villages to be characterized as matrifocal.[45]

Rowena T. Mingo (1915–2004) was born in Eldorado village, West Coast Berbice. "Roe" Mingo, as she was familiarly called, captured the essence of women's lives in the villages, especially during the 1920s-1930s, a heyday in the migratory movement to mining. "The women," she explained, "performed all the work that men could do and they did even more, particularly when the men went to the goldfields." The women included mothers who farmed, and never considered back-dam cultivation as a hindrance to their childcare duties. Infants accompanied their mothers to their farms while the older children attended school. At dawn, women loaded their boats with their children, if any, along with cooking utensils, and sometimes fishing apparatus for a day's work on their backlands' farms. Other family members or villagers were often taken along to augment working parties. The same canals that facilitated travel to the farms were sources of food in the form of "bush fish" which included "hassa," "houri," "sunfish," and "patwa." Women caught fishes in the canal, cooked meals on their farms, and took some of the prepared food back to their homes at the end of the day for other family members. Ms. Mingo explained that the "catching of fish with nets was mostly done by men, but some women learned the skill from their fathers or other males of the older generation."[46] Thus, the women were able to utilize fishing to supplement diets and to sell the surplus catch as fresh, fried, or dried fish.

The farmland at the back of Eldorado village was a major center of activities. The area was located some four and a half miles away from the residential sections with the rice cultivation area even farther away. Women who produced rice used a work-pattern dating back many generations to coincide with a seasonally grown crop, heavily dependent on rain and sun at strategic times for its success. Planting, nurturing, and harvesting seasons required frequent treks to the rice fields. "Roe" Mingo detailed aspects of the harvesting operations such as tying the reaped rice-stalks in bundles, collecting them in oxen-drawn carts, and processing them in a homemade apparatus. In addition, oxen were used to stamp on the rice in husk or "paddy" to separate it from the plants' stalk. The "paddy" was usually placed in a wooden mortar, a hollowed out receptacle, which was made from a tree trunk and pounded with a pestle, a long wooden stick, to loosen the husk from the grain. Next, locally woven flat baskets were utilized by the women and their assistants to winnow the product and fully separate the husk and grain. The processed rice was divided into three portions. One portion, earmarked for home consumption, was placed in baskets and later stored in the "rice-room" of the family house. Another portion of rice was shared with other family members and friends in the village or sent to relatives who lived in non-rice producing areas. The third portion of rice was taken to the village factory and sold. The money gained became part of the family's income.[47]

An inescapable situation that villagers coped with was the anguish of family members or associates who died untimely deaths while migrating and mining. Tragically, "Roe" Mingo identified Samuel Waldron and Abraham Mingo, family members who went to the gold fields to work, but drowned and were buried in the interior. She pointed out that "Tommy Semple drowned attempting to cross" the aforementioned dreaded Itanami Fall. George Dodson, a miner who was absent from home for long periods earned the sobriquet, "Bush Di-Di." Unexpectedly, extended absences due to prospecting sometimes caused tremendous consternation. Some women who were left at home for prolonged periods assumed that their partners who were miners had died and were buried in the interior. The women then formed relationships with new partners and some even bore their children; often adding to those of the previous partners'. Dynamics of village life vibrated, and confusion reigned if and when the former non-deceased partners returned.[48] Nonetheless, the majority of departing male labor migrants devised a number of ways to assist their families to cope in their absences.

Migrating Miners and Their Families: Coping

Facing unforeseen circumstances, migrating miners ensured that the well-being of their families was not compromised with their departure. They usually left varying amounts of nonperishable food staples such as sugar, rice, flour, and beans as well as cash to last their families until they sent remittances. Departing miners believed that they would recover some gold and diamonds before long and send money for their families through their employers' offices, by trusted friends, or through the post office before their return. Extended families usually provided a helping hand during times of adversity, as in the case of the aforementioned Wright family.[49] Overall, the refrain of former miners "we missed our families but we had to do it," exemplify villagers' efforts to cope with the aftermath of enslavement and human and ecological hostilities as well as their determination to make a living for themselves and dependents.[50]

In 1925, Lionel Arthur who was born on June 3, 1909, in Mocha Village, East Bank Demerara ran away to the gold and diamond fields. As a student at Mocha Wesleyan School, Arthur had passed the School-Leaving Examination, but he had no interest in becoming a teacher as his mother and the schoolmaster had planned. In those days such an achievement was quite significant. Nevertheless, Arthur spent twenty-eight years in the interior and raised his family including ten children on the coast. He sent money home by various means, and on intermittent visits home he brought the bulk of his earnings. Arthur first prospected at Buck Canister, located on the left bank of the Mazaruni River. Although he originally worked with small-scale prospecting groups, he also worked for large mining concerns, and in the timber industry.[51]

Mining life inevitably included contact with Guyana's indigenous or Amerindian population who largely inhabit the interior regions. Arthur and his fellow miners slept in hammocks, which were of Amerindian or African origin. This open-ended "sleeping-bag" woven from string, twine, rope, or grassy or fibrous material was secured by tying it between two poles or tree-trunks, at least two feet above ground. According to Arthur, relations with Amerindians were usually cordial. However, on first encountering African-descended miners, Amerindians always appeared reluctant to entertain contact with the strangers. On becoming familiar with the newcomers, Amerindians visited their camps and performed small tasks, such as fetching loads for which they were paid from the miners' supplies of rations (salted fish or meat, dried peas or rice). Customarily, agents attached to local and foreign concerns purchased gold and diamonds from small-scale miners.[52] After they found gold or diamonds, miners could also sell them at the stores and pay their outstanding bills and make new purchases. Firms with headquarters in Georgetown operated many of the stores. Caribbean islanders or locals like J. G. Stephens and A. B. Drakes established similar businesses in the interior. Itinerant hucksters, some of them women, infrequently traded at mining sites. [53]

Leisure time was limited, but during some weekends or every fortnight miners could walk for miles to a river-landing store. Games such as dominoes or cards were played and a drink of rum enjoyed before the return journey to the work-site. The only other contact miners had beyond the frontier was the bimonthly mail delivery. In the mining regions, gun-owners hunted "bush-meat" or wild pigs, deer, agoutis, iguanas, and labbas to supplement their food supplies. Fish was always plentiful and sometimes weighed over thirty pounds. This was "bush life."[54]

Miners anticipated and prepared themselves for the challenges of healthcare in the interior. Their insightful perspectives on the state of healthcare in the villages and the interior convey interesting glimpses of coping mechanisms utilized by these labor migrants. In most coastal villages of the 1920s to 1930s, a physician and/ or surgeon, dispensers, and practitioners of traditional or "bush-medicine" provided healthcare while the more serious cases were treated at hospitals. During that period, in Buxton Village, Drs. Abensetts, Ramdeholl, and Singh provided both government (public) and private medical services. Other healthcare providers included Kemp "Doc" Carter, an apothecary, who was also the sick-nurse and dispenser attached to nearby Cane Grove Hospital. Dispensers such as Morgan, Munroe, Abrams, and Cromwell, from the village, were often assigned for long periods to interior government dispensaries, clinics, or hospitals. Traditional healers or "bush-medicine" practitioners included Agatha Williams; familiarly known as "Mammy Fiffee" (1905–2004) who was trained by her grandmother, Gang-Gang, a formerly enslaved person. "Mammy Fiffee' handed down her skills to a daughter. Bowman Marcus, a Conga "bush-medicine" doctor, and the father of Elfrieda Marcus better known as "Miss Frieda" passed on his skills to her.[55]

Miners never assumed that their well-being in the interior was guaranteed. At the outset of each trip prospectors traveled with patent medicines including Canadian Healing Oil to relieve both internal and external maladies. Items, such as Oil of Clove for the relief of tooth aches; soft-grease to use as a poultice on boils and abscesses; Vick's Vapour Rub for chest and head colds; and Andrew's Liver Salts, that, reputedly, cured all stomach illnesses, were integral parts of prospectors' valuables. Miners carried preventative and curative medicines prepared by traditional healers to ward off and to treat minor illness in the bush. Miners knew that health hazards in the interior were legendary. Nonetheless, it must be remembered that illnesses occurred on the coast and in the interior where, for much too long, afflicted persons received inadequate or no medical, surgical, dental, or psychiatric treatment. National healthcare services developed slowly and therefore fostered self-medication as a coping mechanism.[56]

Coping with the period spanning the global depression, which was certainly not the best of times for miners, their families, or associates was challenging on several fronts. During the 1930s, appeals were made to the governor and larger private businesses by small-scale independent operators for assistance to purchase equipment and supplies as well as help with transportation. By 1934, although weather conditions were adverse, some intrepid small-scale miners outfitted themselves for prospecting trips. Unfortunately, incessant rainfall caused floods, which severely limited their work.[57] Simultaneously, food prices increased steeply. River travel was dangerous and the height of the water obliterated all the markers which guided boat operators in the interior. In fact, "men who had been travelling on the Mazaruni River for years" reported "that they had never seen it in such a condition."[58] As the rains fell unabatedly, floods reached crisis levels in Essequibo County. Throughout the colony, people were starving and fleeing their inundated homes to the safety of government buildings. Consequently, the worsening situation caused the enactment of *Ordinance #1 of 1934* "to regulate the price of food . . . to ensure the food supply for the people of the country."[59]

Meanwhile, fifty small-scale claim-owners including Josephus Headley of West Coast Demerara who had suffered economic hardships submitted a petition to the Legislative Council through the Commissioner of Lands and Mines. The petitioners requested some concessions to mitigate their losses in the interior and in the villages due to the recent floods. Most disconcerting for the petitioners, was the fact that they had already paid all of their mining fees, but the flood had prevented them from working at their sites. On returning home, they had found the same flooded conditions, which prevented them from farming. They were hard-pressed to upkeep their families. As it turned out, their petition was successful and the payments were reimbursed.[60]

Spanning the nineteenth and twentieth centuries were prospectors who would not cope with a mining life. Charles Egerton Denbow Jr., was born in No. 2 Village, Corentyne, Berbice County on February 21, 1877. At school he had displayed academic competency, passed the requisite teaching examinations, and by 1891, he was employed as an assistant headmaster. The next year, considering a teacher's salary inadequate, Denbow became an engineering apprentice at Albion sugar estate. By September 1894, he and an older brother, George L. A. Denbow, a clerk employed by Garnett & Co. left for the Konawaruk goldfields. In July 1895, after completing his second stint, Denbow decided that he would no longer cope with "bush" life. He joined the British Guiana Police Force, rising to the rank of County Sergeant Major by October 20, 1914, the first African Guyanese to do so.[61]

Alan Carter could find no reason to endure the rigors of gold and diamond mining. Carter was born in Buxton Village, East Coast Demerara on July 10, 1923. After making a single trip, he concluded that life in the interior was not for him because he "wanted to live to an old age." Carter reverted to faming and sold yams, sweet potatoes, and cassava at Stabroek market, Georgetown.[62] In the 1950s, David Hatton, another Buxtonian, also made a single trip to prospect in the interior, but left soon after returning for England where he lived until his death in November 2010. Hatton did not find the precariousness of frontier conditions romantic.[63] In 1948, at twenty-eight years old, Howell Wood (1924–2011) and a group of other Buxtonians, Desmond Fiffee, Cyril Abrams, Burchell Booker, Norman Jarvis, and Rickford Weatherspoon began working for the Outlaw Mining Company, a Canadian gold mining firm. This was Wood's way of coping with the economic uncertainties of the aftermath of World War II. Within a few years, he married, started a family, could no longer cope with a mining life, and reverted to carpentry. An enduring benefit from mining for Wood was his ability to acquire resources to construct the home he and his family occupied in Buxton-Friendship Village.

Gold and diamond seeking may seem a litany of seasonal and pendulum migrations, but occasionally, the prospectors paused for enjoyable times. As the announcement in the *Tribune,* April 15, 1934, proclaimed—"DIAMOND SEEKERS DANCE"—to be held "at Jerrickville, Victoria Village, East Coast Demerara . . . on Friday night . . . by diamond seekers." The Jerricks, for whom the venue was named, similar to the Goring and Elias families, were well known for their gold and diamond mining interests.[64] Many villagers participated in those joyous occasions, which celebrated the homecoming of returning mining migrants. Art forms such as music and dancing enabled villagers to cope with the travails of their labor; and the varied vistas of village life. The ways in which the inhabitants conducted their affairs ensured their continuity.

At a critical juncture of the country's socioeconomic development, migration and mining influenced changes and continuity. The multidimensional vistas of life in this colonial society and the ways in which working class persons utilized agriculture and mining to provide for their needs at a particular epoch, ensured the empowerment of their offspring to explore additional vistas. Migrating miners—rural, urban, and overseas workers—were co-participants in establishing the industry and ultimately contributing to national development. Families coped and provided for the future and as the following chapter delineates, villagers cooperated, transferred traditional knowledge, and incorporated additional financial plans and programs to wrestle with ongoing challenges.

CHAPTER 9

Knowledge Transfer and Cooperativism: Agriculture and Mining Eras

Enslaved Africans in Guyana transferred some of their traditional ancestral practices to empower themselves and provided for the future. In particular, their use of rotating savings and credit associations such as *esusu, susu, ajo,* and *awo* enabled self-purchase and fostered their acquisition and retention of property. Early in the nineteenth century, freed persons engaged in agricultural and business activities, saved their money, individually and cooperatively purchased real estate, and formed villages. As property-owners, they constantly faced challenges from colonial bureaucrats who enacted arbitrary tax laws and land use regulations. The measures often seemed designed to stifle the initiatives and progress of the laboring class. Sometimes, economic hardships caused delinquencies in villagers' tax payments. However, when mining emerged later in that century, added income from the industry enabled villagers to make timely payments. Within a few years, villagers' tax payments were again tardy, but a major financial institution in which they saved denied them loans. In the early twentieth century, farmers, miners, and other villagers formed Cooperative Credit Banks to utilize their resources to benefit themselves. The origins and longevity of African-Guyanese villages resulted from the inhabitants' ability to utilize ancestral practices, incorporate modern methods, and apply the outcomes to transform their conditions.[1]

At the outset of freedom, the villagers had displayed an understanding of the ramifications of living in an export-oriented capitalist economy. Expatriate bureaucrats and businesspersons usually lived in the colony only for specific periods before returning to Great Britain—stashing their financial accumulations abroad. The majority of the population comprising of the working class had little or no input into administrative, infrastructure, social, political, and economic decisions. Changes were implemented to benefit colonial bureaucrats, the economic elites, and overseas economies. The onus was on the working class not only to transform their lives, but to improve the quality thereof.

This chapter focuses on ongoing patterns of cooperativism among the rural population that enabled them to withstand challenges as they endeavored to develop themselves and their communities. In an environment where their labor was viewed as valuable only if controlled to produce benefits for others, the laboring class faced myriad roadblocks as they strove to improve their standard of living. Against tremendous odds, village farmers, miners, and others instituted and supported financial institutions, maintained their payments of taxes, retained their properties, and ensured the longevity of their habitats. This analysis focuses on the ways in which villagers conducted financial activities and demonstrated the utility for capital in human and economic development. In the early 1800s, they transferred their African knowledge of saving, in the form of the aforementioned *esusu* and *susu* (box-hand, partner) and so forth and acquired the resources to purchase themselves and real estate. At the dawn of the twentieth century, they incorporated Western financial models such as Cooperative Credit Banks to save their money and took care of their responsibilities.

In the unrelenting environment of colonial Guyana, villagers understood the value of real estate as a means of subverting mechanisms of the plantocracy and mineocracy from engulfing them. The plantocracy, from the late 1830s, were unable to adjust to the financial acumen of the emancipated people who purchased property and asserted their freedom. Planters reduced the wages of the workers and supported ongoing importation of labor migrants to minimize the nascent workforce's viability. Planters influenced legislators to enact arbitrary legislation including unfair tax and draconian vagrancy laws as well as measures to curtail villagers' real estate purchases. Sometimes, legislation was passed to prevent groups of workers from traveling overseas.[2] By the 1890s, when it dawned on the mineocracy that African Guyanese had accumulated resources to become independent miners, which counteracted their intent of dependency for the workers, the mining hierarchy reduced the workers' wages and mostly advocated labor migration.[3] In both instances, the African descended, as though in anticipation of those developments had a number of ways to overcome the roadblocks. Utilizing mental and physical adroitness, they adopted ancestral and Western practices and adapted to changing environs during the agricultural and mining eras.

Applying Knowledge Transfer to Cooperativism in the Agricultural Era

Enslaved Africans were of diverse origins, but they understood the need to overcome tribal, linguistic, and cultural barriers and to utilize ancestral ways of pooling their human and financial resources to face the challenges of surviving in the Americas.[4] By the second half of the eighteenth century, the enslaved in the British Caribbean were legally allowed to engage in independent agricultural and commercial activities. Some of them saved their money by recreating and utilizing ancestral practices common in West and Central Africa and elsewhere on the African continent such as the aforementioned *esusu, susu, ajo,* and *awo* (box-hands, partners, or money-clubs), forms of rotating savings and credit associations. Ongoing importation of African ethnics such as Igbos (Ibos), Ibibios, Asantes, Tivs, Nupes, and Yorubas from different locations of their home continent fostered the continuance of the practices.[5] Africans and their descendants also commonly used the earth and elsewhere as hiding places for their money and other valuables during and after enslavement. In combination, their accumulations from rotating savings and credit associations, mutual aid, and other societies enabled members of the group to acquire considerable amounts of financial resources. The names of the saving mechanisms may vary, but the objective of cooperative accumulation of financial resources is the same in African diaspora communities.[6] Consciousnesses of Africa and traditional practices survived and enabled aspects of self-determination in New World communities such as Guyana.

As early as 1828, a time when slavery still existed, the colonial government decided to establish a savings bank for the use of the enslaved and their offspring. The institution was specially created within the colonial Treasury (and under the control of the Receiver General). Most enslaved and freed persons refused to utilize the bank due to their mistrust of the colonial officials, but they continued to accumulate money through their own devices.[7] After the Abolition of Slavery and during the Apprenticeship period, *Ordinance #71 of 1836,* was enacted to allow the government to establish a savings bank for the laboring class.[8] The bank was established at a time of increasing importation of indentured workers into the colony from Asia, Africa, the Caribbean, and elsewhere.[9] Unlike Africans and their descendants previously enslaved in Guyana, who were not paid for their labor, the indentured workers received wages from the time they began working, enabling them to utilize the banks as well.[10] New African workers among the indentures gravitated to the villages that former enslaved people had established. They intermingled and often formed durable relationships. African cultural knowledge and practices thereby received added infusions and sustenance.[11] Throughout the nineteenth century, Africans and their descendants in Guyana were able to weave a common thread of social, cultural, and economic activities and fostered their survival.

The historical antecedents of such activities and organizations are rooted in an African past and have African diaspora links. Enslaved Africans in eighteenth-and nineteenth-century Guyana established mutual aid organizations that predated and are often compared to the English version of friendly and burial societies that emerged by 1842. The organizations' functions included the collection and distribution of resources in times of need, such as in sickness, for burials, and other financial expenses. A. D. Wells and D. Wells, in *Friendly Societies in the West Indies*, quoting M. J. Herskovits, indicated that the organizations "seem to have been an institution founded upon the traditions brought overseas by the slaves themselves, and certainly antedates any West Indian Friendly Society," of later years. The case of the "*dokpwe*" of Dahomey in mutual aid activities is also cited.[12]

In African diaspora communities, evidence abounds of distinct institutions that are linked to Africa. In late eighteenth-century Jamaica, it was estimated that the enslaved had, approximately, 20 percent of all the money in circulation. African Americans had tens of thousands of dollars to deposit in the Freedmen's Savings Bank after emancipation in 1865.[13] Locally, the utility of these associations and the money generated was indicative of an intrinsic non-Western culture that comingled with elements of European practices, which enabled the African descended to provide for the future despite roadblocks erected by planters and their allies. Persons of African ancestry operated friendly and burial societies alongside other institutions for mutuality and financing. Simultaneously, the provision of parallel financial institutions in their communities by the colonial regime continued. It should not be a surprise then that in tandem with the advent of the seasonal migration of villagers to the gold and diamond fields, the government established an additional financial institution, the Post Office Savings Bank, especially in predominant African Guyanese urban and rural areas.[14]

A correlation existed between the strategically weaving of government financial institutions into the fabric of the lives of the working class when more money circulated in their communities. Therefore, it was not accidental that by the late 1880s when the new government-sponsored Post Office Savings Bank emerged, community members already imbued with the importance of saving, utilized the new institution. The economy of the country had diversified and agriculture along with mining enabled workers access to additional earnings. By the early twentieth century, under the guidance of their own leaders, villagers established what became known as the Cooperative Credit Banks. At a time when their payments of taxes were decreasing, villagers cooperated and initiated another way to meet their responsibilities. One of their most important achievements was the initial acquisition of real estate largely through their savings from agricultural labor. Villagers working in the gold and diamond fields subsequently parlayed some earnings into paying taxes and thus, retained their properties.

Payments of Rates and Taxes: A Segue From Agriculture to Mining

Before the advent of mining, taxation disputes loomed large within the village environment. Government taxation policies were usually viewed by residents with skepticism and as another way to hinder their socioeconomic development. Villagers felt that taxes were disproportionately levied on them compared to plantation owners who benefited from tax concessions that enabled them to offset the cost of and to finance state-subsidized immigration. Villagers knew that immigration served to undercut their bargaining power for increased wages, thwarted their labor initiatives and strike actions, and displaced them as plantation workers.[15] On several occasions, legislation was enacted, which irritated many villagers and their leaders because they deemed the measures as limiting their self-determination. For example, the *Partition Ordinance of 1851,* and *Ordinance #1, 1852,* prohibited the further purchase of land by groups of over twenty persons and slowed villagers' ability to acquire real estate.[16]

Following the disturbances of 1856, during which Portuguese-owned businesses were damaged; legislation enacted to compensate them ensured that payment fell squarely on the shoulders of African Guyanese, many of them villagers, who were protesting government's bias against them and preferential treatment in support of commercial activities by the Portuguese. Villagers viewed an added Registration

Tax as designed to thwart their independence and activism.[17] *Ordinance #33, 1856,* authorized the appointment of village commissioners to oversee public works and sanitation issues. Villagers were compelled to pay rates which were earmarked to offset expenses for maintaining and improving their localities while their authority to hire overseers and other public works employees was restricted.[18]

Viewing these developments as more unwarranted government intrusion, villagers challenged the decisions. A tumultuous period followed as conflict over taxation and property matters between villagers and bureaucrats became increasingly contentious and reached alarming heights. In instances, disputes spurred the offended parties to appeal to the governor who remained pragmatic; forcing a delegation of locals to seek an audience with the British monarch in Great Britain.[19] As conflict escalated over the nonpayment of taxes by some villagers, law enforcement authorities intervened and a few of them were arrested. After court proceedings, some individuals were incarcerated and their properties seized and sold for the infractions. Disputes over taxation simmered because villagers perceived colonial bureaucrats and planters as always working in tandem against their interests. Some villagers felt that resisting capricious tax increases and unwarranted actions by the colonial hierarchy would prevent their marginalization.[20] A key feature of the ongoing taxation dispute was the villagers' resentment of arbitrary impositions by those who appeared determined to worsen their circumstances.

In that vein, some villagers argued that the government should find alternative ways of raising money, other than whimsically increasing their taxes.[21] In the ensuing decades, government administrators sometimes hesitatingly, but often presumptuously, continued to intervene in village affairs. Taxation Ordinances in 1866 and 1873, did not ameliorate relationships between the government and villagers. Conflict exacerbated as some residents remained adamantly opposed to what they considered as government's officiousness. But, with slight regard for the residents, colonial officials enacted *Village Ordinance #4 of 1883,* which reorganized village administration. The Central Board of Villagers, Local Superintendent Office, Village Councils, and elections were abolished. Public works in villages fell solely under the colonial civil engineer's control. Unlike past arrangements where residents participated in the village administration, but received no financial subsidies, the reorganizing provided the engineer's office with substantial grants to cover expenses. The collection of rates and sanitation works shifted to the Central Board of Health. Villagers who were committed to self-determination resented the continuing efforts to relieve them of what they regarded as their responsibilities.[22]

During the decade of the 1880s, migration and mining were increasingly aspects of village life. Although some villagers were farmers and artisans, or pursued other occupations, the coming of mining proved timely because of a depression in the sugar industry. Gold and diamond mining siphoned villagers for mostly seasonal work, as in the case of David Benjamin Collins, a farmer born at Two Friends Village, East Coast Demerara, who became a miner.[23] Mining increased the circulation of money in the villages and enabled more village property-holders to meet financial obligations such as tax payments. Many labor migrants who worked in the industry and sent remittances home demonstrated their commitment to their families. In addition to maintaining the stability of family units, remittances for most recipient households provided a source of supplementary income.[24] The ongoing migration enabled villagers to take care of their responsibilities based on the ebb and flow of income from both agriculture and mining.

In the early 1880s, during the fledgling years of gold production, the payment of taxes by villagers sometimes fell short. After the mid-1880s, as an employment avenue, mining enhanced some villagers' ability to meet their financial obligations, although unforeseen circumstances thwarted the efforts of others. In 1887, a disastrous flood on the East Coast Demerara devastated many villagers' means of livelihood and contributed to a shortfall in tax payments. Of $18,988 collectively assessed on village real estate, approximately, 82 percent or $15,629 was collected. The next year's property assessment totaled $19,227 of which $17,065 or 88 percent was collected. In 1888, the arrears in taxation payments for 1887, were fully paid with the additional income from mining as a key factor.[25] One year later, villagers resumed their complaints against renewed efforts by government officials to exclude them from making decisions on the use of their tax payments.[26] Many villagers and their supporters felt that they should be the ones who managed their affairs because they often performed better than government-sanctioned administrators.

In 1889, a sum of $16,203.98 of the year's tax assessment of $19,381.98 was collected. *The Report of the Inspector of Villages for 1890,* revealed that rates were fairly well-paid, to the extent that Danielstown Village, Essequibo had no arrears and had even overpaid. Notwithstanding, the period of substantial rate-payments between 1890 and 1892, was short-lived. By March 31, 1893, village revenues totaling over $19,000 were collected; leaving a balance of $2,715. Officially, it was construed that the failure to collect the balance of the revenue was due to the absence of many village proprietors who were away working in the gold fields. A few months later, by July 31, 1893, when many of the mining migrants returned home, all payments were fully paid. Unfortunately, the following year inclement weather conditions bedeviled villagers and caused short-falls in the payments of their taxes. Farmers lost much of their crops because of floods. Meanwhile, mining validated the utility of a diversified economy to cushion adversities beyond human control and enabled village councilors to use tax payments for infrastructure works to repair damages caused by flooded conditions.[27]

Increasingly, from 1891, village councilors could expend funds independently. This development emanated from the recently amended constitution, which provided for more local people to become members of the Legislative Council and the Court of Policy.[28] The next year, at the behest of villagers, legislation was enacted to give them the authority to elect their "own councils with almost full municipal powers" in nineteen villages.[29] *Ordinance #11 of 1892,* consolidated and amended laws relating to the administration of the villages. Village Councils, comprising proprietors, now had the power to levy rates and carry out maintenance and public works to improve their localities. Village councils, through the Central Board of Health, were empowered to borrow funds on behalf of villagers. Provisions in the ordinance gave the inhabitants of the communities "a considerable degree of self-government."[30] Local residents controlled their own affairs, earnings from the goldfields increased, and tax payments continued.

Mining, other economic activities, and the visible progress among villagers gained bureaucratic attention once more. The postmaster general noted that as a direct result of increased activities associated with gold mining, extended mail deliveries and better mail boxes were needed in the villages. He urged new facilities for purchasing stamps and providing other services.[31] By 1889, additional financial institutions were established under the aegis of the post master general; some of them in areas with existing banking facilities. Mining was indeed the catalyst.

The Emergence of Post Office Savings Banks in the Mining Era

In the midst of the seasonal migrations to the gold and diamond fields many villagers utilized the newly established government-sponsored Post Office Savings Banks to save their money. Ostensibly, many of the bank's branches were established in areas where large numbers of miners lived and worked. By December 1, 1889, Post Office Savings Banks were established in the villages of Buxton and Mahaica in Demerara and on the Essequibo Islands of Wakenaam and Leguan. In addition, a bank was established at Fort Wellington, Berbice; another place where many African-Guyanese miners lived. From the inception of these repositories, efforts were made to place them where residents of all three counties had access to them. The next year, one bank each was established at Bagotville, West Bank Demerara and Plaisance, East Coast Demerara, followed by a branch at Mahaicony, West Coast Berbice in 1893. Three years later, the *Echo* reported that gold miners were increasing their deposits in the savings bank. When a branch of the bank opened at Beterverwagting Village, East Coast Demerara in 1897, the postmaster general explained that although government savings banks already operated at the nearby villages of Plaisance and Buxton, the increasing population and financial importance of the districts warranted additional branches. All three of the villages, Beterverwagting, Plaisance, and Buxton, were homes to a significant number of gold and diamond miners in the nineteenth and twentieth centuries.[32]

By October 1, 1909, Weldaad, Golden Grove, Belfield, Anns Grove, Providence, Fellowship, Naamryck, and Danielstown, villages with predominant African-Guyanese populations were receiving the services of Post Office Saving Banks. By that date, a total of forty-eight branches were in operation throughout the country. Urban branches established in areas where significant numbers of African Guyanese resided included Albouystown, Carmichael Street, Regent Street, Cornhill Street, and at the General Post Office Building, Georgetown. New Amsterdam, the main city in Berbice County also had a branch. Branches at Morawhanna, Bartica, Arakaka, Potaro, and Tumatumari served interior gold and diamond miners while the branch at Wismar, Upper Demerara River provided services when bauxite operations began.[33]

Lionel Arthur, Darrell Gray, and Sydney Dundas, all former gold and diamond miners, had held deposit accounts at the Providence, Queenstown, and Anns Grove branches, respectively. Although there was a branch in Buxton, Howell Wood preferred to save his money at the main branch in Georgetown. All of the men simultaneously participated in "box-hands" or the aforementioned *esusu*.[34] Messrs. Arthur,

Gray, Dundas, and Wood were all property-holders whose savings enabled them to make timely tax payments and to keep their properties.

In her 1986 study, *The Development and Demise of the Savings Bank System in Guyana*, Donna Danns identified the occupations of depositors in the Post Office Savings Banks between 1895 and 1905. Danns pointed out that the largest occupational category of depositors in the banks comprised laborers (37 percent) who were of African, Asian, or Portuguese origins. Gold and diamond mining workers were included in the category for laborers. Although the occupations of a large number of depositors were not stated, there were so many "gold diggers" that a separate occupational category was created for them. Furthermore, Danns' analysis indicated that the distribution of accounts in various branches confirmed that African-Guyanese villages such as Buxton, Plaisance, and Belfield had a high percentage of depositors.[35]

Throughout the period encompassing 1910 to 1930, branches of the Post Office Savings Bank were established colony-wide. The 1932 *Report of the Post Master General* indicated that "Creoles and others" had saved an additional $38,000 in the Post Office Savings Banks. The category "Creole and Others" included a majority of African-Guyanese depositors.[36] In 1933, envisioning the potential for expanding access to the bank, the Post Master General targeted as prospective depositors, persons in isolated mining and distant communities. Similar to his nineteenth-century predecessor, he felt that "family remittances and other periodical payments...on behalf of miners and others working away from their homes" would be enhanced.[37] By the 1950s, in addition to the urban branches, fifty-six branch post offices and postal agencies in the rural areas transacted business. In 1952, there were 109,617 depositors with total deposits of $15,955,706. The next year depositors numbered 112,829 with a slight decrease in capital to $15,544,972.[38] The government invested the banks' deposits mostly in foreign enterprises. Some of the deposits were also utilized to finance local projects or was made available to commercial institutions. Colonial bureaucrats had decided that it was prudent to establish savings banks for the working class to save their money, but a majority of the depositors were denied loans. This was a significant drawback, especially during times of financial emergencies.[39]

Tax collection from 1894 to 1899, was characterized by fluctuating payments and the continuation of court cases and penalties against those proprietors accused of nonpayment. In the mid-1890s, the resurgence of an economic depression, especially in sugar, an industry which still provided employment for some nearby village residents, affected tax payments. Prolonged rainfall had caused flooded conditions, created havoc with drainage and irrigation on the coast, and thereby limited the ability of villagers to make their tax payments fully. At the same time, the propitiously expanding local mining industry enabled villagers to minimize their shortfall in tax payments.[40] Yet the working class faced myriad challenges as the nineteenth century ended.

Amidst the ongoing conflict that had characterized this colonial society, the African component of the population had instituted "order and progress" in their lives by making a living from a variety of economic pursuits directly and indirectly related to agriculture and mining. Acquisition of real estate had exhausted their savings, they struggled to retain their properties, grappled with economic depressions, and against all odds, individually and cooperatively, villagers continued to accumulate material resources. They utilized their earnings to improve their conditions or invested in future projects. At the dawn of the twentieth century, only 75 percent of villagers paid their taxes.[41] Alarmed that economic hardships and situations over which they had no control, jeopardized tax payments and property ownership, community leaders decided to tackle the situation directly. With legislation to deny these depositors loans from the Post Office Savings Bank as the backdrop, villagers organized and established additional financial institutions, accumulated resources to make tax payments, and retained their properties. Once more, they "made a way out of no way."

DIVERSITY AND MONEY MANAGEMENT: COOPERATIVE CREDIT BANKS IN THE MINING ERA

The Council of Village Chairmen held its first meeting on August 1, 1901, and a priority on its agenda was to caucus on those matters generally affecting the villages. Consequently, village chairmen and overseers were delegated to implore proprietors to pay their taxes. The next fiscal period ending in 1902, ushered in improved tax-collection. Tax arrears from previous years were collected as well. By 1903, proprietors in seven villages had paid their taxes in full. Some villages averaged 96 percent of payments, but no rural community paid less than 80 percent of their assessed taxes. In Den Amstel-Fellowship Village, West Coast Demerara, the council experienced no difficulties with financial matters.[42] Still, many felt that the entire community should devise a way to ensure all proprietors paid their taxes on time to avoid penalties. As a result, decisions made at the Village Council meetings had important short- and long-term consequences for the longevity of these communities. At a conference held in 1904, the then chairman, James McFarlane Corry, introduced and explained to the attendees the utility of establishing their own banking system. One year later, after obtaining additional information on organizing a bank to suit their own needs, the first local Cooperative Credit Bank was established at Victoria Village, East Coast Demerara. The financial institutions, which functioned from 1905, were supported by farmers, miners, and other depositors.[43]

Meanwhile, villagers again experienced a series of uncertainties. Conflict between the working class and the colonial hierarchy erupted into wide-scale riots and public protests in the colony. As previously mentioned, national and natural disasters took their toll in 1905. At the same time, some villagers who were engaged in farming received decreasing or less than expected payments for their produce. Periodic droughts and floods hindered travel to prospecting sites, prevented work, and curtailed gold and diamond production. Mining workers who were affected by this unfavorable situation earned little or no income from the occupation. In several instances, gold and diamond miners who prospected on rented land were delinquent in their payments due to disasters and circumstances beyond their control. In 1906, unusually heavy rainfall lasted for four months and forced many prospectors who usually worked in the North West District to remain on the coast, pursue planting, or some other type of work.[44] Morawhanna, in the interior was inundated with an all-time high of 171.5 inches of rainfall. In many districts, rainfall soared to an average of 50 percent higher than the previous year's. In March 1907, the Meteorological Department recorded the highest rainfall to that date in the colony. However, by 1909, foreclosure of properties due to default in payments was no longer a feature of village life. Droughts occasionally occurred and hindered farming and mining.[45] Nevertheless, during the 1911–1912 periods, villagers' tax payments amounted to over 98 percent of the total. This was attributed to weather conditions favoring agriculture and mining in the years before the droughts.[46]

By 1912, three branches of the Cooperative Credit Bank existed. In addition to the first branch, at Victoria Village, one bank each was established at Buxton Village, East Coast Demerara and Rose Hall Village, Berbice County. At the 1914 Village Chairmen's Conference, the participants agreed that it was time for the government to establish similar institutions.[47] *Ordinance #9, 1914*, established Cooperative Credit and District Banks. During 1915, government officials who had been impressed with the efforts of the villagers, and recognized the utility of the proposed institutions, facilitated the opening of District Credit Banks as well.[48] By the end of that year, three such establishments with 280 shareholders operated in villages. One year later, there were eleven additional banks with 2,228 shareholders; and by 1917, twenty-four banks with 4,078 shareholders existed in several more villages. It is noteworthy that from 1909 to the outbreak of World War 1, an upward spiral in rate collection ensued during a period when depositors were supporting several financial institutions. Later, the difficulty incurred in collecting taxes was attributed to wars and adversities affecting agricultural, mining, and other workers.[49]

By 1920, twenty-seven Cooperative Credit and District Banks were in operation. Persons of African descent held 34,680 shares while East Indians held 20,130, out of a total of 62,170 shares valued at $1 each. Shareholders included agriculturalists, tradespersons, coconut-oil producers, hucksters, and gold and diamond miners.[50] Approximately, one decade later, on December 31, 1930, the number of government cooperative credit banks remained at twenty-seven, although the shareholders had increased to 10,883 persons. Of the 81,909 shares, the African-Guyanese component held 47,566 and East Indians 26,677.[51] In the early years of the cooperative credit banks' operations, shareholders had expressed misgivings about the utility of small loans to take care of a number of necessities. For example, they considered the prices they paid to purchase food, clothes, and other supplies as exorbitant. In June 1919, coinciding with the operations of other existing financial institutions, a trading bank based on the cooperative system was established. Shareholders could send their products to Georgetown for sale through the trading banks. In turn, they could purchase tools, groceries, clothing, footwear, and other items through the banks at cheaper prices than elsewhere.[52]

World War I had adversely affected the positive trend in tax collection. By 1916, the rate of collection stood at approximately 92 percent, a decrease of over 6 percent from the 1911–1912 periods. In 1916, tax collection based on a financial year was replaced by a calendar one; and in 1917, property-holders were required to pay increased taxes. Financial returns from agriculture, gold, and diamonds decreased, but bauxite production was underway and many villagers decided to migrate and to work in that industry. Financial remuneration from bauxite work, no doubt, assisted some proprietors and family members in paying their taxes and payments on properties held steady until the 1920s. During the Great Depression and World War II eras, data on payments returns were mostly unavailable, except for ten villages during 1936–1937, with property-holders in Agricola, Bagotstown, Mocha, and Peters Hall paying 94, 95, 94, and 100 percent of their taxes, respectively. Proprietors with arrears of payments in the three villages paid in full the next year. From 1948 to 1956, payments remained in the 80 percent range.[53]

By the middle of the twentieth century, the average village tax collection had decreased by nearly 20 percent compared with the earlier 1900s period. This was the period when accelerated Caribbean migration overseas, especially to Great Britain occurred. Many would-be internal labor mining migrants left the country. The economy of the country fluctuated and village tax payments were affected. Notwithstanding, some villagers still farmed and gold and diamond seekers continued to acquire properties, and paid their taxes. Officials of the Deeds Registry, Georgetown recorded August 30, 1955, as the date the transport of Lloyd Devonish, a diamond seeker from Fellowship Village, West Coast Demerara, became effective for grant #4281 and lots twenty-three, twenty-four, and ninety-one in the Den Amstel-Fellowship Village District. Similarly, Trenton Ezekiel Pellew, a farmer, purchased property at Clonbrook Village, East Coast Demerara on February 7, 1956.[54] Members of the Pellew family were among the intrepid and legendary gold and diamond miners who combined their earnings for a variety of economic enterprises. Customarily, they purchased properties because accumulation of real estate was a traditional way of investing money, and farming and mining helped homeowners to improve their properties and to pay their mortgages and taxes.[55]

Conclusion

Migration and mining occurred parallel to the agricultural production of the villagers and provided residents with resources to sustain their communities. The Cooperative Credit Banks organized by African Guyanese, similar to the government-sponsored Post Office Savings Banks operated concurrently with *esusus* and susus, and mutual aid and friendly societies of those very depositors whose ancestors had utilized African antecedents captured in their knowledge transfer and cooperativism. Depositors' ability to support multiple financial institutions simultaneously enabled them to retain their property. This largely accounts for the longevity of their habitats.

Ancestors understood the tenets of a capitalist society; utilized their earnings and ancestral practices applicable in mutuality; saved and cooperatively acquired resources and property to transform their situations. Recognizing the inevitability of modernization, exemplified by bureaucratic laws, they updated saving mechanisms, invested the resources further, and improved their quality of life. Ultimately, as the following chapter delineates, gold, diamond, and bauxite mining afforded participants avenues to extend their physical labor into craft skills and entrepreneurial activities. Similar to their nineteenth-century ancestors who transferred their knowledge of traditional practices to foster favorable outcomes, twentieth-century African-Guyanese advantageously utilized a variety of African antecedents.

CHAPTER 10

African Continuities, Jewels, and Economic Linkages to Mining

Africans were present during Europeans' earliest quests for precious metals and gems in what became known as Guyana.[1] The African presence in early mining expeditions supports the deliberateness of European slave trading, that is, to procure workers with particular expertise for planned exploitation of the natural resources of the Americas. In the nineteenth century, long after the initial expeditions and with increasing gold and diamond recoveries, African descended miners predominated in the then British colony. While the enslaved in the Americas invented equipment to ease the drudgery of their work on plantations and farms, their free descendants in colonial Guyana who were jewelers and craftspersons, utilized gold and diamonds to make a living.[2] This chapter focuses on the ways in which persons of African ancestry employed their mining, technological, and entrepreneurial abilities to create other economic opportunities. A critical segment of the chapter's analysis is the intertwining of African cultural practices and nurturing with specific activities and particular craft skills. This topic, the link between people of African ancestry who excavated gold and diamonds and the connection with those retaining expertise in ancestral skills such as goldsmiths and jewelers, has been long neglected in various socioeconomic studies of the country.

The research validates the claim that invisible cargoes in slave ships crossing the Atlantic Ocean included Africans of various abilities and skills.[3] African continuities created economic enterprises, but the unraveling of their identities and attributing particular skills to specific groups is complicated because Europeans conferred nontraditional names on them before, during, and after their transportation across the Atlantic Ocean. The departure points of Africans from the continent were sometimes distant from their natal habitats which further obscure the specificity of identities akin to regional or local cultural practices. Exact origins were unrecorded and Africans' of diverse cultural, tribal, ethnic, and linguistic affiliations commingled during the Transatlantic Ocean crossings. Colonial Guyana's slave registers recorded Africans with a multiplicity of origins and identities who had special technologically related skills and expertise when they were procured. For decades after the British slave trade's demise and the ending of slavery, as already noted, thousands of Africans entered the colony as indentured labor migrants, formed relationships with the nascent community, and revitalized cultural practices.[4] Later, gold and diamond mining enabled those of African descent to display further their capabilities as laborers and abilities as business and craftspersons. Members of the group expanded the jewelry industry, some former prospectors became investors and entrepreneurs in mining, spearheaded other economic undertakings, and increased employment opportunities.

Historiographically, literature focusing on the British Caribbean working class in the years after emancipation tends to pinpoint them as being in the grips of colonial and expatriate domination. The working class and their leaders are often characterized as non catalysts for meaningful change, preferring accommodation within the existing status quo. During the nineteenth and twentieth century, regional and national unrests were sparked by labor exploitation, impoverishment, and discrimination. Periodically, workers protests were followed by some reforms, but often dissatisfaction lingered.[5] As a rule, the causation sparking upheavals in colonial Guyana was the abject socioeconomic conditions of the working class. Continuities and changes in the country during the period are usually couched in terms of upheavals and their aftermath.

In this chapter, a number of counterpoints to accepted notions of marginalization, impoverishment, protests, and disempowerment of the African-Guyanese people are offered. The abilities of those largely associated with the mining industry are analyzed here to allay notions of the inability of this segment of the country's population to make a living continually, improve themselves, and situate their communities for change. In particular, their persistence in utilizing their expertise in a variety of undertakings diversified the country's economy. The energy of this group, persistently directed to forging economic linkages and their progress, is a central feature of the chapter. While the two previous chapters focused on their economic activities locally, this one incorporates miners and jewelers' forays overseas. Analyzing African diasporans in ways other than portraying them as mere victims of colonial and capitalists' labor exploitation and its legacies is a long overdue perspective.

The Unbroken Chain: Africa to the Americas

Until the nineteenth century, inhabitants of extensive areas of Central and West Africa were captured, sold, transported, and enslaved in the Americas.[6] Although their treatment, living, and working conditions were inhumane, some of their traditional cultural practices persisted. In the nineteenth and twentieth centuries, semi-and voluntary migrations and settlement of West and Central Africans in the Americas helped to sustain some specific ancestral skills in many diasporic communities. Several scholars who examined the records of the transatlantic ocean slave trade have offered varying perspectives on the transported as manifested by "clusters" and the contributions of Africans to disparate areas of the Americas. Hence, distinct ethnic affiliations and the intermingling of Africans at ports of departure, on board ships, and in the Americas produced varying representations of practices attributable to several areas of Africa.[7]

In Guyana, registration of the enslaved from 1817 to 1834, identify ports of departure, names, ethnicities, and occupations. Ports of departure, not specific natal origins, included Africa, Guinea, and the Caribbean Islands. Examples of non-European names of the enslaved were Accra, Amba, Conga, Cumba, Cudjoe, Ebo, Ishamael, Juba, Mingo, Quaco, Quasheba, and Yabba, useful in confirming African identities and links. Many of the enslaved had tattoos and other body marks, symbolic of African connections. The Slave Registers identified persons by "country marks on foreheads" and "tattoos on temples" and the nondescript word "Africa" was often recorded in many entries. The registers are particularly informative on the occupations of enslaved persons. Although many of them were farmers, there were numerous industrially skilled persons such as engineers, mill- and wheelwrights, coopers, carpenters, mariners, and blacksmiths. Craftspersons included weavers, tailors, seamstresses, basket- and rope makers, and goldsmiths. The registers identified younger persons who were apprenticed to older and skilled slaves; a practice which ensured some expertise and special skills were passed on generationally.[8] Members of the descendant community improved on these skills further, especially as jewelers.

Eldorado in Guyana and Linkages to the Expanding Jewelery Industry

Gold and diamond mining and its linkage to jewelery making and other African traditions offer insights into antecedents of centuries'-old practices and continuities. Following the ancient traditions of Ethiopians and Egyptians who used enslaved labor extensively in their gold mines, Europeans procured Africans and ensured that precious metals and gems in the Americas were excavated, exported, or traded.[9] All raw materials were not exported from producing countries, and portions of the commodities were utilized locally. By 1864, in the midst of encouraging discoveries of precious metals and gems in the Guianas and surrounding areas, Benjamin Victor Abraham, the owner of a city jewelery establishment, was a major shareholder in one of the country's first gold and diamond mining companies.[10] Retail outlets for jewels confirm that the protracted search for the commodities included parallel plans for export and local utilization.

In the 1720s, long before the discovery of gold in commercial quantities in Guyana, enslaved and free Africans were used "as porters and diggers in mining expeditions."[11] Almost a century later, in 1817, King was a thirty-year old enslaved African-born goldsmith whose owner was Mina Heyliger. Moses, ten years younger, had the same status and skill as King, but he was born locally.[12] Goldsmiths of different ages in the colony, decades before gold was discovered in large quantities, attest to the long standing recognition of a connection between labor, raw material, and outcomes based on expertise and skills. Africans were valuable assets in the exploration, exploitation, and utility of the natural resources of the Americas.

William Bosman's *New and Accurate Description of the Coast of Guinea: Divided into the Gold, the Slave, and the Ivory Coast*, an account of his travels in West Africa during the seventeenth and eighteenth centuries, identified the generational context of African occupations. Bosman indicated that "if the father is a fisherman...and the child a male, he breeds him up to his vocation early."[13] He emphasized "their chief handicraft, with which they are best acquainted being smithery" and with their "tools they can make all sorts of war arms" and "their most artful works are in gold...the thread and contexture of which is so fine."[14] Enslaved Africans in the Americas brought the "mental blueprints" of their heritages that the Middle Passage could not erase" including ancestral skills such as smithing that they retained, retooled, and utilized despite their varied experiences.[15]

The use of jewels and gold ornaments can be traced to the earliest periods of human society and "is probably the oldest art form."[16] From antiquity, African empires including Nubia, Egypt, Ghana, and Mali had "traditions of mining and metallurgy" and "the worlds oldest mine was uncovered at an iron site in Swaziland where it operated more than 45,000 years ago."[17] The gold of "West Africa was traded in Europe as early as the tenth century and probably before...from the kingdoms of Ghana, Mali" and elsewhere.[18] By the fifteenth century the Portuguese, followed by nationals of England, France, Holland, Denmark, and Spain, were exporting gold mostly from West Africa. Upper Niger, Senegal, Gambia, and Ghana yielded significant amounts of gold.[19]

Gold was essential in commercial undertakings and was valued as a measure of wealth. Ghanaians crafted elaborate ornaments, utensils, ceremonial regalia, coins, and jewels from gold. They were miners, entrepreneurs, and traders, and "skilled artisans in Asante and other Akan states became known far afield for the artistic quality of their gold jewelery and brass castings," and the Portuguese who built Ghana's infamous El Mina slave trading fort "were astonished by the resplendent gold ornamentation of local women."[20] In the slave trading era, Ghanaian miners, artisans, and craftspersons were captured and exported like the commodities they had excavated and fashioned. Guyana's records indicate that Africans from the Central and Western areas of the continent, some of whom were goldsmiths, were taken there.[21]

Slave trading was not a random activity. In the ongoing European quests to find precious metals and gems in the Americas, Africans in contiguous regions were sought and procured for specific economic activities such as mining and metalworking.[22] The legend of El Dorado, the fabled city of gold, loomed large in the history of gold explorations. It was a powerful impetus and magnet for the marauding Spanish, French, Dutch, Portuguese, and English who expected to find, virtually "streets paved with gold and studded with diamonds" in the Americas, ever since the days of Sir Walter Raleigh.[23] As early as the sixteenth century, in areas of New Granada (Colombia and Panama) Africans were imported and enslaved to work specifically in gold. Similar to other enslaved Africans, many of them possessed an array of skills and expertise.[24] By the eighteenth century, "Africans designated as 'Minas' were brought to Brazil from gold-producing regions of West Africa, including Greater Senegambia and the Gold Coast" (Ghana) to seek "and develop panning and digging for gold."[25]

The evolution of gold and diamond prospecting in Guyana contrasted with Colombia's and Brazil's. Local forays were intermittent from the eighteenth through nineteenth centuries, when the mining industry was finally established. Colombia's and Brazil's industries were continuous from the sixteenth and eighteenth centuries, respectively. In Guyana several European nations, including the Dutch, French, and British searched for El Dorado, unlike Spanish Colombia and Portuguese Brazil where only a single European nation was involved in the colonial mining operations of each country. Locally, attempts to find gold and diamonds had continued based on some measure of success.[26] Africans in those prospecting trips represented occupational continuities at home and abroad. The acquisition of goldsmiths such as the aforementioned King and the nurturing of the younger Moses, confirm the continuities of traditional skills which had survived brutal capture and confinement, the nightmare of the transatlantic slave trade, the horrors of enslavement, and the tumult of emancipation. African diaspora gold and diamond miners engendered the jewelery industry and other economic linkages related to mining.

A direct correlation existed between the presence of goldsmiths and the specificities of slave trading. African labor was not only procured to recover gold, but it was also anticipated that the skilled were needed to transform the raw material into items of utility. As noted, Africans were craftspersons before enslavement and multi-skilled persons in colonial Guyana's Slave Registers certify connections with ancestors of similar abilities and pinpoint the transfer of expertise to descendants. While an argument can be made for "deep culture," the technological skills and expertise found among diaspora-born craftspersons prove that Africans had much more than the physical ability to perform arduous tasks. Occupational skills passed from one generation to the next because of "the association between creative contributions to science and technology and the involvement" of craftspersons "in the activity of the work."[27]

Gold and diamond miners in the colony were in an appropriate situation to foster the rekindling of techniques, coming as they did from the same communities as the jewelers. Skills in metallurgy and goldsmithing, nurtured through enslavement and emancipation, were given additional outlets with the advent of the precious metals and gems industries. Metallurgy and jewelery-making as aspects of the country's history exhibited continuities with Africa and are linked to the colony's economic diversity. For instance, of the twenty enslaved persons owned by James Millar in 1832, fourteen of them worked in his foundry. Skills in fashioning ores and metals account for the high visibility of Africans and their descendants as engineers, mechanics, blacksmiths, coppersmiths, tinsmiths, coopers, and millwrights in the country. Metalworking and technological skills, which are akin to the work of jewelers, require the ability to recognize particular metals and to design them in specific ways by using fire.[28] After enslavement ended, skilled craftspersons were often required for work on the sugar estates, thereby maintaining an outlet for their dexterity.

Guyana's post-slavery censuses identified the skills and occupations of Africans and their descendants. Similar to the Slave Registers, occupations recorded in the censuses often reflected a connection between parents, offspring, community elders, younger and older men and women.[29] As noted elsewhere, during the nineteenth century, with the infusion of more African-descended workers due to the coming of indentured migrants, ancestral traditional practices were reinvigorated. Birth and Death Registers show that newcomers and those long resident in the country commingled. From the 1830s, vestiges of Africa continued and assumed renewed expression with the influx of the new Africans and the resulting fusion, especially in villages. Some skills practiced in Africa were transferred, sustained, and recreated.[30] Villages afforded their inhabitants environments of relative isolation to interact and also availed the residents opportunities for nurturing, self-determination, and empowerment. The enclaves were pivots whereby the inhabitants revolved around agriculture, mining, and other economic pursuits. Thus, the impetus for them to participate fully in all aspects of technology related to ores, minerals, and metals was omnipresent.

By 1850, the tradition of jewelery-making by members of the African-Guyanese community benefited from the Caribbean-wide gold rush to Venezuela. In addition, an intra-Guiana migratory movement and gold-rushes in Dutch Guiana (Suriname), French Guiana, and British Guiana (Guyana) accelerated from the middle of the nineteenth century, contemporaneous with prospecting ventures to Venezuela and Colombia by Caribbean nationals. This movement lasted for decades. Edward Baird, born in Hopetown Village, West Coast Berbice and who later farmed at Belair Village, Corentyne worked as a gold miner in Dutch and French Guiana. Later, Ewart Sydney Rawlins, born on September 21, 1889, in Ithaca Village, West Coast Berbice became a diamond seeker at Mabaruma, North West District in 1919. The next year, Rawlins was a gold miner in Venezuela.[31] In addition to new employment vistas, prospecting for gold and diamonds made available the commodities for jewelers. Migration within the Atlantic world to nearby gold producing areas had also promoted jewel-making opportunities for other Caribbean people. William Bourne and Christopher Hodge, African-descent natives of Trinidad and Tobago and St. Thomas, Virgin Islands, respectively, countries with no known deposits of gold, were goldsmiths in their homelands.[32] The jewelery made by these skilled persons played a key role in fostering the continuity of particular crafts, which helped to maintain aspects of Africa in the Americas.

Locally, by the 1860s, both mining and jewelery-making were fait accompli and Benjamin Abraham and others sold gold chains, rings, and other items in their establishments. The aforementioned exhibitions that were also intercolonial and international affairs enabled African Guyanese to display their expertise. The exhibitions ensured the survival of particular skills aligned to metal workers, smiths, and engineers. At the 1866–1867 exhibition, William Romeo, a blacksmith from Clonbrook Village exhibited and demonstrated a steam engine he had invented to process the community's agricultural products. Persons "engaged in the physical labor of production invented or recreated machinery and utilized the products in a variety of ways."[33]

Along with the emergence of gold and diamond mining, the multi-skilled activities of those born during the late nineteenth century are noteworthy. Isaac Alfred Baird born on March 4, 1885, at Two Friends Village, East Coast Demerara became a shoemaker, saddler, gold and diamond miner, and a businessman. Archibald Fitzdonald Jackman was born at Paradise Village, East Coast Demerara on February 6, 1886. In the early 1900s, as a resident of Nabaclis Village, he was identified as a jeweler, goldsmith, and watchmaker. Edward Ebenezer Collins who was born in New Amsterdam, Berbice in 1890, had an even more varied experience. Collins was educated in Litchfield Village, West Coast Berbice. While living at nearby No. 40 Village, he first married Cecelia Saul and later, Elizabeth Booker and became the father of three sons and six daughters In his long life, Collins was a policeman, rice farmer, prospected for gold and diamonds and as a labor migrant, he worked in Surinam's forestry industry. Joseph Assanah, born December 15, 1900, at Victoria Village, East Coast Demerara, came from a long line of fishermen and he too followed that occupation. Yet for a period of time, Assanah, the father of nine sons and three daughters, worked as a motor mechanic in the Mazaruni diamond fields. He was one of the inventors of "the smith lifting propeller."[34]

African Diaspora Continuities and the Jewelery Industry

Precious metals and gemstones produced profits for many persons, diversified the economy, and generated employment through its spawning of other industries. The commodities have a direct connection to the jewelery industry including the design and manufacture of rings, chains, bracelets, earrings, brooches, trophies, and other items. Varying quantities of gold were sold on the coast by prospectors who brought the items with them on their trips out of the interior. Outlets in Georgetown and the villages of Plaisance, Buxton, Fellowship, Den Amstel, and Tushen were supplied with the gold. Some personnel at goldsmith shops in Lombard Street between the Commissary's Office and La Penitence and in the Charlestown district bought some of the raw gold. The goldsmiths or proprietors operating the businesses included African Guyanese.[35]

Members of this group had invested time, energy, and capital, and were among the first persons to prove "the existence of payable deposits of gold and diamonds."[36] They deserve credit for their production. Other segments of the population contributed to the industry, but it was these landowners, artisans, and laborers who were at the forefront in the development of mining.[37] Some among the rapidly increasing East Indian population were involved in the industry, but until 1928, legal measures were in place to prevent them from working in the gold and diamond fields. By that year, it was felt that Regulation 106, the "Restriction as to the Registration of East Indian Immigrants" working in the diamond and gold fields "should be deleted," and recommendation was made to modify the Immigration Ordinance or Regulations to reflect the change.[38]

East Indians are prominent in accounts of the development of the country's jewelery industry—an economic linkage to gold and diamond mining— but, the visibility of African Guyanese is obfuscated. This omission has minimized their roles at the forefront of all utilization of the commodities.[39] The occurrence of gold mining in the colony was indicative of the "transfer of an activity" and "an entire cultural system from production to" utility as jewelery, trophies, coins, and as storage of wealth.[40] African diaspora workers were much more than mere diggers of gold and diamonds. In Guyana, their goldsmith skills honed from antiquity in distant Africa were manifested more prominently with the coming of the precious metals and gems industries. Skilled Africans enslaved in Guyana, those who were freed, and their descendants had had opportunities to participate in mining and craftsworks long before East Indian immigrants arrived in the country. Although one group of persons does not have a monopoly on specific expertise and India like Africa is renowned for precious metals and gems, the subcontinent and continent both have traditions of craftspersons who worked in the commodities from ancient times.[41]

In 1891, *The Census Report for British Guiana* indicated the inclusiveness of craftspersons in the jewelery industry.[42] This census, the first one enumerated since the establishment of gold and diamond mining, contained a single occupational category for those who were goldsmiths, jewelers, and watchmakers.[43] African Guyanese, among other craftspersons, were well represented in the category.[44] The ability of Africans in the Americas to retain, to recreate, and to utilize inherent skills, practiced in their ancestral homelands from antiquity is irrefutable. Diaspora-born persons, who similar to their ancestors, not only labored in the gold and diamond mining, but also traditionally honed skills as goldsmiths, jewelers, engravers, and occupations requiring technological expertise, confirm linkages over time and space.[45]

The gold and diamonds, which formed the backbone of the jewelery industry, did not simply walk out of the earth's bowels or swim out of the maws of the rivers unto jewelers' work-benches for fashioning and designing. The rivers of the country remain littered with the bodies of countless African-descended persons while its earth entombs numberless more who ventured into the gold and diamond fields.[46] The compatriots of the deceased workers created linkages and established and expanded enterprises, such as the jewelery industry, in which they played a pioneering role. Availability of raw materials, labor, ancestral traits, creativity, financial accumulations, and entrepreneurship, linked and intertwined to characterize the jewelery industry. The nexus between African continuities and nurturing explain the intertwining link in labor activity and outlet to practice craft skills associated with the product itself.[47]

Generational links created by the miners ensured the longevity of the industry. Well-known mining families such as the Jerricks, Semples, Pellews, and Wades ensured the continuity of the occupation by passing on their expertise to their offspring and other family members. In 1936, Walter Bagot followed the footsteps of his father Albert "Pepper" Bagot who disliked cane-cutting and opted for mining. In the 1890s, Walter's grandfather, Phillip Bagot pioneered the family's involvement in mining when he prospected in Omai, Essequibo. This practice enabled experienced family members to initiate the unfledged into the occupation and, thereby ensured continuity.[48] The generational and nurturing aspects were critical factors in the interrelationship between the actual recovery and eventual utility of the products and sustaining African continuities in diaspora communities that exist to this day.

Consider the case of Clifford Hector who was born in 1927, at Victoria Village, East Coast Demerara. Hector became a jeweler's apprentice in 1941, under the tutelage of Samuel Felix who was born in the 1880s. By 1947, Hector began training apprentices including Maurice July, born in Mocha Village, East Bank Demerara and the nephew of Felix. In turn, July continued to train a younger generation of jewelers including his son. Over the decades, Hector, a father of eight trained many jewelers including his two sons, who preferred to work in undertakings other than jewelery making.[49] Until the mid-1950s some apprentices and jewelers like Messrs. Hector and July worked for or were attached to A. C. Young, the proprietor of *Colonial Jewelery Establishment*, America Street, Georgetown. Unfortunately, a year after he bought the business from A. C Young, Samuel Felix became a road fatality. Ray Felix, the nephew of the deceased, then assumed control of, and operated the business along with Clifford Hector. Although the business eventually relocated, was subsequently sold, and finally closed; Messrs Hector and July continued to practice their craft. To ensure that succeeding generations built on the foundations of their ancestors, both jewelers passed on the traditional skill to their descendants[50] (See Figures 10.1 and 10.2).

Figure 10.1 Clifford Hector, jeweler.

Figure 10.2 Maurice July, jeweler and apprentice son.

Characterizing themselves as jewelers, rather than goldsmiths, because they work in gold, silver, and diamonds, these craftspersons also set semiprecious stones and applied engravings to a variety of items including ornaments, coins, medals, trophies, and jewelery. As offshoots of diamond mining, jewelery-making and designing are also linked to toolmakers, advertisers, retailers, wholesalers, and local and international traders. Clifford Hector travelled periodically to the United States to conduct business with merchants and traders in the Gold and Diamond Districts of Manhattan, New York. The district is located mostly in the vicinity of 40th to 50th Streets and between Fifth and Sixth Avenues (Avenue of the Americas).[51]

By the late nineteenth century, as indicated earlier, jewelery businesses operated in wide swaths of the capital city such as such Lombard, Regent, Leopold, Robb, and America streets. The goldsmiths shops mentioned in *The Report of the Gold Industry, 1895* were precursors to the one operated by Gaskin and Jackson's Ltd. after the mid-twentieth century. This business was established by Herman Gaskin in May 1966, after his local apprenticeship and training overseas.[52] At fourteen years of age Gaskin became a jeweler's apprentice, under the tutelage of the Vangenderens, a family noted for the craft. In 1956, he left the colony for the United Kingdom to train further at the Central School of Arts and Crafts, London. On his return, Gaskin first established the jewelery business in Camp Street, Georgetown. By the time the business relocated to Robb Street, a few years later, the staff which were originally four employees, numbered ten. In the 1990s the company relocated to King Street. It also conducted operations at its factory at Land of Canaan, East Bank Demerara with twenty employees. Among its senior employees was a Jamaican craftsman who had worked there for over twenty years. An apprentice from the United Kingdom was undergoing training.[53]

When Eric Jackson returned home, Gaskin invited him to join the business. Jackson had started his career as an apprentice jeweler with Humphrey and Company in 1955. Three years later he left for the United Kingdom to train at the same institution as Gaskin. Later, Jackson worked for eight years in Switzerland before returning to join the business as its works manager. Both Gaskin and Jackson purchased gold, silver, and diamonds from licensed traders and dealers and catered to local and overseas customers. Jewelery such as bracelets, chains, earrings and various kinds of medals, buttons, pins, badges, graduation rings, and trophies were made. Eventually, Gaskin's and Jackson's became "Jewellers for the Guyana Government."[54] Gold and diamonds not only afforded opportunities for craftspersons, but also provided the means for them to become entrepreneurs; thereby creating additional employment.

Goldsmiths and jewelery enterprises were not only linked to the mining of precious metal and gems, but also to the diamond–cutting industry. This industry associated with the refining of gemstones fostered skills and technologies, generated employment, and provided revenue for the country. The dental profession is also linked to the occupation of mining. Gold is utilized in decorating, capping, and filling teeth to enhance personal appearance and for health reasons. Messrs. Grant and Rodney, who operated a dental establishment on the East Coast Demerara, used Guyana's gold, and as skilled craftspersons they provided employment opportunities for others. The use of gold for teeth alteration is another form of African cultural retention or continuity.[55]

Economic and Diasporic Linkages: Creating Employment Opportunities

As early as 1890, it was noted that gold and diamond mining had expanded employment opportunities by creating linkages to jobs in the maritime industry for boat-builders and others. The tradition continued into the twentieth century as epitomized by Edward Nassy (1881–1935), one of the country's most noted boat-builders and the proprietor of Nassy Boat and Launch Works in LaPenitance. Nassy was born at Aurora Village, Essequibo County and lived there until 1895. He relocated to Georgetown and was introduced to a well-known boat-builder named William Schipers. While under the tutelage of Schipers, Nassy decided to accompany a cousin into the interior goldfields. Prospecting was discouraging and he returned to the city and boatbuilding. After training as a marine designer, he established a boat-building business. Nassy spearheaded the construction of government-operated launches such as the *Paku* and *Seba,* which plied local rivers and were used by miners on their way to the gold and diamond fields. He supervised the construction of the cruising yacht *Hazle Grey*, famous for its record-breaking sailings in Trinidad and Tobago. Nassy made a name for himself in boat-racing; competing with vessels he had constructed and winning prizes for his skills and dexterity. At the time of his death, the business was one of the best-equipped and most up-to-date in the city.[56] J. A. Eytle of Brickdam, Georgetown was not a boat-builder, but he belonged to the cadre of gold and diamond mining entrepreneurs who transported their workers to and from their prospecting sites in vessels they owned.[57]

In a diasporic context, entrepreneurial, maritime, and gold and diamond prospecting as economic linkages synchronized in Guyana. In the nineteenth century, African Americans had been active in "gold rush" California, Australia, Guyana, and elsewhere in the mining diaspora. Although African descended sailors had served on United States Navy and Merchant Marine vessels, and worked on ships under the captaincy of Paul Cuffee and others, diaspora-wide, Marcus Garvey's Black Star Lines monopolizes discussions on their entrepreneurial and technological positions in the maritime Atlantic world. Bridging the 1800s and 1900s, emigrationists and colonizers to disparate parts of the diaspora are highly publicized, while largely unheralded is the action of a multifaceted group of African Americans who in 1922, intertwined entrepreneurial, maritime, and gold and diamond mining in a significant way.[58]

The unfolding of what was envisioned as "one of the most" interesting "chapters in the history of" African "people of the New World" began "with the departure on August 23, from the Port of New York, of the steamship *Anna May* bound for the diamond fields of British Guiana on a prospecting expedition"[59] As reported in the *Afro-American*, the project included several African American bankers, investors, and businessmen. "Chas E. Mitchell and Col. Jas M. Canty of Charlestown, South Carolina, A. Beresford Gale of Philadelphia and" one "Crawford of New York, who was also the operating manager with offices at 80 Wall Street," Manhattan were the financiers.[60] This group owned the *Anna May*, which was ideally suited for the waterways to the diamond fields of Guiana. The commanding officer of the vessel was Capt. Lionel O'Neil, a veteran of the British Merchant Marine. Major

William H. York accompanied the expedition as the owners of the vessel representative. "Enroute to Guiana, the steamship was scheduled to call at Charleston, South Carolina, Savannah, Georgia, Jacksonville, Florida, and several West Indian ports;" African diaspora people were involved at all levels of mining in Guyana.[61]

The industry created economic linkages, expanded a variety of businesses, and influenced increased employment opportunities.[62] Businesspersons advocated the need for even more goods and services in the mining regions. They also commented on the increasing revenue government derived from the sale of consumer items in the interior as well as from the transportation of food, equipment, and sundry supplies. In addition, gold mining along with legislation enacted, ostensibly to facilitate access to government land for prospecting, influenced a burgeoning number of peasant industries and created a variety of employment opportunities.[63] Other offshoots of gold and diamond mining were the purchasing of buildings or shops by employees of that industry who also made small investments in diverse industrial undertakings.[64] These developments represented continuities consistent among community members.

Enslaved and freed Africans were involved in retail and wholesale businesses.[65] In the 1840s, soon after emancipation, members of this community had encountered a number of challenges as businesspersons. Portuguese and Madeirans were increasingly imported and the new migrants were selectively favored with financial patronage by the colonial authorities to assist them in business ventures. Members of the African descended business community were not granted such favors by colonial bureaucrats. They were increasingly denied access to credit and free or cheap land, and legislation was enacted to tax them heavily.[66] As late as the 1870s, despite myriad roadblocks, the aspirations of Africans and their descendants in commercial enterprises continued with relative success; some like Charles Hicks Augustus Denbow, Senior, ran a drug "store" business (apothecary/chemist shop) on the Corentyne Coast.[67]

Discriminatory actions affected members of this community and limited expansion of their businesses but, their ventures in these enterprises were never totally abandoned. Harsh as the human and ecological environments were, those persons were not all victims of their circumstances because they took action to improve their social and economic conditions. Some of the more successful miners were involved in building construction, transportation services, and sugar, coconut, rice, and diary production as entrepreneurs and investors. Others were merchants, traders, and financers of prospecting groups. These linkages, directly and indirectly related to mining, facilitated the spread of technologies, created training and employment opportunities, and increased incomes in both urban and rural areas for a cross section of the Guyanese population.[68]

Among the entrepreneurs and investors involved in gold and diamond mining who achieved social and economic mobility and created employment opportunities were Maclean Ogle and the previously mentioned, Andrew Benjamin Brown. Ogle of Buxton Village, East Coast Demerara was a son of the dispenser at La Bonne Intention sugar estate. He became a teacher and later a lawyer and legislative council representative. Brown was born in 1857, in Den Amstel Village, West Coast Demerara. He became a teacher, lawyer, politician, and held legislative office.

Brown was the co-owner of Plantation Middlesex, Canal No. 2 Polder, West Bank Demerara where coffee and cocoa were cultivated and he invested in the coconut industry. He contributed enormously to the country—administratively, legally, politically, educationally, and economically and left a long-lasting legacy for which he was honored nationally in 1922. Ogle and Brown financed groups of villagers to prospect for gold and diamonds. To his credit, Brown undertook a visit to the interior gold and diamond fields so that he could experience the rigors of mining.[69]

George "Premier" Arthur, Josephus Headley, "Tiger" Jarvis, Frederick Hope, Frederick "Bandola" Willman, and E. J. Levans were among those who invested their profits from gold and diamond mining and financed mining groups in the interior. Other former miners invested their profits in a variety of ways and created employment opportunities. Jacob Gershom Stephens, born in Buxton-Friendship, East Coast Demerara in 1879, was a former miner who became a diamond merchant and business proprietor. Stephens employed miners as well as diamond buyers in the interior and on the coast. He was a president of the British Guiana Miners Association. In January 1956, spectacular caches of diamonds were found in vicinities of the Cuyuni River. E. J. Levans and J. G. Stephens provided transportation, food, and equipment to and at the sites.[70]

Grandison Nathaniel Collins, a resident of Ann's Grove Village, East Coast Demerara was originally a farmer. After spending considerable time as a gold and diamond miner prospecting in Mazaruni, Puruni, and Cuyuni, Essequibo County, Collins became a merchant. "Sharkey" Andries another former miner became a building contractor and owned a large-scale farming enterprise, which provided employment for many persons in his natal village of Buxton, East Coast Demerara.[71] West Coast Berbician, Alexander Barrington Gillis who was born in 1864, and died in 1935, was among the first African-Guyanese entrepreneurs legally permitted to outfit their own self-help and independent gold and diamond mining prospecting groups from 1895. Gillis used the profits he acquired from mining to establish a building construction business which employed several persons. They included George Mc D. Hackett and Aaron Hope who were apprentices/journeymen in Gillis' business. Hackett, eventually a master builder, owned his own construction firm. He served as president of the Master Builders Association of Guyana. Aaron Hope migrated to Aruba and worked with the Lago Oil Refinery as a building superintendent. Years later, he returned home and continued building construction. Hope also manufactured tiles and landscape ornaments.[72] Other former miners used their profits to make aesthetic contributions. Martin "Fish" Hamer bought an organ for Buxton Wesleyan/Methodist Church.[73] Entrepreneurs whose vision and expertise were not constrained by bureaucratic and environmental conditions originated from gold and diamond mining.

Two such entrepreneurs, who diversified their investments and epitomized the astuteness of establishing economic enterprises in both rural and urban settings, were William Alexander "Pigeon Island" Brown and Frederick Lugard Mahaica. One of the earliest mining millionaires, "Pigeon Island" Brown, was a legend in the country. In 1910, on account of spectacular discoveries of gold by prospectors under his direction, he was nicknamed "Pigeon Island" Brown for posterity. Brown was named for the island in the Cuyuni/Arangoy River, Essequibo, where he and his mining employees struck it rich. Brown was the owner of this exceedingly profitable claim.[74] Approximately, two decades later, a local newspaper, *The Daily Argosy*, reported on a surprising story in England's *Sunday Despatch*, concerning "Pigeon Island" Brown, a millionaire from British Guiana who had purchased gold prospecting machinery in the United Kingdom.[75] Brown was identified as "a picturesque figure" and a "West Indian negro grandson of a slave, but the possessor of a sense of gold-divining that verges on the uncanny."[76]

Earlier in the twentieth century when Brown had perused copious geological materials on the interior region, he concluded that there was the likelihood of immense deposits of gold on Pigeon Island. Many of his contemporaries were skeptical, but Brown was not dissuaded from investing all that he had to buy the rights from the government to work on the island. Brown, who did not usually display his feelings, could not believe "he struck it rich—rich beyond his own dreams."[77] The deposits were so significant that he eventually had several hundred men working for him. As an entrepreneur and investor, he used his first profits to purchase several flat bottomed boats to carry his workers up the river. His conveyances included a motor-launch, which was the first of its type ever seen on the upper reaches of the Essequibo River.[78] Brown's prolific gold discoveries enhanced economic linkages in boat building and transportation services and created employment opportunities for many other persons.

On account of his wealth, Brown spent over 20,000 Pounds (sterling) to purchase gold mining equipment during the UK business visit. Astutely, Brown then informed the newspaper reporter that he was expanding his prospecting operations because the deposits on the island could not last forever. By August 1934, approximately a year after Brown's UK visit, prophetically, as he had envisioned, a short-lived gold-boom occurred in the Cuyuni area where Pigeon Island is located. A local newspaper characterized the gold discoveries as a probable return of the "Pigeon Island" days of the 1910–1914 eras. Hundreds of prospectors and speculators who were eager to be part of the gold-rush, hurried to the district every day.[79] Ironically, a few years later, at sixty years of age in 1937, "Pigeon Island" Brown's powers in divinity transcended gold recovery to include forays into human spirituality.[80]

Diamond magnate Frederick Lugard Mahaica, who is revered in Guyana's mining history, was born in 1922, and eventually became a millionaire. Fred Mahaica, as he was fondly known, spent decades during his youth acquiring the prowess to recover gemstones. Initially, during the 1930s, year after year in company with and under the tutelage of his father, they braved inclement weather, trudged through the jungles, and crossed many rivers searching for precious stones. Sometimes, because they could not find any diamonds independently, the Mahaicas had to seek employment with foreign diamond prospecting companies. Throughout the 1940s they both utilized rudimentary methods prospecting for the stones. By the 1950s, the younger Mahaica's avid readings on all aspects of diamond mining and his investment in the purchase of a suction pump proved fortuitous. Installing pile-driving equipment and using other machinery to break boulders, Mahaica increased his production of diamonds. Before long the increasing amounts of diamonds that Mahaica produced made him a millionaire. Along the way, two characteristics of the younger Mahaica surfaced: the patience to understand that the experience he was gaining would auger well for him later; and his unsatiated quest for knowledge about diamonds.[81]

Young Mahaica perused many geological volumes to learn how to identify stones quickly. Some of the mining publications he read were borrowed from engineers attached to the companies for which he worked, but the majority of the reading material was bought by Mahaica himself. He believed that gaining more expertise would enable him to formulate new techniques to recover the precious stones and to minimize the likelihood of damaging them. Mahaica's methods in prospecting for diamonds proved very productive. Sometimes, Mahaica deposited gemstones at local banks worth a million dollars or more. He also traded on the international diamond market.[82] On account of his success, Mahaica, like Brown, employed hundreds of miners including family members and villagers from Mahaicony, East Coast Demerara to work on claims he owned at Kurupung, Essequibo. One of Mahaica's most profitable claims was given the location name, "Mahaica Creek" in his honor. Mahaica insisted that "there were good diamonds in Guyana…we have gemstones of better quality than most of the rest of the world, equaled only by those in Central Africa."[83]

Mahaica diversified his business to include other kinds of economic enterprises. He constructed a $750,000 rice mill in his natal village. He created employment opportunities with the mill as well as through the construction of a feed factory and many other buildings. Mahaica's other economic undertakings included a diary farm at Kurupung and the ownership of several prize-winning race horses. Most of Mahaica's employees became property-owners themselves. In the capital city, Georgetown, Mahaica conducted much of his business from his head-office located on Middle Street, again generating more employment opportunities. Among the projects he envisioned for the years after the 1950s, were the erection of a hotel and a number of factories. Mahaica also "helped many students with their education through scholarships."[84] When he died in December 1991, those who eulogized him referred to the ways in which he created employment opportunities for others and justifiably credited diamonds for Mahaica's contributions in diversifying Guyana's economy.[85]

African diaspora mining migrants who parlayed their earnings into profitable enterprises were not confined to Guyana. Barbadian Fitzgerald Ince captured the sheer magnitude of migration, mining, and the African diaspora when he migrated to the Orinoco region of Brazil, which borders Guyana to join an 1890s gold-rush. In the 1920s, he returned home with an enormous amount of gold and purchased a vast area of land on which he produced bananas and sugar cane as well as a type of grass suitable for furniture manufacturing. Generating employment opportunities and more income by the businesses, Ince bequeathed much property and wealth to his descendants.[86]

For thousands of years, humans have been fascinated with gold. No doubt because of "its great natural beauty and universal durability" as well as its aesthetic, ornamental, decorative, commercial, investment, and industrial uses.[87] In its long history, gold assumed different meanings, largely due to the interests of persons involved. Geochemists, mining engineers, metallurgists, artists, jewelers, numismatists, and miners value gold for its many properties, utility, storage capacity, and in representing wealth. Gold can be forged into bars, ingots, and coins. "Gold is a store of" and "unit of value, a medium of exchange and a source of capital" and as currency, it has purchasing power. Economists view gold as "a valuable standard against which wealth is measured and" as "an imperishable medium for balancing international accounts."[88] Similarly, diamonds are valuable because of their industrial, commercial, ornamental, and other uses. Thus, the worldwide value and utility of gold and diamonds has served as motivating forces to attract workers, investors, and entrepreneurs worldwide.

Economic Linkages: The Case of the Bauxite Belt

Unlike gold and diamonds, the mining of bauxite offered no possibility for workers to become independent operators. Bauxite mining was monopolized exclusively by a few foreign-owned companies. In the development of mining in Guyana, the claim exists that "capitalists who dominated bauxite production in the colony" never encouraged or "fostered forward and backward linkages to the rest of the economy."[89] No factories were ever constructed in the country to produce pots, pans, and other utensils or automobiles and aircraft parts that are manufactured from bauxite. Following this trajectory, studies on the bauxite industry hold its hierarchy responsible for the lack of economic linkages and diverse employment opportunities.[90] Allegedly, for most of the decades the Canadian owned Demerara Bauxite Company operated in Guyana, it failed "to act as an engine of economic development...for the whole economy," exported all of the bauxite it produced, and tended to rely overwhelmingly on imported technical services and industrial inputs.[91] Ultimately, some studies noted, bauxite operated only as an extractive industry and "an enclave without forward or backward linkages to the rest of the economy...failing to motivate the developing of a manufacturing industry."[92] While the above perspectives have some merit, entrepreneurial and other economic activities by workers and residents of the bauxite belt were not unknown.

Management of bauxite conglomerates were mainly focused on extracting the ore and creating economic linkages abroad, but expansion of related local business activities occurred. Entrepreneurial ventures in the bauxite mining community included a variety of commercial enterprises operated by small-scale shopkeepers and vendors. The two markets at Mackenzie and Wismar and outlets along the Wismar-Christianburg Road attested to increased business opportunities with the coming of bauxite mining. Part of the discussion in Chapter 6 includes an outline of the expansion of businesses independent of the bauxite company. Similarly, the working class in the bauxite belt utilized the traditions of their compatriots in other areas of the country and created business enterprises in the face of expatriate monopolies. They situated their community for change.

The localities of gold and diamond mining had created the need for expanded water transportation. The providers of such services were not only the government and larger private entities, but also independent boat-builders and operators who created economic linkages. A variety of river and oceanic transportation services were provided by locals to facilitate the production of gold, diamonds, and bauxite. These economic linkages were not usually large-scale nor did they always confirm to a particular mode. Entrepreneurs and self-employed business persons operated in tandem with what has been characterized as a monopolistic undertaking—bauxite mining—and an example of an independent operator is significant because of diaspora ramifications.

James Edmund Orna, familiarly know as "Dutchie" exemplified the case of an entrepreneur who created economic linkages to bauxite production, and provided employment opportunities. Orna was an African diaspora labor migrant, and a native of the former Dutch Guiana (Suriname), for which his alias "Dutchie" was bestowed. He was born on March 14, 1914, in Nikerie, a town on the Corentyne River, across from Guyana. Educated during his formative years in the Dutch-speaking country, Orna as a young man migrated to Guyana, where English is the official language. He applied himself to acquiring both electrical and mechanical engineering skills. At a time when "working at Mackenzie", as an employee of the Demerara Bauxite Company was highly desirable, Orna used his electrical and mechanical skills to establish independently his own boat transportation service.[93]

Although other vessels had provided transportation services between both banks of the Demerara River, "Dutchie's" operations were unique because he was the first boat-proprietor to provide twenty-four hour services. This type of transportation service was indispensable for bauxite production at Mackenzie because many of the employees worked a shift-system, encompassing a twenty-four hour routine. Workers especially needed cross-river transportation at all times. Utilizing his independent economic enterprise linked to the larger undertaking of bauxite mining, Orna trained and employed several assistants and served as the magnet for his three brothers to migrate from Suriname to Guyana. Orna married a local woman, Hilda Alman, with whom he had eight children. He died at ninety-three years of age on April 7, 2007. Orna, his siblings, offspring, and their descendants exemplified African diaspora labor migrants who seized opportunities and contributed to the development and longevity of Guyana's mining industry.[94]

After World War II, migration to the United Kingdom and North America which enveloped the Caribbean region included current and potential mining workers. In the 1970s, on the verge of Suriname's independence from the Netherlands, the Dutch administration gave nationals of its soon-to-be-former colony, Dutch Guiana, a final opportunity to enter Holland without any immigration restrictions. A mass exodus of "Dutch" nationals from Linden/Mackenzie and its environs to the European nation ensued to precede the 1975 deadline. As the end of the twentieth century neared, a downturn in bauxite production influenced a reverse migration of Caribbean and other nationals out of Guyana. Natives of St.Vincent, St. Lucia, and Barbados, for example, returned to those countries, some with their Guyanese-born offspring. Lack of markets, burgeoning shipping costs, tariffs, and fees imposed by several developed countries after the Guyana government nationalized the bauxite industry, adversely impacted production, profits, and employment. By the mid twentieth century, bauxite exports had been the most lucrative in the mining sector, but at the dawn of the twenty-first century the situation reversed. In 2009, according to an International Monetary Fund Report, the value of gold exported from the country was 86% percent while that of bauxite was 14% percent of the sector.[95] Ultimately, El Dorado in Guyana created opportunities for migration, mining, and the African diaspora to contribute to the country's development.

Conclusion

Enslavement and its effects were not mere interludes in the lives of Africans and their descendants in Guyana. Following the 1833 Abolition Act, the newly freed had to figure out how to maneuver in the half-freedom of the Apprenticeship period. Remembering the lessons enslavement had taught, the former bondspeople challenged the limitations which half-freedom imposed on them; and questioned their work requirements and the wages they earned. Upheavals that followed led to a shorter-than-legislated Apprenticeship period. Within a few months of the enactment of the 1838 Emancipation Act, the recently freed, pooled savings they had accumulated mostly from agricultural production and commercial transactions while they were enslaved, and wages earned during their Apprenticeships and bought real estate. Among the purchasers were multi-skilled persons, who in the succeeding years, produced and sold sugar, coffee, coconuts, ground provisions, fruits, vegetables, and dairy and livestock. to make a living and provide for the future. Rural residents who were also fishermen and women and fishmongers, and artisans, engendered self-employment, organized strikes to protest plantation conditions, and sometimes migrated. Throughout the post-emancipation decades, although migration offered opportunities to explore varied vistas, African Guyanese faced ongoing challenges from the plantocracy, mineocracy, bureaucrats, and their allies. Even worse, adverse ecological conditions caused by floods and drought periodically created havoc in their living and working locations. Still, as a group, from the overwhelmingly agricultural era to the diversified economic period fostered by their labor in the mining industry, they survived in the social and ecologically hostile environments of nineteenth- and twentieth-century Guyana.

In the pre-mining era from the 1830s to 1880s, African Guyanese forged ahead while articulating and actively protesting the disadvantages they faced and the advantages taken of them. Post-emancipation conditions were often as abysmal as enslavement. Nonetheless, they created mechanisms for the improvement of their situations. Some of them utilized educational opportunities and made important contributions to their communities. From the 1890s, after importing East Indians, Chinese, Portuguese, Madeirans, and free and liberated Africans to shore up the country's perennial short labor supply, the colonial hierarchy decided to import even more African-descent labor. As part of an industrial age when diversified economies enabled nations to prosper, bureaucrats and entrepreneurs focused on importing this type of labor because of its immeasurable contributions to the colony's export-oriented economy. In the late nineteenth century, a time when racist views pervaded about the inferiority of some races—mostly non-Europeans—it was the assumption that only African-descended persons could work under abominable and deadly conditions which made them preferable.

Migration provided a significant mechanism of adaptation in the Caribbean. Migratory movement in and out of the region is well-documented. Aspects of internal migration such as rural to urban movement and *vice versa* have also received attention. In this case, not only rural to rural but also African diaspora migration to gold, diamonds, and bauxite mining are unique. The finding of this study show: most miners were agricultural in orientation, versatile, and quickly adapted to mining; many originated locally, from the Caribbean, South and North America, and Africa; miners diversified the economy and contributed to national development; they created employment and expanded economic activities; their high mortality rates did not staunch the flow of chain migration to mining; workers determined according to market and environmental conditions whether to work in gold, diamonds, or bauxite; this shift in the direction of migratory movement provided the panacea for a partial solution to their untimely deaths; after World War II, educational opportunities, overseas migration to the United Kingdom, the United States, and Canada pointed the way to other vistas for those of African descent to explore.

The dynamics of a colonial policy geared to remit profits from the Empire to improve the standard of living in Great Britain, had serious consequences for those of African descent in the then British colony. While still enduring the hardships associated with a sugar economy, they travelled *en masse* to the gold and diamond fields, died in unprecedented numbers, and thereby diminished the colony's labor supply. Therefore, efforts to diversify the economy hinged on the availability and longevity of this highly productive labor force in agriculture and mining. But, although frontier conditions in the gold and diamond mining regions urgently needed infrastructure development, colonial officials and entrepreneurs pursued piece-meal and cost-minimizing strategies and favored replenishing the labor supply by importing additional migrants. Nonetheless, life-taking raging rivers studded with rocks, fallen trees, and waterfalls and unpathed jungles with wild animals and ferocious reptiles added to the deadly perils mining workers faced.

These workers had grown tired of ongoing conflicts, which had characterized coastal plantation work and they desired less micromanaging. They were not unaware of their worth, but the self-sacrifice inherent in their seasonal migrations under hazardous conditions can be described as experiencing pain to gain progress. Behind all of the pendulum movements between the interior and coast, was a vision that revealed itself by the mid-twentieth century. Many miners, including bauxite workers, equipped their offspring to function in other occupations by providing them with access to higher education. After all the back-breaking and nerve-racking conditions endured by their agricultural ancestor, their migrating mining-descendants envisioned improvements in the quality of life of their progeny.

Women endured the vicissitudes of life in coastal, interior, and riverain areas. In analyzing the motivations of this segment of the Guyanese population, their visions, disappointments, and fulfillments afforded instructive insights into laudable coping mechanisms. Their "travails and triumphs" fostered an understanding of the intricacies of continuities and changes through economic ventures. Men and women were central in family, communal, and economic activities, and together they played pivotal roles in the country's development. Challenges associated with enslavement, emancipation, globalization, modernization, "order and progress" and authoritarianism bedeviled them, but they created ways to prosperity.

By the1940s, earnings from gold and diamond mining contracted while bauxite production expanded. This was only a pause in the prosperity of gold mining which was forecasted more than a hundred years earlier. After an interlude spanning the 1950s to the 1980s, production of the precious metal accelerated. The period from the 1880s to early 1900s was one of spectacular gold discoveries. By the end of the nineteenth century largely due to capital intensive investment in interior infrastructure development, gold output increased tremendously. Over a century ago, predictions were made about the potential wealth of this industry but substantial investments took a long time. Still, mining workers persisted, produced profits, and caused bureaucrats and investors to deal with infrastructure issues. By the 1950s, the political scene was in transition and the mining industry faced reorganizing. In the Cold War era, developing countries such as Guyana, which have historically been producers of raw material for the industrialized and developed nations faced added dilemmas. As a supplier of raw materials for overseas markets and a major importer of manufactured goods, industrialization in Guyana was limited. In the post-independence period from the 1960s, governments' efforts to nationalize foreign companies which controlled bauxite production netted mixed results at best. The longevity of Guyana's precious metal and gems industries rested squarely on the minds and might of the workers. They faced the initial challenges under adverse conditions, and directed their labor elsewhere when they deemed it propitious.

Africans in Guyana never lost the ability to evolve from one level to the next. They survived enslavement and improved their lot during the nineteenth and twentieth centuries. From the inception of mining, many African diaspora workers migrated to the gold and diamond fields even after bauxite production began. Gold and diamond miners persistence vouchsafed that deposits of those commodities were far from exhausted. Although the gold and diamond regions were areas of almost overwhelming environmental hostilities, compared with coastal plantation work it appears that these labor migrants viewed interior mining as the lesser of two evils. Many of the workers had grown tired of ongoing conflicts which characterized coastal plantation work.

People of African descent were among the earliest involved in the establishment of the gold and diamond industries of Colombia, Brazil, French Guiana, and Suriname, but this study is possibly the first in the Americas to link the activity of precious metal and gems' mining to a craft—jewelery-making and other economic enterprises. The activities of skilled persons from enslavement through succeeding generations revealed remaining links to their ancestral homeland and nurturing in the Americas. Early jewelry and related businesses were established and owned by individuals from this segment of the country's population. Using their expertise and skills some of them became investors and entrepreneurs, and were instrumental in expanding economic opportunities.

While it is true that the colonial hierarchy went to extraordinary lengths to control the lives and labor of the workers, bureaucrats relative lack of effectiveness should be measured against the accomplishments of the working class. This group added another perspective with which to view enslavement, emancipation, globalization, modernization, and "order and progress." On a world-wide scale, environmental disasters still bedevil mining, but high mortality rates and migration did not sound the death knell for gold operations in Guyana that has prolonged into the twenty-first century. New life was breathed into the gold industry by the end of the previous century with infusions of more modern machinery and aggressive investing that increasingly helped to realize the predictions of huge profits forecasted during the nineteenth century. Diamond mining limps along. As the twenty-first century unfolds, some foreign mining conglomerates are switching their production between gold and bauxite, following a pattern established by African-descended miners. The legacies of migration and mining have come full circle; and *The International Monetary Fund Staff Country Reports-Guyana: Recent Economic Developments, Selected Issues, Statistical Appendices*, 1990s-2010, indicate that profits from gold remain significant.

The findings indicate that the unwavering authoritarian mindset of the power structure was unnecessary to arouse the Guyanese working people. To a degree, the efforts, achievements, and contributions of mining workers have been overshadowed or even ignored because of the abiding interests in the country's social, political, and economic ills and agriculture. The research also indicates that proceeds from mining were used by workers in positive ways that have heretofore been only scantily acknowledged. Overall, the book confirms that African diaspora labor migrants traversed Guyana and engaged in back-breaking pioneering efforts which contributed significantly to the triple mining industry's impact on national and international development. Many mining workers who survived the challenges of their employment provided for their own future and created opportunities for their offspring to function effectively in various areas and levels of society.

Appendix A

Table A1 Purchasers of plantation New Orange Nassau (Buxton), Demerara, amounts paid ($)

From plantation Lusignan:		From plantation Good Hope:	
Name	**Amount $**	**Name**	**Amount $**
Charles M'Kenzie	350	David London	270
Howe M'Kenzie	250	Bat Thornton	210
Nicolas M'Kenzie	300	M'Donald Hannibal	200
John Scott	300	Cornett Phiffie	300
King Daniel	300	Jack Robinson	230
Abercrombie	200	Howard Thornton	210
Alexander M'Kenzie	300	Albert Brutus	200
Corobal M'Kenzie	160	Simon Will(s)	200
Richard M'Kay	100	Pero Will(s)	130
Friendship Peter	250	Rasmas Hay	200
Sammy M'Kay	150	James Stuart	300
Columbus M'Kay	160	David Stuart	200
Charles Bati (s)field	300	George Sanco	300
David Spencer	200	Calais Quammina	300
Present M'Kenzie	160	Punch Stuart	300
Edward Francis	100	Bob Stuart	220
Whiliminner	150	Edward Stuart	200
Simon M'Kenzie	100	Tom Griffith	400
Thomsin Stuart	130	Alexander Ross	270
James Jones	127	Providence Stuart	200
Corporal M'Kenzie	39	Castello Stuart	290
		Walter Young	300
		Harry Pollard	200
		Perot Jemmie	200
		Harry Pollard	200
		Perot Jemmie	200
		Coboas Stuart	200
		Roger Brian	160

From plantation Nonpareil:

Name	Amount $	Name	Amount $
Duncan Adams	264	Cudgo Adams	200
Dundas Baily	300	Bob Quammie	235
Frank Baily	255	John Williams	100
Colin Baily	253	Hercules Bergess	100
Assi Baily	280	Mary Diamond	220
Cadgo Adams	215	Luben M(A)ccra	200
Murry Diamond	234	Jacob Gill	260
Robert Baily	261	Dr. M'Beth	160
Quammina Baily	250	Sandy Duke	273
Murry Cummings	250	Prince Paul	214
Hoxborough David	240	Glasscow Hopkinson	300

From Plantation Enterprise:

Name	Amount $	Name	Amount $
Glasscow Sarry	235	Valentine Brutus	120
Simon Scott	214	Matthew Jones	225
Saul Hopkinson	235	Bonopart Cato	200
Shappo Glasscow	239	Isaac Farley	200
Aukeva Sam	330	Charles Massiah	130
John Castelo	300	James London	160
Kingsgate Sifox	316	Billy Chester	241
Prince M'Kaa	206	Adam Newtown	200
Charles Cuffy	200	Caesar Moffat	200
Ben Hopkinson	200	Christmas Cuffy	200
John Smith	220	Glasscow Cove	300
Billy Manuel	236	Tom Gibson	230
John Fallett	200	Duncan Jarvis	200
Charles Fillington	100	Captain Stepney	200
Granville Greenfield	235	Martin Deal	250
John M'Cra	250	Novell Jones	140
Alexander Sam	230	Nero Castilo	240
Captain Hanily	230	Roderick Porter	362
Charles Enterprise	290	Nelson Rogers	200
Will Enterprize	230	Hanibal Estman	200
Punch Enterprize	256	Lamb Rogers	200
Jack Lustall	100	Alexander Smith	200
Farmer Lesprance	200	Corn Scott	150
John Mike	1,000	Primo Osanna	207
Ben Porter	200	Primo Simon	153
Friday Campbell	235	Alman Mack	110
Lewis Bacche	100	Michael Burion	57
Trotman Enterprize	235	Cyrus Calvander	56
Adam Stoan	200	Stephen Marrant	150
Stephan Morian	200	York Trim	200
Abraham Fox	200	Downes Corian	100
Total 141. A True Copy	District B., May 11, 1840	A. M. Lyons, Stipendiary Magist	

Source: *CO 111/171, Original Correspondence British Guiana 1840 April-June*. Despatch 64 dated May 15, 1840.

Table A2 Persons for transports of lots, Winkle Village, 1890

Names	Names
Susan Aaron	Frederick Headecker
Peter Adolf	Sally Headecker
Mrs. Lima Bell	Mary Henry
Flora Beresford	Lydia Houston
John Bernard	Harriet January
Lyander Bernard	George Joseph
Miss Binning	John Joseph
Mrs. Britton	Nathaniel Joris
Henry Carter	Georgiana Lonefield
C. B. Carto	Adelaide Louis
Maria Class	Rosalie Lunchon
Mrs. Cummings	Martin laRose
Frederick Cyrus	Julia Lancaster
Mrs. Darrell	Louis Manners
Edward Denba	H. Mento
Madelene Denbar	Mrs. Myre
Grace Dewall	Mrs. Paul
Richard Downer	William Pitt
Apollo Duke	Johanna Prince
Charlotte Duke	Julia Rollins
Elizabeth Dykerman	Priscilla Rose
Eliza Ferdinand	Susan Rose
Rose Ferrell	Michael Smith
Robert Forbes	C. G. Stubbs
Charlotte Frederick	William Thorne
Rosaline Fredericks	Mrs. West
Ann F. Gonzuel	Anna Westcott
George N. Harris	Rebecca Woolf
	R. A. Williams

Source: CO 114/46, Sessional Papers, British Guiana September 30, 1890, 38–39.

Appendix B

Selected Mining Ordinances and Regulations, 1880s–1940s

Gold mining Ordinance #16 of 1880	Grants of land and royalty on gold & silver
Ordinance #10 of 1886	To establish Government Land Department
Gold mining Ordinance #16 of 1886	Repeal of Section on Royalty, Power of Governor and the Court of Policy to make regulations
Mining Ordinance #4 of 1887	
Ordinance #18 of 1887	Supervision of Crown Lands, Forests, Rivers & Creeks . . . vesting of minerals rights in the Crown
Mining Ordinance #5 of 1889	Special provisions for the unlawful possession of gold
Ordinance #9 of 1890 *repealed by 6 of 1936	To establish the Institute of Mines and Forests of British Guiana
Mining Precious Stones Regulations, 1891	
Mining Regulations 1892	
Mining Ordinance #8 of 1894	Definition of "possession of gold" to prevent gold stealing.
Mining Precious Stones Regulations 1896	
Mining Ordinance #15 of 1896	Conferring magisterial powers on government officers.
Mining Regulations 1899	
Dredging Regulations 1901	
Mining Ordinance #3 of 1901	Definition of valuable minerals; Rights to base metals on private lands; Dredging concessions; Permission to occupy and test the value of Crown Lands.
Ordinance #25 of 1902	To establish the Department of Lands and Mines from the Dept of Lands and the Dept of Mines.
Mining Ordinance #1 of 1903	
Mining Regulations 1903	
Mining Precious Stones Regulations 1903	
Mining Regulations 1905	
Mining Precious Stones Regulations 1905	
Mining Ordinance #9 of 1911	Payment of duty on profits of gold
Mining Ordinance #27 of 1916	Redefinition of "valuable materials"
Ordinance #6 of 1918	To facilitate geological and mineralogical surveys.
Mining Ordinance #47 of 1918	Substituting tax on profits for royalty on gold won by dredging
Mining Ordinance #34 of 1920	
Ordinance #10 of 1924 *repealed 1940	Mining Districts Omnibus Licenses
Ordinance #22 of 1924	Mining Districts Sunday Trading
Mining Regulations 1924	Employment of persons in mining districts

Continued

Continued

Mining Ordinance #28 of 1925	Definition of claim license for mining Governor-in-Council to make regulations
Mining Ordinance #14 of 1929	
Bauxite Mining Regulations, 1930	Area by concession; notice to grant lease; license to mine on private and colony lands
Mining Ordinance #20 of 1931	Right of governor to refuse to issue lease or concession
Mining Ordinance #25 of 1933	Payment of royalty; repeal of "Duty on gold won by milling and dredging" Validation of mining licenses
Mining Ordinance #of 1939	Powers of the Commissioner or warden to order work to cease.
Mining Ordinance #26 of 1939	Issue of license to mine; approval for transfer of license.
Mining Ordinance #18 of 1942	Definition of raw gold
Mining Ordinance #1 of 1943	Approval of secretary of state for exclusive permissions.
Mining Ordinance #18 of 1943	Power of the governor to close areas to prospecting and mining.
Mining Ordinance #1 of 1945	Refusal of concession or lease.
Mining Ordinance #17 of 1946	

Source: *British Guiana. Laws and Statutes. Index to the Mining Regulations, 1924* (Georgetown: The Argosy Ltd., 1927); and *Sherwood A. J. Lowe, Historical Analysis of Mining Policy, Law and Administration in Guyana: 1831–2002* (Georgetown: Pavnik Press, 2003), 155–157.

Appendix C

Interviews with Former Miners and Others*, 1995–2010

Name	Place of Interview	Date
George E. R. Adams*	Victoria Village, Guyana	November 23, 1996
Linton "Jack" Adams*	Den Amstel Village, Guyana	December 27, 1996
Lionel Arthur*	Mocha Village, Guyana	November 17, 1996
Franklin Assanah*	Victoria Village, Guyana	October 26.1996
Henley "Piggy" Assanah*	Victoria Village, Guyana	October 26.1996
Walter Bagot*	Den Amstel Village, Guyana	December 29, 1996
Octavius Barnwell*	Mocha Village, Guyana	October 17, 1996
Alan Carter*	Stabroek Market, Guyana	December 31, 1996
James Culley*	Buxton Village, Guyana	October 24, 1996
Dr. Charles E. Denbow*	Bowie, MD	September 28, 2007
Arnold B. Drakes*	Queenstown Village, Guyana	November 21, 1996
Richard Dodson*	Blankenburg Village, Guyana	December 29, 1996
Sydney Dundas	Anns Grove Village, Guyana	December 31, 1996
Keith Easton	Lanham, MD	December 8, 2007
Darrell Gray*	Queenstown Village, Guyana	January 3,1997
Lenox Griffith	Florida (Telephone)	July 1, 2006
David Hatton*	Edmonton, London, England	November 24, 1997
Leon Headley*	Georgetown, Guyana	November 5, 1996
Clifford Hector	Hadfield St, George Town	December 20, 1996
Dr. Walter B. Hope	Howard University, USA	July 12,2006
Maurice July	Peters Hall Village, Guyana	December 22, 1996
Edward "Guitar" Levans*	Vergenoegen, Guyana	October 30, 1996
Pollydore Miggins*	Den Amstel Village, Guyana	December 20, 2002
Rowena "Roe" Mingo*	Brooklyn, New York, USA	November 16, 1997
Philomena Murray*	Zorg-en-Vlygt, Guyana	November 19, 1996
Frank Proctor	Upper Tulse Hill, UK	June 21, 2010
Benjamin Thomas*	Buxton Village, Guyana	November 3 1996
Harold Thompson	Hackney, London, England	May 30, 2010
A. Wright-Toussaint*	Mocha Village, Guyana	October 17, 1996
Winston Valentine*	Golden Grove/Nabaclis	October 26, 1996
Joseph Watson*	Washington, DC, USA	February 5, 1995
Eunice Williams	Brooklyn, NY, USA	February 28, 2006
Howell Wood*	Buxton-Friendship, Guyana	October 15, 1996

* Deceased.

Notes

Introduction

1. Sir Walter Raleigh, *The Discovery of The Large, Rich, and Beautiful Empire of Guiana* (London: Hakluyt, 1845. Reprint, 1596 ed.), 1 and C. A. Harris and J. A. J. Villiers, *Storm Van's Gravesande: The Rise of British Guiana* (Nendeln, Liechtenstein: Kraus Reprint Ltd., 1967. Reprint, 1911 ed.), 61–62.
2. *British Guiana: Report of the Labour Commission, 1890* (Georgetown: C. K. Jardine, 1890), i–xix and Norman E. Cameron, *The Evolution of the Negro* (Westport: Negro Universities Press, 1970. Reprint, 1929 ed.), 20, 27.
3. Ashton Chase, *A History of Trade Unionism in Guyana, 1900–1964* (Georgetown: New Guyana Co. Ltd., 1964); Walter Rodney, *A History of the Guyanese Working-People, 1881–1905* (Baltimore: Johns Hopkins University Press, 1982); Brian Moore, *Race, Power, and Social Segmentation in Colonial Society: Guyana After Slavery, 1839–1891* (New York: Gordon Breach Science Publishers, 1987); Maurice St. Pierre, *Anatomy of Resistance: AntiColonialism in Guyana, 1823–1966* (London: Macmillan Ltd., 1999); Kean Gibson, *The Cycle of Racial Oppression in Guyana* (Lanham: University Press of America, 2003); Odida T. Quamina, *The Mineworkers of Guyana: The Making of A Working Class* (Atlantic Highlands, NJ: Zed Books, 1987); and Sherwood A. J. Lowe, *Historical Analysis of Mining Policy, Law & Administration in Guyana: 1831–2002* (Georgetown: Pavnik Press, 2003).
4. Modernization and "order and progress" used here to identify group transformation with traditional and Western inputs. Cameron, *The Evolution of the Negro*, 6–10, 44–62, 27–32; Alan Young, *The Approaches to Local Self-Government in British Guiana* (London: Longman and Green, 1958), 119–120, 138, 200, 203; James G. Rose, "African-Guyanese Business," *Emancipation Vol.2* (1994), 35–37. See Tulio Halperin Donghi, *The Contemporary History of Latin America*. Edited and Translated by John C. Chasteen (Durham: Duke University Press, 1993), 127–140, 158–207 and O. Nigel Bolland, *The Politics of Labour in the British Caribbean: The Social Origins of Authoritarianism and Democracy in the Labour Movement* (Kingston: Ian Randle Publishers, 2001) for discussions on authoritarianism and dependency.
5. Alan Lancaster, "Proposals for Hinterland Settlement and Development of British Guiana, 1884–1890," *History Gazette 59* (1993), 2–23.
6. Ibid.
7. Harry Madramootoo, *The Story of Rice in Guyana* (Georgetown: Guyana Rice Corporation, 1973); Mohammed Shahabudeen, *From Plantocracy To Nationalism: A Profile of Sugar in Guyana* (Turkeyen: University of Guyana, 1983); Walton Look-Lai, *Indentured Labor and Caribbean Sugar: Chinese and Indian Migrants to the British West Indies, 1838–1918* (Baltimore: The Johns Hopkins University Press, 1993); K. O. Laurence, *A Question of Labour: Indentured Immigration Into Trinidad and British*

Guiana, 1875–1917 (New York: St. Martin's Press, 1994); David Northrup, *Indentured Labor in the Age of Imperialism, 1834–1922* (New York: Cambridge University Press, 1995); and Clem Seecharan, *Sweetening "Bitter Sugar": Jock Campbell: The Booker Reformer in British Guiana, 1934–1966* (Kingston: Ian Randle, 2004).

8. George Lanning and Marti Mueller, *Africa Undermined: A History of the Mining Companies and the Underdevelopment of Africa* (New York: Penguin Books, 1979), 19–60; Raymond E. Dumett, *El Dorado in West Africa: The Gold Mining Frontier, African Labor, and Colonial Capitalism in the Gold Coast, 1875–1900* (Athens: Ohio University Press, 1998), 1–50, 84–109.
9. *Report of the Labor Commission, 1890,* Circular 2, July 23, 1890; *British Guiana: Report of the Gold Industry and the Crown Lands Commission, 1895* (Georgetown: C. K. Jardine, 1895), 1–3; *British Guiana: Report of the Commissioner of Mines, 1895–1896,* 3–4; "Mining Industries," *British Guiana Exhibition Wembley, 1924,* 75.
10. H. A. Will, "Colonial Policy and Economic Development in the British West Indies, 1895–1903," *The Economic History Review* (1970), 129–147; Ann Spackman, "The Role of Private Companies in the Politics of Empire: A Case Study of Bauxite and Diamond Companies in Guyana in the Early 1920s," *Social and Economic Studies* 24:3 (1975), 341–378; and Stephen Labaton, "DeBeers Agrees to Guilty Plea to Re-enter the U.S. Market," *New York Times,* July 10, 2004, C1, 3.
11. Barbara P. Josiah, "A. B. Brown, Forbes Burnham, Herbert N. Critchlow, Claude H. A. Denbow," in Colin Palmer, ed., *Encyclopedia of African-American Culture and History: The Black Experience in the Americas,* 2nd. ed. (Thompson Gale/Macmillan: Farmington Hills, MI, 2006 ed.); Josiah, "Organizing Within the Diaspora: Claude H. A. Denbow, Howard University & the League of Coloured Peoples," *Journal of African American History* 95:2 (2010), 229–247; and Josiah, "From Shaw University to 'the Booker T. Washington of British Guiana:' Hon. E. F. Fredericks and Contributions of HBCUs to the African Diaspora." Paper presented at the *Association for the Study of African American Life and History, 95th Annual Meeting,* Raleigh, NC, October 3, 2010.
12. Ian Molho, "Theories of Migration: A Review," *Scottish Journal of Political Economy* 33:4 (1986), 396–419; Donald Larson and Yair Mundlak, "On the Intersectoral Migration of Agricultural Labor," *Cultural Development and Economic Change* 45:2 (1997), 295–319; and Lok Siu, "Migration Stories: Serial Migration and the Production of Home and Identity in Transnationalism," in Andrew R. Wilson, ed., *The Chinese in the Caribbean* (Princeton: Markus Wiener Pub., 2004), 159–190.
13. Dawn Marshall, "A History of West Indian Migrations: Overseas Opportunities and 'Safety-Valve' Policies," in Barry Levine, ed, *The Caribbean Exodus* (New York: Praeger Publishers, 1987), 19; Ransford W. Palmer, *Pilgrims From the Sun: West Indian Migration to America* (New York: Twayne Publishers, 1995), 6–87.
14. Ruth Simms Hamilton, Kimberly Eison Simmons, Raymond Familusi, and Michael Hanson, "Rethinking the African Diaspora: Global Dynamics," in Ruth Simms Hamilton, ed., *Routes of Passage: Rethinking the African Diaspora Vol.1. Part 1* (East Lansing: Michigan State University Press, 2007), 12, 17.

1 African Diaspora Migrant Miners and Guyana's El Dorado

1. Leslie Peter Cummings, *Geography of Guyana* (London: Collins Press, 1975, ed.), 6–12 and *Gazetteer of Guyana* (Georgetown: Government of Guyana, 1974), i.

2. Sir Walter Raleigh, *The Discovery of the Large, Rich and Beautiful Empire of Guiana* (London: Hakluyt Society, 1845. Reprint, 1596 ed.), 1; *Imperial Institute: The Mining Laws of the British Empire and Foreign Countries*, Vol. IX, British Guiana (London: H. M. S. O., 1927), 49, gives 1702 as the year gold mining began; *British Guiana Mining Industries* (Georgetown: Commissioner of Lands and Mines, 1918), 3–15; *Mining Industries. British Guiana Exhibition Wembley 1924*, 72–83. See V. T. Harlow, ed., *The Discovery of the Large and Beautiful Empire of Guiana by Sir Walter Raleigh* (London: The Argonaut Press, 1928, ed.), 61–62; Cornelius C. H. Goslinga, *The Dutch in the Caribbean and the Guianas, 1680–1791* (Dover: Van Gorcum and Co., 1985), 441–443.
3. James Rodway, *History of British Guiana from the Year 1668 to the Present Time*. 3 Volumes. (Georgetown: J. Thompson, 1891–1894), 217; C. A. Harris and J. A. J. Villiers, *Storm Vans Gravesande: The Rise of British Guiana* (Nendeln/Liechtenstein: Kraus Ltd., 1967 ed. Reproduced from Hakluyt Society, 1911 ed.), 61–62; Goslinga, *The Dutch in the Caribbean*, 442–443.
4. Raleigh, *The Discovery of the Large*, 1.
5. Rodway, *History of British Guiana*, 182–185, 217–219.
6. Rodway, *History of British Guiana*, and Harris and Villiers, *Storm Vans Gravesande*, 63–64.
7. Rodway, *History of British Guiana*, 172–178; Harris and Villiers, *Storm Vans Gravesande*, 119; D. Graham Burnett, "Robert Schomburgk in British Guiana," *Journal of Caribbean Studies* 15:1–2 (Fall 2000) 11–37 and D. Graham Burnett, *Masters of All They Surveyed: Exploration, Geography, and A British El Dorado* (Chicago: The University of Chicago Press, 2000), 1–253; and Marc Herman, "Searching for El Dorado," *Civilization* (July 1997), 42–51.
8. Harris and Villiers, *Storm Vans Gravesande*, 63–64.
9. Burnett, *Masters of All They Surveyed*, 17–20.
10. Rodway, *History of British Guiana*, 172–178; Harris and Villiers, *Storm Vans Gravesande*, 119; Burnett, "Robert Schomburgk in British Guiana," 11–37; and Burnett, *Masters of All They Surveyed*, 198–253.
11. Rodway, *History of British Guiana*, 178–186; Harris and Villiers, *Storm Vans Gravesande*, 120–121; and Hope Hunter, "Gold in British Guiana, Its History and Prospects," *Timehri 6* (1882), 264. See Jacqueline Braveboy-Wagner, *The Venezuela-Guyana Boundary Dispute: British Colonial Legacy in Latin America* (Boulder: Westview Press, 1984).
12. Rodway, *History of British Guiana*, 186–189, and Harris and Villiers, *Storm Vans Gravesande*, 121–123.
13. Rodway, *History of British Guiana*, 170–184, and H. J. Perkins, *Notes on British Guiana and Its Gold Industry* (London: Sampson Low, Marston, and Company, 1896), 10–11.
14. Ibid., Rodway, *History of British Guiana*, 186, the group was operated by E. V. Abrahams.
15. Rodway, *History of British Guiana*, 182–185, 217–219.
16. Rodway, *History of British Guiana*, 217–219, and Norman E. Cameron, *The Evolution of the Negro* (Westport: Negro Universities Press, 1970. Reprint, 1929 ed.), 61–62.
17. *British Guiana: Report of the Crown Surveyor, 1882* (Georgetown: Royal Gazette, 1883), 4–5. *The Gold Mining Ordinance #16, 1880* and *Imperial Institute: The Mining Laws of the British Empire*, 3–15.
18. First gold export recorded in, *The Report of the Crown Surveyor*, 1882, 5. See Rodway, *History of British Guiana*, 217–219 and Perkins, *Notes on British Guiana and Its Gold Industry*, 11, 37–39.

19. *British Guiana: Report of the Commissioner of Mines, 1895–1896,* 18–22; *Mining Industries: British Guiana Exhibition at Wembley, 1918,* 77–83; *British Guiana: Report of the British Guiana Commission, 1927* (London: H. M. Stationery Office, 1927), 24–25; *Handbook of Natural Resources of British Guiana* (Georgetown: The Daily Chronicle Ltd., 1946), 23–24; and Sherwood A. J. Lowe, *Historical Analysis of Mining Policy, Law & Administration in Guyana: 1831–2002* (Georgetown: Pavnik Press, 2003), 4, 19–30, 155–157.
20. *British Guiana: Report on the Blue Book for Statistics, 1889,* 6; Hunter, "Gold in British Guiana," 265; Rodway, *History of British Guiana,* 218. See Lowe, *Historical Analysis of Mining Policy.* Appendix 1–3, 155–157.
21. Harris and Villiers, *Storm Vans Gravesande,* 127–130; Rodway, *History of British Guiana,* 220–226, 267; Burnett, *Masters of All They Surveyed,* 198–253; and Lowe, *Historical Analysis of Mining Policy,* 30–35.
22. British Parliamentary Papers (BPP), 1890, 21–22; Hunter, "Gold in British Guiana," 264–265; and *British Guiana Exhibition at Wembley,* 74–75.
23. Rodway, *History of British Guiana,* 267.
24. BPP, 1890, 6, 21–22 and Rodway, *History of British Guiana,* 267.
25. J. B. Harrison, *British Guiana and Its Resources* (London: The West India Committee, 1907), 9 and *Mining Industries,* 1918, 3–10.
26. BPP, 1890, 56; Alan Lancaster, "Proposals for Hinterland Settlement and Development of British Guiana, 1884–1890," *History Gazette* 59 (1993), 2–23. See Walter Rodney, *A History of the Guyanese Working People, 1881–1905* (Baltimore: Johns Hopkins University Press, 1982), 58, 97–99, 101–102, 144, 170; and Lowe, *Historical Analysis of Mining Policy,* 23–30.
27. G. M. Stockley, "The Geology of British Guiana and the Development of Its Natural Resources," *Bulletin 25* (1955). See Douglas Smith, "The Origins of Porknocking in Guyana," *McGill Studies in Caribbean Anthropology* (1969), 81–92; Alan Lancaster, "An Unconquered Wilderness: A Historical Analysis of the Failure to Open Up the Interior of British Guiana, 1838–1919," MA Thesis, University of Guyana, 1979, 91; and Lowe, *Historical Analysis of Mining Policy,* 42.
28. *British Guiana: Report of the Labor Commission,* 1890 (Georgetown: C. K. Jardine, 1890), i–xix for support of migration; Perkins, *Notes on British Guiana and Its Gold Industry,* 23, *The Institute of Mines and Forests of British Guiana,* established in 1890 to regulate recruiting.
29. Walter A. Rodney, "Upper Guinea and the Origins of Africans Enslaved in the New World," *The Journal of Negro History LIV* (1969), 327–345; Joseph E. Inikori, "The Origin of the Diaspora: The Slave Trade From Africa," *Tarikh* 5:4 (1978), 1–19; Johannes M. Postma, *The Dutch in the Atlantic Ocean Slave Trade, 1680–1815* (Cambridge: Cambridge University Press, 1990, 56–118; B. W. Higman, *Slave Populations of the British Caribbean, 1807–1834* (Baltimore: Johns Hopkins University Press, 1984); Winston F. McGowan, "The African Slave Trade to Guyana," in Winston F. McGowan, James G. Rose, and David A. Granger, eds., *Themes in African Guyanese History* (Georgetown: Free Press, 1998), 19–38; and Timothy R. Buckner, "The Slave Trade Apex in the Eighteenth Century," and Aribidsi A. Usman, "The Nineteenth-Century Black Atlantic," in Toyin Falola and Kevin D. Roberts, eds., *The Atlantic World, 1450–2000* (Bloomington: Indiana University Press, 2008), 96–113 and 114–134.
30. The National Archives of the UK (TNA) T71/391–446 Colonial Slave Registry, 10, Demerary, Essequibo, and Berbice. Slave Registers, 1817–1832: Records of the Treasury (RT); (TNA) Colonial Office Original Correspondence (COOC) 111/161,

British Guiana, 1837–1838, Orders in Council Correspondence Concerning the Abolition of Slavery. See David Eltis, "The Traffic in Slaves Between the British West Indian Colonies, 1807–1833," *Economic History Review*, 2:25 (1972) 55–65; Eric E. Williams, *From Columbus To Castro: The History of the Caribbean, 1492–1969* (New York: Vintage Books, 1984, ed.), 304–305; A. Meredith John, "Communications, The Smuggled Slaves of Trinidad, 1813," *The History Journal*, 31:2 (1988), 365–375; Guy Granum, *Tracing Your West Indian Ancestors: Sources in the Public Record Office* (London: Public Record Office, 1995), 4–5, 23, 42–44, 62, 88; and Paul Farnsworth, "From Past to the Present: An Exploration of the Formation of African-Bahamian Identity during Enslavement," in Jay B. Haviser, ed., *African Sites Archaeology in the Caribbean* (Princeton: Markus Wiener Publishers and Kingston, Jamaica: Ian Randle Publishers, 1999), 94–130, on those legally transported from the Bahamas to Guyana.

31. Rawle E. Farley, "The Shadow and the Substance," *Caribbean Quarterly* 4:2 (1955), 132–153. See A. R. F. Webber, *Centenary History and Handbook of British Guiana* (Georgetown: The Argosy Co. Ltd., 1931), 158–174; Vere T. Daly, *A Short History of the Guyanese People* (London: Macmillan Education Ltd., 1975), 168–171; and Maurice St. Pierre, *Anatomy of Resistance: AntiColonialism in Guyana, 1823–1966* (London: Macmillan Education Ltd., 1999), 22–25.
32. James G. Rose, "The Strikes of 1842 and 1848," in Winston F. McGowan, James G. Rose, and David A. Granger, eds., *Themes in African Guyanese History* (Georgetown: Free Press, 1998), 158–200.
33. BPP Relating to Immigration, 339. No. 85. Reports and Correspondence. Copy of Despatches from Governor Light to Lord Russell, November 18, 1839. Letter dated November 30, 1839 signed by plantation purchasers from Edward Carberry, *Inducements to the Coloured People*, 24. COOR 111/171, 1840:2. Despatch 48, April 9, 1840, "Affairs of British Guiana. Statement of Transports of Land in favour of Purchasers (formerly Apprenticed Labourers)," August 1, 1838–April 8, 1840 and COOC 111/253, 1848. See no. 6464, *The Last Will and Testament of John Williams*, who died November 17, 1885, one of the original purchasers of land in Parika Village, East Bank, Essequibo; Rodway, *The History of British Guiana, Vol. 2*, 1893, 1–308; James Rodway, *Handbook of British Guiana* (Georgetown: The Committee, 1893), 27–28; J. Graham Cruickshank, "The Beginnings of Our Villages," *Timehri* 7:3 (1921) 65–76; Cameron, *The Evolution of the Negro*, 108–113; Rawle E. Farley, "The Rise of Village Settlements in British Guiana," *Caribbean Quarterly* 3 (1953), 101–109; Alan Young, *The Approaches to Local Self-Government in British Guiana* (London: Longman, Green & Co., Ltd., 1958), 8–23; Barbara P. Josiah, "After Emancipation: Aspects of Village Life in Guyana, 1869–1911," *The Journal of Negro (African American) History, 82* (1997), 105–121; and Carl B. Greenidge, *Empowering Peasantry in a Caribbean Context. The Case of Land Settlement Schemes in Guyana, 1865–1985* (Mona, Kingston: University of the West Indies Press, 2001), 17–21.
34. Colonial Slave Registry, 10; Cameron, *The Evolution of the Negro,* 6–11 on economic activities of the enslaved and recently freed. See M. J. and F. S. Herskovits, *Trinidad Village* (New York: Alfred Knopf, 1947), 257–265; A. F. and D. Wells, *Friendly Societies in the West Indies, Report on a Survey* (London: H. M. S. O., 1953), 10–21; Shirley Ardener, "Study of Rotating Credit Associations," *Royal Anthropological Institute Journal 94* (1964), 201–229; Aubrey W. Bonnett, *Institutional Adaptation of West Indian Immigrants to America: An Analysis of Rotating Credit Associations* (Washington, DC: University Press of America, 1981), 4–7, 27, 33–39; J. D. Elder, *African Survivals in Trinidad and Tobago* (London: Karia Press, 1988), 99–100;

Donna F. Danns, *The Development and Demise of the Savings Bank System in Guyana* (Georgetown: Bank of Guyana, 1986), 7–9, 18; and Barbara P. Josiah, "Creating Worlds: A Study of Mutuality and Financing among African Guyanese, 1800s–1950s," *The Journal of Caribbean History*, 38:1 (2004) 106–127, for financial activities.

35. Cameron, *The Evolution of the Negro,* 6–11 and Josiah, "Creating Worlds."
36. TNA: COOC 111/171, Light to Russell: Despatch 75: Enclosure 64, June 4, 1840 Berbice Transports. Cruickshank, "The Beginnings of Our Villages," 65–76; Cameron, *The Evolution of the Negro*, 108–113.
37. *Report of the Titles and Land Commissioners On Claims to Land in the County of Berbice* (Georgetown: C. K. Jardine, 1893), 328–331.
38. Ibid., and Cameron, *The Evolution of the Negro*, 108–113; Young, *The Approaches to Local Self-Government*, 11–23.
39. TNA: COOC 111/171, Despatch 48, April 9, 1840 "Statement of Transports," *Report of the Titles and Land Commissioners On Claims to Land in the County of Essequibo* (Georgetown: C. K. Jardine, 1893), 313–316; Young, *The Approaches to Local Self-Government*, 10–11; *Queenstown 1841 (Westfield, Mocha, Dagaraad)150 Anniversary Publication* (New York, 1992); *Queenstown 2000 (Sankofa "GO BACK AND FETCH IT")* (New York,The Queenstown Essequibo Association,2000); *Queenstown 2004 Anniversary Publication* (New York: The Queenstown New York Association, 2004); and "Overseas-Based Villagers at Queenstown Celebrations," *Stabroek News*, August, 4, 2004, 1–2.
40. *Reports of the Title and Land Commissioner, Essequibo*, 313–316.
41. Colonial Slave Registry 10, on economic activities of enslaved farmers, tailors, seamstresses, bakers etc that fostered self-purchase and Cameron, *The Evolution of the Negro*, 6–11.
42. TNA: COOC 111/1842 Vol. 2. Despatch 98, Petition from the African Association of British Guiana.
43. TNA: COOC 111/190. Despatch from Light to Stanley dated April 2 and 4, 1842: Enclosures 3.2 and 4.2 on land acquisitions and cultivation at Victoria, Queenstown, Cove & John, and Den Amstel. See Cameron, *The Evolution of the Negro,* 6–118; Cruickshank, "The Beginnings of Our Villages," 65–76; Young, *The Local Approaches to Self-Government*, 8–23; and Michael Moohr, "The Economic Impact of Slave Emancipation in British Guiana, 1832–1852," *Economic History Review* 25 (1972), 588–607.
44. Robin D. G. Kelley, *Freedom Dreams: The Black Radical Imagination* (Boston: Beacon Press, 2002), 3.
45. "Overseas-Based Villagers at Queenstown Celebrations". See *Cimbux-Buxton Pride Vol. 1 #1*, (April 1991; *County High School Alumni; First Anniversary Publication* (1997)), *2nd Publication* (1999); Eusi Kwayana, "Village Notes," *Cimbux Inc. 10th Buxtonian Reunion October 11, 2003*; *Cimbux-Buxton Pride Special Post 10th Reunion Ed.* (2003); Josiah, "Creating Worlds," 106–127; *Buxton Pride, Newsletter, Cimbux Inc.* (2004; *11th Cimbux Inc. Committee For The Improvement of Buxton, Inc., Biennial Reunion* (2005); *The New Victoria: A Souvenir Magazine 170th Year (Redevelopment Through Revival of Our Traditions and Culture) Vol. 1* (September 2009); *170th Anniversary Celebration of the Purchase of Victoria Village*, November 29, 2009, St. Stephens Lutheran Church Auditorium, 2806 Newkirk Avenue & E 48 St. Brooklyn, NY 11226; *Purchase & Pride, 170th Anniversary, Buxton Village, Guyana, 1840–2010. Celebrating Our Heritage* (Commemoration Publication, 2010, *Cimbux, Inc.*, "Saturday Evening Memorial Dance"... Bladensburg, MD... May

28, 2011; and *Cimbux Inc.*, 22nd Annual Souse Party"...Greene's Residence, Washington, DC...July 9, 2011.). The descendant community from Victoria, Buxton, Queenstown, Den Amstel, Belladrum, Danielstown etc., in the UK, the US, Canada, and Guyana hold numerous activities, raise funds, and support their compatriots at home to maintain their villages and African cultural heritage.

46. Young, *The Approaches to Local Self-Government*, 8–23 and Josiah, "After Emancipation,"105–121.
47. TNA: T71/757–765 Colonial Slave Registry, 10. The Slave Compensation Commission. Compensation-Valuer's Returns, Register of Claims, Claims and Certificates' Originals, Certificates for compensation, and Lists of Awards, RT; US Department of State, *Royal Gazette of British Guiana,* 29, July 24, 1834 on regulations for laborers, Special Justices of the Peace, and Stipendiary (Special) Magistrates. Consular Despatches Before 1905, Demerara, Record Group (RG) 59, National Archives, College Park, MD (NACP); Dwarka Nath, *A History of Indians in Guyana*, (London: By the author, 1950, 83, 104, 830; and Robin Blackburn, *The Overthrow of Colonial Slavery, 1776–1848* (London: Versco, 1988), 457–458 for owners payments. See M. J. Steele, "A Philosophy of Fear: The World View of the Jamaican Plantocracy in a Comparative Perspective," *The Journal of Caribbean History,* 27:1 (1993), 1–20; Douglas Hall, "The Flight From the Estates Reconsidered: The British West Indies, 1838–1842," in Hilary Beckles and Verene Shepherd, eds. *Caribbean Freedom:Economy and Society From Emancipation to the Present.* (Kingston, Jamaica: Ian Randle Publishers and Princeton: Markus Wiener Publishers, 1993), 55–63; Michael Craton, "Continuity Not Change: The Incidence of Unrest Among Ex-Slaves in the British West Indies, 1838–1876," in Hilary Beckles and Verene Shepherd, eds., *Caribbean Freedom: Economy and Society from Emancipation to the Present* (Kingston, Jamaica: Ian Randle Publishers and Princeton: Markus Weiner Publishers, 1993), 192–206; and Demetrius Eudell, *The Political Languages of Emancipation in the British Caribbean and the U.S. South* (Chapel Hill: U. N. C. Press, 2002), 8–157.
48. Cameron, *The Evolution of the Negro*, 6–118; Young, *The Approaches to Local Self-Government*, 7–38; and Rodney, *A History of the Guyanese Working People*, 32, 60.
49. Rodney, *A History of the Guyanese Working People*, 32–40; K. O. Laurence, *A Question of Labour: Indentured Immigration Into Trinidad and British Guiana, 1875–1921* (New York: St. Martin's Press, 1994), 1–7, 17, 186, 293–296, 510; David Northrup, *Indentured Labor in the Age of Imperialism, 1834–1922* (New York: Cambridge University Press, 1995), 2, 4–10, 16–28; Greenidge, *Empowering A Peasantry In A Caribbean Context*, 13; Rose, "The Strikes of 1842 and 1848," 158–200; O. Nigel Bolland, *The Politics of Labour in the British Caribbean: The Social Origins of Authoritarianism and Democracy in the Labour Movement* (Kingston, Jamaica: Ian Randle Publishers, 2001), ix–155.
50. Greenidge, *Empowering A Peasantry In A Caribbean Context*, 13 and Rose, "The Strikes of 1842 and 1848," 158–200. See Eudell, *The Political Languages of Emancipation*, 8–157.
51. TNA: COOC 111/161, Correspondence on the Condition of Hill Coolies in British Guiana, Correspondence on indentured laborers and domestic servants who migrated to British Guiana since August 1834. From the Official Returns of the Special Justices, September 1838; TNA: COOC 111/174, Foreign Office to James Stephen, July 7, 1840: Despatch May 29, 1840 on 500 African Americans from Baltimore emigrating to Trinidad and British Guiana; Edward Carberry, *Inducements To The Colored People of the United States To Emigrate To British Guiana* (Boston: Kidder and

Wright, 1840), 6–24; J. Graham Cruickshank, "African Immigrants After Freedom," *Timehri* 3:6 (1919), 74–85; G. W. Roberts, "Summary Statistics on Indentured and Associated Migration Affecting the West Indies, 1834–1918," *Population Studies* 20 (1966), 125–134; and Barbara P. Josiah, "Caribbean Migration To Guyana: Identity, Patterns, and Contributions From Slavery to WW I," in Emory J. Tolbert, ed., *Perspectives on the African Diaspora, Volume Two: Since 1800* (Boston: Houghton Mifflin, 1999), 285–304.

52. TNA: COOC 111/171, "Transfer of Captured Africans from Surinam and the Bahamas,'" April 8, 1840; and "On Emigration of Captured Africans from the Bahamas to Sierra Leone," June 25, 1840; and TNA: COOC 111/183 "Return of Immigrants and Captured and Liberated Africans, Rio de Janeiro," August 31, 1841. See J. Graham Cruickshank, "Among the Aku (Yoruba) in Canal #1, West Coast Demerara," *Timehri* 3:4 (1917), 70–82; Johnson U. J. Asiegbu, *Slavery and the Politics of Liberation, 1787–1861: A Study of Liberated African Emigration and British Anti-Slavery Policy* (New York: Africana Publishing Corp., 1969), 39–44, 72–80, 104–146; Walter Rodney, ed., *Guyanese Sugar Plantations in the Late Nineteenth Century: A Contemporary Description From the Argosy* (Georgetown: Release Publishers, 1979), 87; Blackburn, *The Overthrow of Colonial Slavery*, 419–472; Northrup, *Indentured Labor*, 1–7, 17.

53. See Ordinance #1, 1868 enacted October 1, 1868 to register births and deaths, *The Laws of British Guiana, 1803–1921*, Vol.2, 1866–1884 (London: Waterlow & Sons Ltd., 1923. *Register of Births, Demerara, Essequibo, and Berbice, 1869*; and *Index of Births, Demerara, Essequibo, and Berbice, 1869* (Georgetown: General Registrar Office, 1870).

54. Barbara P. Josiah, "After Emancipation," 105–121.

55. TNA: COOC 111/171 Light to Russell: Despatch 79 June 6, 1840 Enclosure 69 *The Voluntary Subscription Immigration Society*... the *Brig Porpoise*... Passengers... Immigrants, Baltimore, US, 316–318.

56. *Register of Births and Register of Deaths, Demerara, Essequibo, and Berbice, 1900* (Georgetown: General Registrar Office, 1901); *Reports of the Institute of Mines and Forests of British Guiana, 1890–1920; The Gold Mining Gazette*, on West Indians, French and Dutch (Suriname) Guianese, Africans, and African Americans. See Walter A. Rodney, "Barbadian Immigration into British Guiana, 1863–1924," Paper, *Ninth Annual Conference of Caribbean Historians*, Cave Hill, University of the West Indies, 1977, 19–25; Bonham Richardson, *Caribbean Migrants: Environment and Human Survival in St. Kitts and Nevis* (Knoxville: University of Tennessee Press, 1983), 18, 74, 80–88; Annette Palmer and Clement London, "Migratory Labor in the English Caribbean 1940–1945," *Journal of Caribbean Studies* (1981), 199–215; and Bolland, *The Politics of Labour*, 156–158.

57. *Report of the Labor Commission, 1890*, i–xix and Circular #2, July 23, 1890; *Results of the Decennial Census of the Population of British Guiana*, 1841–1891, Microfilms, National Archives and Records Administration: Washington, DC (NARA); *British Guiana: Report on the Results of the Census of 1891* (Georgetown: The Argosy Co. Ltd., 1892), viii; E. D. Rowland "The Census of British Guiana, 1891," *Timehri 4* (1892), 40–68; *Report of the Institute of Mines, 1890s–1920s*. See *The Gold Mining Gazette* on overseas recruitment of African diaspora workers and the role of steamer and shipping agents; R. R. Kuczynski, *Demographic Survey of the British Colonial Empire. Volume III. West Indian and American Territories* (London: Oxford University Press, 1953), 17, 165–190.

58. Ruth Simms Hamilton, ed., *Routes of Passage: Rethinking the African Diaspora* (East Lansing: Michigan State University Press, 2007), 1–32.

59. Rose, "The Strikes of 1841 and 1848," 158–200 and Kimani Neheusi, "The Causes of the Protest of 1905," in James G. Rose, Winston F. McGowan, and David A. Granger, eds., *Themes in African Guyanese History* (Georgetown: Free Press, 1998), 251–276.
60. Colonial Slave Registry 10 the enslaved were mechanics, woodcutters, shingle-makers etc. Hugh Ross' twelve enslaved workers in Berbice were cotton pickers, cleaners, and weavers. Cameron, *The Evolution of the Negro*, 6–23, 26; Moohr, "The Economic Impact of Slave Emancipation," 595–598; and Clive Lloyd Matthews, basket-maker and vendor, Haimaruni (Lowood) Village, Demerara River, interview with author, Stabroek Market, Georgetown, December 14, 1996, on this ancestral practice. See Rodney, *A History of the Guyanese*, 84; and Mohammed Shahabudeen, *From Plantocracy to Nationalization: A Profile of Sugar in Guyana* (Turkeyen: University of Guyana, 1983) on East Indians' and Africans' contributions. Judith A. Carney, *Black Rice: The African Origins of Rice Cultivation in the Americas* (Cambridge: Harvard University Press, 2001); and "With Grains in Her Hair: Rice in Colonial Brazil," *Slavery and Abolition*, 25:1 (April 2004), 1–27, on African rice producers in the diaspora.
61. *Report of the Commissioner of Essequibo and Pomeroon River Districts, 1901–1902*, 6; Report, 1905, 6; *Report 1908–1909*, 4; *Report of the Institute of Mines, 1913*, 12, on deceased Amerindian workers.
62. *Census Report, 1891*, viii; Rowland, "The Census," 40–68; Kuczynski, *Demographic Survey*, 17, 165–190.
63. *Census Report, 1891*, viii; Rowland, "The Census," 40–68 and Kuczynski, *Demographic Survey*, 17.
64. Michael Moohr, "The Discovery of Gold and the Development of Peasant Industries in Guyana, 1884–1914: A Study in the Political Economy of Change," in *Caribbean Studies* 15:2 (1975), 57–72; Rodney, *A History of the Guyanese Working People*, 33; and Brian L. Moore, "Colonialism, Cricket Culture, and Afro-Creole Identity in the Caribbean after Emancipation: The Case of Guyana," *The Journal of Caribbean History* 33:1–2 (1999), 54–73.
65. Rodney, *A History of the Guyanese Working People*, 108–109.
66. *Census Report, 1891*; Rowland, "The Census," 40–68; Kuczynski, *Demographic Survey*, 17; Henry Kirke, *Twenty-Five Years in British Guiana* (Westport: Negro Universities Press, 1970. Reprint London: Sampson Low, Marston, and Company, 1898, ed.), 207–215; Rodney, *A History of the Guyanese Working People*, 33, 84, 109; Walton Look-Lai, *Indentured Labor and Caribbean Sugar: Chinese and Indian Migrants to the British West Indies, 1838–1918* (Baltimore: The Johns Hopkins University Press, 1993), 52, 57, 59–61; Andrew R. Wilson, ed., *The Chinese in the Caribbean* (Princeton: Markus Wiener Publishers, 2004), vii–xxiii; and Lisa Yun, *The Coolie Speaks. Chinese Indenture Laborers and African Slaves of Cuba* (Philadelphia PA: Temple University Press, 2008), 5–70.
67. *Report of the Institute of Mines, 1913*, on one Chinese laborer who died of dysentery, "Chinese as Gold-Miners: Number Steadily Increasing," Georgetown, *The Daily Argosy*, October 1, 1935, 5; and Marlene Kwok-Crawford, "Chinese Immigration into Guyana," Georgetown, *Stabroek News*, February 3, 1994.
68. *Census Report, 1891*, viii; Rowland, "The Census," 40–68; Kuczynski, *Demographic Survey*, 17. See Lomash Roopnarine, "East Indian Indentured Emigration to the Caribbean: Beyond the Push and Pull Model," *Caribbean Studies* 31:2 (July–December 2003).
69. *British Guiana: Report of the Diamond and Gold Industries, 1928* (Georgetown: The Argosy Co. Ltd., 1928), 28–30, Regulation 106, Mining Regulations, 1924, ending East Indian restrictions.

70. *Report of the Labor Commission, 1890,* i–xix. Rodway, *Handbook of British Guiana,* 60; William Francis and John Mullin, eds., *Handbook of British Guiana, 1922* (Georgetown: The Argosy Co. Ltd., 1923), 115–116; H. A. Will, "Colonial Policy and Economic Development in the British West Indies, 1890–1903," in *Economic History Review* 23 (1970), 129–147, for physical descriptions of the groups; Basdeo Mangru, "Abolishing the Return Passage Entitlement Under Indenture: Guianese Planter Pressure and East Indian Government Response, 1838–1917," *Caribbean Quarterly* 32:3–4 (1984), 1–13; and James Rose, "The Repatriation Controversy and the Beginning of An East Indian Village System," in *Guyana Historical Journal 1* (1989), 51–67, on incentives for East Indians in coastal agriculture, restrictions, and planters' efforts to discourage repatriation by land colonization schemes and free or cheap land. See Eusi and Tchaiko Kwayana, *Scars of Bondage: A First Study of the Slave Colonial Experience of Africans in Guyana* (Georgetown: Free Press, 2002), 39–44; Greenidge, *Empowering a Peasantry in a Caribbean Context*, 23, 47.
71. Rowland, "The Census," 40–68. See, Clem Seecharan, *"Tigers in the Stars:" The Anatomy of Indian Achievement in British Guiana, 1919–1929* (London: Macmillan: Education Ltd., 1997) for contributions of East Indians to Guyana's development (1929).
72. *The Gold Mining Gazette, June–Sept. 1890*; Samuel Pollard, Henry Walcott, George McCray, James Allen, Joseph Johnson, Den Amstel Villagers recruited for H. July Co. and Star Mining Co. *Report of the Commissioner of Mines, 1895–96,* 2.
73. *Colonial Slave Registry 10. Demerary and Essequibo Slave Registry, 1817,* Vol.1 #548 dated May 31, 1817, for enslaved workers on the Demerara Ferry. *50th Anniversary THD, Fifty Years of Service of the Transport and Harbours Department, January 1931–December 1980* (Georgetown: Guyana National Printers Ltd., 1980), 13–25; Howell Wood, interview with author, Buxton Village, East Coast Demerara, October 15, 1996 and Lionel Arthur, interview with author, Mocha Village, East Bank Demerara November 17, 1996.
74. *British Guiana: Administrative Report, 1890,* 6; Kirke, *Twenty-Five Years,* 135; Terrence Simmons, "Bartica," *Guyana Chronicle,* June 1992; and Lowe, *Historical Analysis of Mining Policy,* 45, note 101.
75. Wood, interview with author.
76. Cummings, *Geography of Guyana,* 11.
77. St. Pierre, *Anatomy of Resistance: AntiColonialism in Guyana,* 2–3.
78. Simms Hamilton, ed., *Routes of Passage: Rethinking the African Diaspora,* 24.
79. Cameron, *The Evolution of the Negro,* 6–10.
80. TNA: COOC 111/553 Despatch 400, December 7, 1906 from governor to the Colonial Office: Enclosure, "The Peoples Association of British Guiana," dated February 1906, on taxation and hardships. Young, *The Approaches to Local Self-Government,* 24–64; Rodney, *A History of the Guyanese Working People,* 1–30, 60–71; David Granger, "The African Village Movement," in *The African-Guyanese Achievement: 155th Anniversary of African Slave Emancipation 1* (1993), 25–26; Brian Moore, "The Social and Economic Subordination of the Guyanese Creoles after Emancipation," in Winston F. McGowan, James G. Rose, and David A. Granger, eds., *Themes in African Guyanese History* (Georgetown: Free Press), 141–157; Rose, "The Strikes of 1842 and 1848," 158–200; Rodney, *A History of the Guyanese Working People*; 72–82; Neheusi, "The Causes of the Protest of 1905," 251–276; Bolland, *The Politics of Labour,* ix–158; Blackburn, *The Overthrow of Colonial Slavery,* 461–464; and Kwayana and Kwayana, *Scars of Bondage,* 19–26.
81. *United States Naval Records Collection,* Office of the Naval Records and Library, RG 45 and *Bureau of Naval Personnel,* RG 24, NARA. Other Guyanese sailors in the

Civil War Navy were Thomas Holloway, Leon Cepeda, William Phillips, Thomas Tillman, George Wallace, Henry Brown, Frederick Beck, and James Larkin); TNA: COOC 111/554, Enclosure "Administration of the Colony" from governor to the Colonial Office dated May 18, 1906, on Guyanese leaving the country. See Barbara P. Josiah, "The Other Migration: African Diaspora Sailors from Latin America and the Caribbean in the U.S. Civil War Navy," Paper, *Association for the Study of the Worldwide African Diaspora Conference*, Rio de Janeiro, October 3–9, 2005 and the *Washington Area Seminar on the History of Latin America*, The Catholic University of America, Washington, DC November 4–5, 2005.

82. Cameron, *The Evolution of the Negro*, 27; *Who is Who in British Guiana, 1938–1940* (Georgetown: The Daily Chronicle Ltd., 1940), 29–30, for John Alexander Barbour-James in West Africa. See Robert T. Vinson, "In the Time of the Americans: Garveyism in Segregationist South Africa, 1920–1940," Unpublished PhD., Howard University, 2001 and Nemata A. Blyden, *West Indians in West Africa: The African Diaspora in Reverse* (Rochester: University of Rochester Press, 2000).
83. Neheusi, "The Causes of the Protest of 1905," 260–264.
84. Ibid.
85. *British Guiana: Royal Commission Report, 1897*, Appendix C Part II, British Guiana; Rodney, *A History of the Guyanese Working People*, 99; and Lancaster, "Proposals for Hinterland Settlement and Development of British Guiana, 1884–1900."
86. Moore, "The Social and Economic Subordination," 152; Neheusi, "The Causes of the Protests of 1905," 257; and Bolland, *The Politics of Labour*, ix–158.
87. *Report of the Labor Commission, 1890*; Perkins, *Notes on British Guiana and Its Gold Industry*, 37–39. See Rodney, "Barbadian Immigration to British Guiana," 22–25.
88. *Report of the Commissioner of Mines, 1895–1896*, 2–3; and Cameron, *The Evolution of the Negro*, 41–61.
89. Kirke, *Twenty-Five Years*, 126; Basdeo Mangru, *Indenture and Abolition: Sacrifice and Survival on the Guyana Sugar Plantations* (Toronto: Tsar, 1993), 59–80, 98–135; Rodney, *A History of the Guyanese Working People*, 86, 174–189, on East Indians having India as a point of reference and measures to mitigate the harsh conditions of indentureship. See Rose, "The Reparation Controversy and the Beginning of An East Indian Village System," 51–67. Monica Schuler, "Kru Migration to British and French Guiana, 1841–1857," in Paul Lovejoy, ed., *Africans in Bondage: Studies in Slavery and the Slave Trade* (Madison: University of Wisconsin Press, 1986) points to the rare exception of an 1853 visit by Kru Chief Jumbo to investigate local conditions of Kru labor migrants.
90. TNA: T71/437CO 114/46. *British Guiana: Court of Policy Decision, 30 Sept. 1890*, 38–39, New Amsterdam: Office of the Registrar. List of names, Appendix B. Colonial Slave Registry, Berbice, 1817, 425–430, #225, 310 enslaved persons (83) children, Winkle Dept., Berbice.
91. Walter Nugent, *Crossings: The Great TransAtlantic Migrations, 1870–1914* (Bloomington: Indiana University Press, 1992) and Thomas Holloway, *Immigrants on the Land: Coffee and Society in Sao Paulo, 1886–1934* (Chapel Hill: University of North Carolina Press, 1980) on opportunities created by European migrants in Latin America.
92. Walter Rodney, *How Europe Underdeveloped Africa* (Washington, DC: Howard University Press, 1982. Reprint, 1972 ed.) and E. Bradford Burns, *The Poverty of Progress: Latin America in the Nineteenth Century* (Berkeley: University of California Press, 1980), studies offering comprehensive discussions on colonization and underdevelopment.

2 Migration and Mining Strategies in a Colonial Society

1. *British Guiana: Register of Deaths, Essequibo District* 3, 4, 16, 17, 18 (Georgetown: General Register Office, 1891–1950s). See *Report on the Gold, Diamond, and Forests Industries of British Guiana: Institute of Mines and Forests of British Guiana, 1908–1922.* On labor shortages, *British Guiana: Report of the Labor Commission, 1890* (Georgetown: C. K. Jardine, 1890), i–xix, 1–66.
2. *Report of the Labor Commission, 1890,* 45–49 and *British Guiana: Report of the Gold Industry and Crown Lands Commission, 1895* (Georgetown: C. K. Jardine, 1895), 1–5, 1–162.
3. H. W. Brands, *The Age of Gold: The California Gold Rush and the New American Dream* (New York: Doubleday, 2002), 334–358, 361–491. See *Panamanian Passages,* The Smithsonian Latino Center, Exhibition, Concourse, S. Dillon Ripley Center, Smithsonian Institution, Washington, DC. October 2009–May 2010.
4. Peter Duignan and L. H. Gann, eds., *Colonialism in Africa, 1870–1960. Vol.4.* The Economics of Colonialism (Cambridge: Cambridge University Press, 1975), 3–32; P. T. Bauer, *Equality in the Third World and Economic Delusion* (Cambridge: Cambridge University Press, 1981), 1–5, 6–85, 163–184; Dinesh D'Souza, "Three Cheers for Colonialism," *The Chronicle of Higher Education*, May 10, 2002, B7–9; and Robert Johnson, *British Imperialism* (New York: Palgrave Macmillan, 2003), 1–12, 107–119, 204–217. See Jan Nederveen Pieterse, *White on Black: Images of Africa and Black in Western Popular Culture* (New Haven: Yale University Press, 1992), 20, 22; and Jonathan P. Spiro, *Defending the Master Race: Conservation, Eugenics, and the Legacy of Madison Grant* (Burlington: University of Vermont Press, 2009), 2–140, on race matters.
5. Walter Rodney, *How Europe Underdeveloped Africa* (Washington, DC: Howard University Press, 1982. Reprint 1972, ed.), 3–288; Basil Davidson, *Africa in History* (New York: Collier Macmillan Publishing Co., 1974. Reprint, 1968, ed.), 160–298; Samuel Agonda Ochola, *Minerals in African Underdevelopment* (London: Bogle L'Overture Publications, 1975), 1–117; George Lanning and Marti Mueller, *Africa Undermined: A History of the Mining Companies and Underdevelopment of Africa* (New York: Penguin Books, 1979), 19–60; E. Bradford Burns, *The Poverty of Progress: Latin America in the Nineteenth Century* (Berkeley: University of California Press, 1980), 5–17, 132–154; Caroline Elkins, *Imperial Reckoning: The Untold Story of Britain's Gulag in Kenya* (New York: Henry Holt & Co., 2005), xi–xvi, 1–61, 354–374; and Adam Hochschild, *King Leopold's Ghost; A Story of Greed, Terror, and Heroism in Colonial Africa* (Boston: Houghton Mifflin, 2006, ed.), 292–306.
6. Ann Spackman, "The Role of Private Companies in the Politics of Empire: A Case Study of Bauxite and Diamond Companies in Guyana in the Early 1920s," *Social and Economic Studies* 24:3 (1975), 341–378; Lanning and Mueller, *Africa Undermined,* 19–60; Odida T. Quamina, *The Mineworkers of Guyana: The Making of a Working Class* (Atlantic Highlands, NJ: Zed Books, 1987),14, 17–18; Alan Cowell, "A New Generation of DeBeers: Diamonds and It Seems the Oppenheimers Are Forever," *New York Times*, April 25, 2003, W1, 7; and Stephen Labaton, "DeBeers Agrees to Guilty Plea To Re-enter the U.S. Market," *New York Times*, July 10, 2004, C1, 3.
7. Howard University (HU): British Parliamentary Papers (BPP) British Guiana, 1890, 323. *The Gold Mining Gazette*, July, 1890. Ordinance #9, 1890, to incorporate the Institute of Mines and Forests of British Guiana, October 4, 1890 in *The Laws of British Guiana* Vol. 3, 1884–1891 (London: Waterlow & Sons Ltd., 1923), 482.

8. H. I. Perkins, *Notes on British Guiana and Its Gold Industry* (London: Sampson Low, Marston & Co., Ltd., 1896), 23.
9. *Report of the Labour Commission, 1890*, 1–66. See O. Nigel Bolland, *The Politics of Labour in the British Caribbean: The Social Origins of Authoritarianism and Democracy in the Labour Movement* (Kingston: Ian Randle Publishers, 2001), ix–xiii, 1–19, 80–81,117–119, 181–191, 336–355.
10. TNA: COOC 111/190, British Guiana, 1842. Despatch 98 "Administration of the Colony" Enclosure: "The African Association of British Guiana"; *Report of the Labour Commission, 1890*, evidence, B. Howell-Jones, 45–49; *The Gold Mining Gazette*, July 1890 and February 25, 1897. See Maurice St. Pierre, *Anatomy of Resistance: AntiColonialism in Guyana, 1823–1966* (London: Macmillan Education Ltd., 1999), 22–50.
11. *Report of the Labour Commission, 1890*, i–xix, 45–49.
12. "Hon. Andrew Benjamin Brown—Symbol," *The Daily Chronicle*, January 9, 1939, 1; "OBITUARY. Hon. Andrew Benjamin Brown. Churchman, Lawyer, and Politician. Hundreds Pay Tribute at Funeral," *The Daily Argosy* January 10, 1939, 5. See Edith Campbell-Brown, *The Life Story of Andrew Benjamin Brown* (Georgetown: B. G. Lithographic Co., 1947); and Vere T. Daly, *A Short History of the Guyanese People* (London: Macmillan Education, Ltd., 1975), 270–274.
13. TNA: COOC 111/652, governor to the Colonial Office "Aspirations of the Negro Progress Convention," April 7, 1924. Enclosure: Memorandum of Reasoned Statement Submitted by the Negro Progress Convention on Behalf of the Negro Population of British Guiana to Governor, Sir Graeme Thomson for Transmission to…Principal Secretary of State for the Colonies, dated February 20, 1924. See Barbara P. Josiah, "Andrew Benjamin Brown," in Colin Palmer, ed., *Encyclopedia of African-American Culture and History: The Black Experience in the Americas* (Thomas Gale/Macmillan Reference, Second Ed., 2005), 338–339.
14. *The Gold Mining Gazette* for information on recruiting, registration, and so on. See *The Echo* December 2, 1893 and Perkins, *Notes on British Guiana*, 23.
15. Ibid. See Selwyn R. Cudjoe, *Caribbean Visionary: A. R. F. Webber and the Making of the Guyanese Nation* (Jackson: University of Mississippi Press, 2009), 3–4.
16. *Report of the Labour Commission, 1890*, i–xix and Perkins, *Notes on British Guiana*, 23.
17. Ibid.
18. HU: BPP, 1890, 323. Appendix A1.
19. *Report of the Labour Commission, 1890*, Minutes, 111-V, See Annexure A-E and VI-IX.
20. *Report of the Labour Commission, 1890*, evidence, J. B. Mayers, owner, Wales Sugar Estate, West Bank Demerara, 1–5, E. C. Luard, manager, Peters Hall Sugar Plantation, East Bank Demerara, legislator, 6–9, T. J. Wakefield, proprietor, investor, 15–17; A. P. Bugle, saw-miller, 17–18; F. A. Mason, manager, Demerara Railroad Company, 19–21, C. Morrison, attorney and manager, Taymouth Manor and L'Union Plantations, Essequibo, 32–34, John Cummings, attorney, Blairmont Sugar Estate, Berbice, 34–36, and G. M. Bethune, manager Enmore Sugar Plantation, East Coast Demerara, 21–22, 68–69, favored African descent labor migrants.
21. *Report of the Labour Commission, 1890*, evidence, J. L. Stoute, manager, Reliance Sugar Plantation, Essequibo.
22. Ibid. James Andrew, attorney and manager, Leonora Sugar Plantation, West Coast Demerara, evidence, 39–41 and Cummings, Blairmont.

23. *Report of the Labour Commission Report 1890*, 1–2, Minutes 111-V and XX-XX2. See Spiro, *Defending the Master Race*, 3–140 on Francis Galton, Charles Darwin, Houston Stuart Chamberlain Count de Gobineau, Comte de Georges Vacher, and Madison Grant as scientific racists.
24. *Report of the Labour Commission, 1890*, i–xix, 1–66; TNA: COOC 111/612 Despatch 213, June 8, 1917, from governor to the Colonial Office "Application from Demerara Bauxite Company for Crown Lands;" *Report of the Institute of Mines, 1919–1920*, 6–7, on labor migrants moving from sugar to mining. See Walter Rodney, "Barbadian Immigration Into British Guiana, 1863–1924," Paper presented at *the 9th Annual Conference of the Association of Caribbean Historians*, Cave Hill Campus, University of the West Indies, April 1977.
25. *British Guiana: Report of the Diamond and Gold Industries* (Georgetown: The Argosy Company, 1928), 30 on removal of *Regulation 106. Mining Regulations, 1924, Restriction on East Indians. The Royal Commission, 1939*, evidence, Dr George Giglioli and Legislator Seaford, *The Daily Chronicle*, February 10, 17, 1939, 12, 6. See Henry Kirke, *Twenty-Five Years in British Guiana*. Reprint 1898, ed. (Westport: Negro University Press, 1970, ed.),126; Basdeo Mangru, "Abolishing the Return Passage Entitlement Under Indenture: Guianese Planter Pressure and East Indian Government Response, 1838–1917," *Caribbean Quarterly* 32, 3&4 (1984), 1–13; James Rose, "The Repatriation Controversy and the Beginnings of An East Indian Village System," *Guyana Historical Journal* 1 (1989), 51–67); and Carl B. Greenidge, *Empowering Peasantry in a Caribbean Context: The Case of Land Settlement Schemes in Guyana, 1865–1985* (Kingston: The University of the West Indies Press, 2001),26–27, 41–42, 49.
26. *Report of the Labour Commission Report 1890*, 1–2, Minutes 111-V and XX-XX2 and *British Guiana: Reports of the Census, 1881 and 1891*. See Basdeo Mangru, *Indenture and Abolition: Sacrifice and Survival On the Guyanese Sugar Plantations* (Toronto: Tsar Publications, 1993), 98–135 and David Northrup, *Indentured Labor in the Age of Imperialism, 1834–1922* (New York: Cambridge University Press, 1995), 117–118.
27. Mangru, *Indenture and Abolition*, 98–135.
28. Yen Ching-Hwang, *Coolies and Mandarins: China's Protection of Overseas Chinese During the Late Ch'ing Period, 1851–1911* (Kent Ridge: Singapore University Press, 1985), vii–viii, xiii–xvi, 99–347; Michael Gonzales, *Plantation Agriculture and Social Control in Northern Peru, 1875–1933* (Austin: The University of Texas Press, 1985), 73–195; Andrew R. Wilson, ed., *The Chinese in the Caribbean* (Princeton: Markus Wiener Publishers, 2004); and Lisa Yun, *The Coolie Speaks: Chinese Indentured Laborers and African Slaves in Cuba* (Philadelphia PA: Temple University Press, 2008), 5–8, 34–71.
29. *Report of the Labour Commission, 1890*, evidence, George Garnett, agent, proprietor, legislator, gold mining and forestry investor, 13–17.
30. *Report of the Labour Commission, 1890*, Garnett, 13–17 and Andrew, 39–41.
31. Ibid. Timber grant owner, J. D. Paterson, evidence, 54–55; J. M. Fleming, manager Diamond Sugar Plantation, East Bank Demerara, 26–28, wanted Africans.
32. Ruth Simms Hamilton, ed., *Routes of Passage: Rethinking the African Diaspora Vol. 1. Part 1* (East Lansing: Michigan State University Press, 2007), 12–17.
33. *British Guiana: Report on the Census*, 1911–1946, identify large numbers of foreign-born of African descent in the colony. George E. Brooks, *The Kru Mariner in the Nineteenth Century: An Historical Compendium* (Newark, DE, 1972), 1–112; Diane Frost, *Work and Community Among West African Migrant Workers Since the Nineteenth*

Century (Liverpool: University of Liverpool Press, 1999), 1–75; and Dwayne E. Williams, "Rethinking the African Diaspora: A Comparative Look at Race and Identity in a Transatlantic Community, 1878–1921," in Darlene Clark Hine and Jacqueline McLeod, eds., *Crossing Boundaries: Comparative History of Black People in Diaspora* (Bloomington: University of Indiana Press, 2001), 105–120, on migratory movement of Liberians, Sierra Leoneans, and Cape Verdeans in the 19th and 20th centuries. See Simms Hamilton, *Routes of Passage*, 16, for perspective on migratory movements in Latin America and the Caribbean.

34. TNA: T71/391–446. *Colonial Slave Registry 10, Demerary, Essequibo, and Berbice. Slave Registers, 1817–1832*, Records of the Treasury; Daniel C. Littlefield, *Rice and Slaves: Ethnicity and the Slave Trade in Colonial South Carolina* (Baton Rouge: Louisiana State University Press, 1981), 74–114, 137 for silversmiths and jewelers among enslaved; C. A. Harris and J. A. J. Villiers, *Storm Van's Gravesande: The Rise of British Guiana*. Hakluyt Society, 1911, ed. (Nende/Lechtenstein: Kraus Reprint Ltd., 1967), 61–62 for Africans in early explorations; Jean Libby, "Technological and Cultural Transfer of African Ironmaking into the Americas and the Relationship to Slave Resistance," Paper presented at Louisiana State University, 1992, slavedata@ afrigeneas.com; Judith A. Carney, *Black Rice: The African Origins of Rice Cultivation in the Americas* (Cambridge: Harvard University Press, 2001), 1–8, 12–38, 43–159 and Carney, "With Grains in Her Hair': Rice in Colonial Brazil," *Slavery and Abolition*, 25:1 (April 2004), 1–27 on enslaved Africans' in rice and the implements the invented for the industry; Candice Goucher, "African-Caribbean Metal Technology: Forging Cultural Survivals in the New World," in Jay B. Haviser, ed., *African Sites Archaeology in the Caribbean* (Kingston, Jamaica: Ian Randle Pub., and Princeton: Markus Weiner Pub., 1999), 143–156; Gwendolyn Midlo Hall, *Slavery and African Ethnicities in the Americas: Restoring the Links* (Chapel Hill: University of North Carolina Press, 2005), xiii–172.
35. *Report of the Labour Commission, 1890*, evidence, Garnett, 14; and *British Guiana: Administrative Report of the Essequibo River District, 1890*, 17 February 1891, 1.
36. Ibid., Garnet and evidence, Wakefield, 15.
37. James G. Rose, "The Strikes of 1842 and 1848" in Winston F. McGowan, James G. Rose, and David A. Granger, eds., *Themes in African-Guyanese History* (Georgetown: Free Press, 1998), 158–200.
38. *Report of the Labour Commission, 1890*, Wakefield, 15 and Howell-Jones, 45–49.
39. *Report of the Labour Commission, 1890*, evidence, David Gibson, 37–39. See Douglas Hay and Paul Craven, eds., *Masters, Servants, and Magistrates in Britain and the Empire, 1562–1955* (Chapel Hill: University of North Carolina Press, 2004) for widespread measures to control labor.
40. COOC 111/190 "The African Association of British Guiana;" COOC 111/553, 1907 " The People's Association of British Guiana;" COOC 111/652 "Aspirations of the Negro Progress Convention." Demetrius L. Eudell, *The Political Languages of Emancipation in the British Caribbean and the U.S. South* (Chapel Hill: University of North Carolina Press, 2002), 8–57; Eusi Kwayana and Tchaiko Kwayana, *Scars of Bondage: A First Study of the Slave Colonial Experience of Africans in Guyana* (Georgetown: Free Press, 2002), 1–36; and Kean Gibson, *The Cycle of Racial Oppression in Guyana* (Lanham: University Press of America, 2003), 7–21.
41. *Report of the Labour Commission, 1890*, evidence, Bethune, 21–22, Cummings, 34–36, R. F. Rose, owner, Henrietta Sugar Plantation, Leguan Island, Essequibo River, 41–43, H. Murray, manager Port Mourant Sugar Plantation, Corentyne, Berbice, 52–54.

42. Alan Lancaster, "Proposals for Hinterland Settlement and Development of British Guiana, 1884–1900," in *History Gazette 59* (1993), 2–23.
43. *Report of the Labour Commission, 1890*, evidence, Rose, 41–43.
44. Ibid. Mayers, 4, Luard, 8, Stoute, 12, and Murray, 53. See Walter Rodney, *Guyanese Sugar Plantations in the Late Nineteenth Century: A Contemporary Description from the Argosy* (Georgetown: Release Publishers, 1979), 2–18, 39–44 and Barbara P. Josiah, "After Emancipation: Aspects of Village Life in Guyana," in *The Journal of Negro (African American) History, 82* (1997), 105–121.
45. *Report of the Labour Commission, 1890*, evidence, Reverend D. J. Reynolds, 23–26.
46. *Report of the Labour Commission, 1890*, evidence, Reverend F. C. Glasgow, 63–66. See Eudell, *The Political Languages of Emancipation*, 11, 73–79, 84–88, 96–97, 124–127, 157, comparative situations in Jamaica and the US South.
47. *Report of the Labour Commission, 1890*, evidence, Glasgow.
48. *Report of the Labour Commission, 1890*, 1–2.
49. *Report of the West India Commission, 1897 Appendix C. Part 11, British Guiana*. Rev. D. J. Reynolds, evidence, 95–96.
50. *Report of the West India Commission, 1897. Appendix C. Part 11, British Guiana*, 162. Letter dated February 4, 1897, from Ebenezer Alex Trotz, 29 Durban St. Georgetown and 182 others on hardships and tragedies in interior mining--and solutions. COOC 111/550 Despatch 107: Enclosure: "Administration of the Colony. "Monster" petition from the Inhabitants" Signed by 5,750 people on the neglect of recommendations of Royal Commission, 1897." See Chapter 4; District Death Registers for high mortality.
51. Black Stalin, "More Come," Record album, *REBELLION*, Trinidad and Tobago Carnival, 1896. See HU: BPP, Colonies West Indies, 1896, Appendix A, 256–257, on physical ability of workers.
52. *British Guiana: Report of the Commissioner of Mines, 1891*, 10.
53. Perkins, *Notes on British Guiana*, 23. *Ordinance #25, 1902,* established the Department of Lands and Mines in Report of the Commissioner of Lands and Mines, 1903/1904. Ordinance #20, 1914 see *Report of the Commissioner of Lands and Mines, 1914/1915*, 26–27; and *Report of The Institute of Mines, 1913/1914*, 8; *Report of the Institute of Mines, 1914/1915*, 4; and *Report of the Institute of Mines, 1915/1916*, 1–2.
54. *Report of the Institute of Mines, 1914–1915*, 4 and *Report of the Institute of Mines, 1915–1916*, 1–2. Ordinance #6, 1936, officially ended the work of the Institute of Mines. See Report of the Commissioner of Lands and Mines, 1936.
55. "Appeal Court To Solve Labour Bureau Problem," *The Daily Chronicle*, November 24, 1934, 4 and "Puruni Prospecting Expedition," *The Daily Chronicle*, July, 28, 1937, 1.
56. *The Mining Gazette*, January-March, 1890; June-August, 1890; and April-June 1892.
57. See *Reports of the Institute of Mines; Reports of the Commissioner of Lands and Mines; The Census Reports of British Guiana; The Echo*; and the *Daily Argosy* for designations.
58. HU: BPP, 1890, *British Guiana Legislation #23* and *Report of the Labour Commission, 1890.* Response to Circular July 2, 1890, dated July 23, 1890, 1. See COOC 111/505 Despatch dated September 29, 1898; COOR 111/506 Despatch 368 dated November 19, 1898, concerning *Ordinance #10, 1898, British Guiana. Immigrants Transported from Barbados and St. Vincent,* and *The Laws od British Guiana Vols 1–6, 1803–1921*(London: Waterlow & Sons Ltd., 1923.

59. *Report of Commissioner of Lands and Mines 1913–1914.*
60. Odida T. Quamina, The*Mineworkers of Guyana: The Making of a Working Class* (Atlantic Highlands, NJ: Zed Books, 1987), 16–24.

3 Mining Factors in a Diversified Economy

1. *British Guiana: Report of the Crown Surveyor, 1882* (Georgetown: Royal Gazette Office, 1883), 4–5; TNA: COOC 111/439, Despatches 230–299, May-July, 1887; *British Guiana: Report of the Lands and Mines, 1918*, other companies were, Kanaimapoo Gold Mining Co, Winter Gold Mining Co. (B. G. Gold Mines), Aremu Mining Co. Ltd., B. G. Dredging Syndicate, Sir Walter Raleigh Gold Mining Co., B. G. Gold Mines Ltd., Demerara Exploration Syndicate., B. G. Gold Concessions Co. Ltd, and Guiana Gold Co.; *Mineral Resources in British Guiana* (Georgetown: Daily Chronicle, 1935), 5, 23–24.
2. *British Guiana: Report of the Commissioner of Mines, 1895–1896*, 2–3; *Report of the Gold, Diamond, and Forest Industries: Institutes of Mines and Forests, British Guiana, 1909–1910.* Appendix Viii (the Institute of Mines); W. A. Swain, "The Gold and Diamond Mining Industry in Guyana, Unpublished Master's Thesis, University of London, 1980; 33–35; and *Imperial Institute: The Mining Laws of the British Empire and Foreign Countries* Vol.ix, British Guiana (London: H. M. S. O., 1927).
3. *Report of the Commissioner of Lands and Mines*, 1902–1956; *British Guiana: Colonial Annual Report*, 1946–1957; and Swain, "The Gold and Diamond Mining Industry, 30; 33–35.
4. H. A. Will, "Colonial Policy and Economic Development in the British West Indies, 1895–1903," *The Economic History Review 23* (April 1970), 129–147 and Alan Lancaster, "Proposals for Hinterland Settlement and Development of British Guiana, 1884–1900," *History Gazette 59* (1993), 2–23.
5. TNA: COOC 111/543 Despatch 22, 6 May 1904, "Arrest of Gold Diggers near the B. G. Frontier... Venezuelan Outrage on Boundary;" Ann Spackman, "The Role of Private Companies in the Politics of Empire: A Case Study of Bauxite and Diamond Companies in Guyana in the Early 1920s," *Social and Economic Studies 24:3* (1975), 341–378; George Lanning and Marti Mueller, *Africa Undermined: A History of the Mining Companies and Underdevelopment of Africa* (NY: Penguin Books, 1979), 19–60 (4).
6. *Report of the Commissioner of Lands and Mines*, 1904–1905, 11–12; Walter Rodney, *A History of the Guyanese Working-Class, 1881–1905* (Baltimore: The Johns Hopkins University Press, 1982), 190–216; and Kimani Neheusi, "The Causes of the Protest of 1905," in Winston F. McGowan, James G. Rose, and David A. Granger, ed., *Themes in African-Guyanese History* (Georgetown, Free Press, 1998), 251–276.
7. Rosemarijn, Hoefte, "Cherchez la femme and Other Quests in the Historiography of the Dutch Caribbean in the Post-Emancipation Era," in Juanita DeBarros, Audra Diptee, and David V. Trotman, eds., *Beyond Fragmentation: Perspectives on Caribbean History* (Princeton: Markus Wiener Publishers, 2006), 173; and Ruth Simms Hamilton, ed., *Routes of Passage: Rethinking the African Diaspora, Vol. 1, Part 1* (East Lansing: Michigan State University Press, 2007), 14–19.
8. *The Echo,* December 2, 1893, 6; *Report of the Institute of Mines*, 1890–1924; and Walter Rodney, "Barbadian Immigration into British Guiana, 1863–1924." Paper presented at *the 9th Annual Conference of Caribbean Historians*, Cave Hill, Barbados, University of the West Indies, April 1977.

9. *British Guiana: Report of the Immigration Agent General, 1920,* 7; "West Indians in Panama Want to Come and Settle Here," *The Daily Chronicle,* January 16, 1934, 4; "The British Guiana Colonization Project Attract Leaders," *New York Amsterdam News* , June 22, 1935, on African American migration; "West Indians for British Guiana," *The West Indian Review,* (July 1935),12; "Introducing Jamaicans," *The Daily Argosy,* April 29, 1937, 7; U. S. Department of State: British Guiana. Letter dated August 29, 1940, from the Law Offices of Brown, Brown Cyrus and Greene, Chicago, IL, File 844B82/20, RG 59,(MD: NACP).
10. "The Labour Situation," *The Daily Argosy,* August 29, 1922, 4.
11. Ibid. See "Colonisation by West Indians," in *The Daily Argosy* August 24, 1933, 4; and Hon. Joseph Eleazar, evidence, *The Royal Commission Enquiry, 1939*, 9.
12. *The West Indian Census, 1946: British Guiana*, xxix-xli; US Department of State: British Guiana. Letter dated February 26, 1948. File 844B.5544F/2–2-2648, RG 59, NACP: MD); and R. R. Kuczynski, *Demographic Survey of the British Colonial Empire: Volume 111. West Indian and American Territories* (London: Oxford University Press, 1953), 16–20, 139–190.
13. *The West Indian Census, 1946: British Guiana*, xxix-xli and J. H. Cruickshank, "Jamaica Comes to Britain," *The Glasgow Herald*, December 1954, 18, 4, TNA: COOC 1031/1569, British Guiana (WIS 121/194/01).
14. *The West Indian Census, 1946: British Guiana*, xxix-xli and *Report on the British Guiana and British Honduras Settlement Commission, 1948* (London: H. M. S. O., 1948), 1–189.
15. Michael Moohr, "The Discovery of Gold and the Development of Peasant Industries in Guyana, 1884–1914: A Study in the Political Economy of Change," *Caribbean Studies 15:2* (1975), 57–72.
16. Ibid. Moohr. See Howard University (HU): British Parliamentary Papers (BPP), 1877–1899, 256–257 and H. J. Perkins, *Notes on British Guiana and Its Gold Industry* (London: Sampson, Low, Marston & Co. Ltd., 1896), 37–39.
17. *British Guiana: Administrative Report, 1890–1900*, Table F; *Report of the Comptroller of the Customs, 1898–1899*, 11–13, tables F and G; *Report of the Blue Book, 1899–1900*, Sec. WA47–48,318; HU: BPP, 1898–1899, 11–13, tables F and G.
18. *Report of the Commissioner of Mines, 1890–1891*, 12 and Perkins, *Notes on British Guiana*, 36–40. See, *Products of British Guiana—Exports and Their Value, 1913–1923, The British Empire Exhibition, 1924*, and H. J. C Conolly, *Economic Geological Survey*... *of the Mazaruni and Puruni Diamond Fields, British Guiana, March-Dec., 1925* (London: The Crown Agents for the Colonies, 1927), 100.
19. *Report of the Commissioner of Mines, 1898–1899*, 20; Perkins, *Notes on British Guiana*, 36–39; and Conolly, *Economic Geological Survey*, 100–103.
20. *Mineral Resources in British Guiana*, 6. See Sherwood A. J. Lowe, *Historical Analysis of Mining Policy, Law, and Administration in Guyana: 1831–2002* (Georgetown: Pavnik Press, 2003), 66–68.
21. *Report of the Comptroller of Customs, 1901–1902*, Table L; *Report of the Comptroller of Customs, 1909–1910*, 9, 12, Table P. See Rodney, *A History of the Guyanese Working People*, 238.
22. *British Guiana: Report of the Gold Industry & Crown Lands* (Georgetown, 1895), 1.
23. *Report of the Gold Industry*, 1895, 1–166; *Report of the Commissioner of Lands and Mines, 1918*, 11; Moohr, "The Discovery of Gold," 65–72; and Lancaster, "An Unconquered Wilderness," 94–99.
24. TNA: COOC 111/504. Despatch 235, July 20, 1898. Enclosure: "Gold Industry Request of Petitioners."

25. Will, "Colonial Policy and Economic Development," 129–147. See, Robert Kubicek, *The Administration of Imperialism: Joseph Chamberlain at the Colonial Office* (Durham: Duke University Press, 1969), 3–11, 24, 128; Peter T. Marsh, *Joseph Chamberlain: Enterprise in Politics* (New Haven, Yale University Press, 1994), 332–358; and Niall Ferguson, *Empire: The Rise and Demise of the British World Order and the Lessons For Global Power* (New York: Basic Books, 2003), 25.
26. Ibid. Robert W. Boyle, *Gold: The History and Genesis of Deposits* (New York: Van Nostrand Reinhold, 1987), 3–6.
27. Marcello M. Veiga, "Artisanal Gold Mining Activities in Suriname." *Report. United Nations Industrial Development, December 1997,* 1–18.
28. Ibid., Veiga, 3. See Marieke Heemskerk, "Gender and Gold Mining: The Case of the Maroons of Suriname," Working Paper #269, July 2000, Women and International Development, International Center, Michigan State University, East Lansing, MI, 4–5. Maroons include the Ndjuka/Djukas, Saramakas, Alukus, Paramakas, Matawais, and Kwintis.
29. Veiga, "Artisanal Gold Mining Activities in Suriname," 4.
30. Ibid., 3.
31. *Report of the Commissioner of Mines, 1895–1896*, 24–25.
32. *British Guiana: Report of the Labour Commission, 1890* (Georgetown: C. K. Jardine, 1890), 54; *Report of the Institute of Mines, 1909–1910*, 8; *British Guiana: Report of the Balata Committee* (Georgetown: The Argosy Ltd., 1912), 15; Norman E. Cameron, *The Evolution of the Negro* (Westport: Negro Universities Press, Reprint 1929, ed., 1970, ed.), 27, 95; Douglas Smith, "The Origins of Porknocking in Guyana" in McGill Studies in Caribbean Anthropology (1969), 88; and Gamaliel Herry, Stone Mountain, Georgia, March 25, 2006, interview by author (telephone), his uncle left Grenada before the 1950s to prospect for gold in Colombia.
33. *Report of the Commissioner of Lands and Mines, 1907–1908*, 7–10.
34. *Report of the Commissioner of Lands and Mines, 1907–1908; Report of the Institute of Mines, 1909–1910, 6–7.*
35. *Report of the Institute of Mines, 1909–1910.* Appendix Viii, Licenses July 1, 1909–June 30, 1910.
36. Ibid. *British Guiana: Report of the Diamond and Gold Commission, 1928* (Georgetown: The Argosy, Ltd., 1928), 15. A privilege is the certificate required to prospect for gold and diamonds. See Veiga, "Artisanal Gold Mining Activities in Suriname;" Heemskerk. "Gender and Gold Mining: The Case of Suriname;" 13–18; Mackay, "Mining in Suriname: Multinationals, the State and the Maroon Community of Nieuw Koffiekamp:" and "Small-Scale Gold Mining in Suriname" for women miners.
37. *The Daily Chronicle*, August 15, 1930, 4.
38. Ibid.
39. Ibid. *Report of the Institute of Mines, 1909–1910.* Appendix Viii, Licenses July 1, 1909–June 30, 1910. See TNA: COOC 111/734/1. Despatch 484, December 24, 1935. Enclosure: "The Diamond Industry."
40. "Women in the News, Mrs. P. J. Willems, Manager, Willems Trading Co.," *The Daily Chronicle,* May 13, 1934, 9; "A Gold Expedition that was Abandoned, $5,000 Claim Against Boundary Commission," *The Daily Chronicle*, June 14, 1939, 7; *The Daily Argosy*, July 13, 1935, 4.
41. "Bouts in the Goldfields. Novel Contest at Tiboku Road," *The Daily Argosy, 24 December 1939*, 7.
42. *The West Indian Census, 1946. British Guiana.* Appendix xlvii, Industry, Status, Occupations, 22–29, Table O. Principal Occupations, Males and Females, 1946.

43. Odida T. Quamina, *The Making of a Working Class: The Mineworkers of Guyana* (Atlantic Highlands, NJ, 1987), 28–29, 54 and *The Social Organization of Plantation Mackenzie*, 3; and Chris Curtis, ed. *Demerara Doctor, An Early Success Against Malaria: The Autobiography of a Self-Taught Physician, George Giglioli, 1897–1975* (London, 2006), 47–108.
44. Jaclyn J. Grier and Laurie Mercier, eds., *Mining Women: Gender in the Development of a Global Industry* (New York: Palgrave Macmillan, 2006), 1–16.
45. Grier and Mercier, *Mining Women*, 6, 14. See, Kwabena O. Akurang-Parry, "Making a Difference in Colonial Interventionism in Gold Mining in Wassa Fiase, Gold Coast (Ghana) The Activism of Two Women, 1874–1893," in *Mining Women: Gender in the Development of a Global Industry* (New York, NY, 2006), 40–57; Raymond E. Dummett, "Joseph Chamberlain, Imperial Financial Railway Policy in British West Africa in the Late Nineteenth Century," in *The English Historical Review 90* (1975), 287–321; Kwabena O. Akurang-Parry, "To Wassa Fiase for Gold: Rethinking Colonial Rule, Eldorado, Antislavery, and Chieftaincy in the Gold Coast (Ghana), 1874–1895," in *History in Africa 30* (2003), 11–36.
46. James Rodway, *Handbook of British Guiana* (Georgetown, British Guiana,1893); Perkins, *Notes on British Guiana, 35–37*; *Report of the Comptroller of Customs*, 1901–1910, 7–16, 61 and Tables K, Q and 1910–1911, 9,11,31; *Report of the Commissioner of Lands and Mines*, 1901–1902, 14–15, 157; 1902–1903, 1; 1903–1904, 1,4, 20, 164; 1904–1905, 1, 1906–1907, 7,12,18, 1907–1908, 18, 1909–1910, Appendix 18; *Report of the Auditor General*, 1901–1902, 4; Conolly. E*conomic Geological Survey*, 15–20; *British Guiana Mining Industries.Exhibition Committee* (Georgetown: Commissioner of Lands and Mines, 1918), 5–8; *British Guiana Exhibition Wembley, 1924*, 75, 80; *British Guiana: Geological Survey Bulletin, 21–27*, 1948–1958; Arthur B. Cummins and Ivan A. Given, eds., *Society of Mining Engineers Handbook Volume 1* (New York: Society of Mining Engineers... Inc., 1973), 9.2–9.8, 45.
47. *Report of the Commissioner of Lands and Mines, 1909–1910*, Appendix 18 and *Handbook of Natural Resources of British Guiana* (Georgetown, British Guiana, 1946), 27–31.
48. Cummins and Given, eds., *Society of Mining Engineers*, 9.2–9.8, 45.
49. Ibid., Cummins and Given. Cyanide replaced mercury/quicksilver by the 1950s.
50. Ibid., Cummins and Given.
51. *Report of the Commissioner of Lands and Mines, 1913–1914*, 14–16, *Report of the Commissioner of Lands and Mines, 1914–1915*, 17–18.
52. *Report of the Commissioner of Lands and Mines, 1913–1914*, 14–16.
53. Ibid.
54. *Report of the Commissioner of Lands and Mines, 1903–1904*, 11, *British Guiana: Administrative Report North West District, 1903*, 5; and *The Echo*, 7 March 1903, 3, 6.
55. Ibid.
56. Ibid.
57. TNA: COOC 111/538 British Guiana, 1903. Despatch 378 "Approval to Hatton Diamond Syndicate," September 17, 1903, Governor to Colonial Office; TNA: COOC 111/538 British Guiana, 1905. Despatch 354 "B. G. Gold Concessions Co Acquired Full Rights from Hatton Garden Syndicate," December 5, 1905; and *Report of the Commissioner of Lands and Mines, 1904–1905*, 8, 12.
58. *Report of the Labour Commission, 1890* and *Report of the Commissioner of Lands and Mines, 1905–1906*, 12–13. On uneven treatment by administrators see Allan Young, *The Approaches To Local Self-Government in British Guiana* (London: Longman and

Green, 1958), 51–54; James Rose, "The Repatriation Controversy and the Beginnings of An East Indian Village System," in *Guyana Historical Journal* 1 (1989), 50–67; Eusi and Tchaiko Kwayana, *Scars of Bondage: A First Study of the Colonial Experience of Africans in Guyana* (Georgetown, 2002), 42–44; Carl B. Greenidge, *Empowering A Peasantry in a Caribbean Context: The Case of Land Settlement Schemes in Guyana, 1865–1985* (Mona, Jamaica, 2001).

59. Jacqueline Braveboy-Wagner, *The Venezuela-Guyana Border Dispute: British Colonial Legacy in Latin America* (Boulder, CO: Westview Press, 1984) and D. Graham Burnett, *Masters of All They Surveyed: Exploration, Geography, and A British El Dorado* (Chicago, 2000), 199–265.
60. *Report of the Commissioner of Lands and Mines, 1905–1906*, 12–13.
61. Ibid., and *Report of the Commissioner of Lands and Mines, 1906–1907*, 9, 12.
62. *Report of the Commissioner of Lands and Mines, 1911–1912*, 12–13; *Report of the Commissioner of Lands and Mines, 1912–1913*, 13; *Report of the Commissioner of Lands and Mines, 1914–1915*, 16–18; *Report of Commissioner of Lands and Mines, April-December, 1915*, 12–13; *Report of the Commissioner of Lands and Mines, 1916*, 12–13; *Report of the Commissioner of Lands and Mines, 1917*, 12, 18, Appendix 1; *Report of the Commissioner of Lands and Mines, 1918*, 16; *Report of the Comptroller of the Customs, 1910–1911*, Table L; *Report of the Comptroller, 1918–1919*, XIII, *Report of the Comptroller, 1924*, 14 and Table J, *British Guiana Mining Industries, 1918*, 11–13; *British Guiana Exhibition Wembley, 1924*, 81, 83; Conolly, *Economic Geological Survey*, 101–102. Diamonds increased from $10 per carat in 1914–1915, to $27.60 per carat in 1919; decreased to $25.80 in 1920 and to $15.40 by 1921.
63. *Report of the Commissioner of Lands and Mines, 1917*, 12, 18, Appendix 1; *Report of the Commissioner of Lands and Mines, 1918*, 16.
64. Maurice St Pierre, 'Industrial Unrest in Mackenzie, Guyana," *McGill Studies in Caribbean Anthropology* (1969), 65–80 and "The Sociology of Decolonization: The Case of Guyana Bauxite," *New World Quarterly 5: 4* (1972), 50–62; Cedric Grant, *Company Towns in the Caribbean: A Preliminary Analysis of Christianburg-Wismar-Mckenzie* (Georgetown: Government Printers, 1971), 1–6; Mohammed Shahabudeen, *Nationalisation of Guyana's Bauxite: The Case of Alcan* (Georgetown, Ministry of Information, 1981), 125–139; and Quamina, *The Mineworkers of Guiana*, 13–21, 99–100.
65. Spackman, "The Role of Private Companies in the Politics of Empire, 341–378 and Lanning and Mueller, *Africa Undermined*, 19–60.
66. *Report of the Commissioner Lands and Mines, 1914–1915*, 17–18; TNA: COOC 111/652 British Guiana, 1924.Secret Despatch dated April 11, 1924, "Enquiry. DeBeers Consolidated Mines"; Spackman, "The Role of Private Companies," 341–378; Lanning and Mueller, *Africa Undermined*, 19–59.
67. TNA: COOC 111/652 British Guiana, 1924. Secret Despatch; Spackman, "The Role of Private Companies," 341–378; and Lanning and Mueller, *Africa Undermined*, 19–59.
68. Spackman, "The Role of Private Companies," 341–378.
69. Spackman, "The Role of Private Companies," 341–363 and Lanning and Mueller, *Africa Undermined*, 19–60. See *Michael Weinstein, The World of Jewel Stones* (New York: Sheridan House, Inc., 1958), 34–36, 39; and Sherwood A. J. Lowe, *Historical Analysis of Mining Policy, Law & Administration in Guyana: 1831–2002* (Georgetown: Pavnik Press, 2003), 88–93.
70. *Report of the Lands and Mines, 1927*, 12–17; Smith, "The Origins of Porknocking," 81–92; Spackman, "The Role of Private Companies in the Politics of Empire," 341–378.

71. Spackman, "The Role of Private Companies," 341–363.
72. Hon. Joseph Eleazar, evidence, *The Royal Commission Enquiry, 1939.*
73. *The Daily Chronicle*, April 11, 1930, 5. See Alan Cowell, "A New Generation of DeBeers: Diamonds and It Seems the Oppenheimers Are Forever," *The New York Times* April 25, 2003, W1, 7; Stephen Labaton, "DeBeers Agrees to Guilty Plea To Re-enter the U.S. Market;" *The New York Times*, July 10, 2004, C1, 3; Margaret Webb Pressler, "DeBeers Pleads to Price-Fixing, Firm Pays $10 million, Can Fully Re-enter U.S.," *The Washington Post*, July 14, 2004, E1–2.
74. *Report of the Lands and Mines, 1934*, 5–6. Petition dated February 14, 1934.
75. Ibid., and Shahabudeen, *Nationalisation of Guyana's Bauxite*, 90–102, 128–139; Carlos Lamur, *The American Takeover: Industrial Emergence and ALCOA's Expansion in Guyana and Suriname, 1914–1921* (Dordrecht, Amsterdam, 1985), 58–73; and Steven Holloway, *Aluminum Multinationals and the Bauxite Cartel* (New York, NY, 1988), 40–41. See Spackman, "The Role of Private Companies;" and Lanning and Mueller, *Africa Undermined*, 19–60.
76. "Pork Knockers Protests, *The Daily Chronicle*, January 20, 1937, 1; "Miners' Protest Against 460,000 Acres Concession," *The Daily Chronicle*, January 24, 1937, 24; "B. G. Consolidated Goldfields Will Encourage Pork-Knockers. Sir Alfred Sherlock (Chairman of Bookers Bros. McConnell) Explains" *The Daily Chronicle*, April 5 1937, 1; and "Bookers Puruni Concessions to be Reduced," *The Daily Argosy*, April 4, 1937, 4.
77. *The Daily Argosy*, April 28–29, 1934, 6.
78. Smith, "The Origins of Porknocking in Guyana," 81–92.
79. *Report of the Census*, 1921, 1931; *The West Indian Census*, 1946; Quamina, *The Mineworkers of Guyana*, 13–16; Grant, *Company Towns in the Caribbean*, 1–84; Walter A. Rodney, "Barbadian Immigration into British Guiana, 1863–1924," Paper presented, *9th Annual Conference of Caribbean Historians*, Cave Hill, University of the West Indies, 1977; Wilfred David, *The Economic Development of Guyana, 1953–1964* (Oxford, 1984), 187–222; Shahabudeen, *Nationalisation of Guyana's Bauxite*, 64–145; Fitzroy A. Baptiste, "The Exploitation of Caribbean Bauxite and Petroleum, 1914–1945," in *Social and Economic Studies 37: 1&2* (1988), 107–127.
80. *Report of the Commissioner of Lands and Mines, 1918*, 19. See appendices of *Reports of the Commissioner of Lands and Mines, 1922–1937*, for number of workers in gold, diamonds, and bauxite.
81. *Report on the Lands and Mines, 1913–1914*, 15; *Report of the Commissioner of Lands and Mines, 1914–1915*, 19, 26, *Ordinance #20 of 1914; Report of the Commissioner of Lands and Mines, 1915 (April-December)*, 15; *Institute of Mines*, 1913–1914, 8, 11; *Administrative Report, 1917*, 242.
82. TNA: COOC 111/608 British Guiana, 1917. Despatch 28, 29, December 29, 1916, February 17, 1917, "Bauxite Shipments to USA" "Export of Bauxite to USA; TNA: COOC 111/612 British Guiana, 1917. Despatch 213 June 8, 1917, "Application from Demerara Bauxite Co. for leases of Crown Lands;" *Report of the Commissioner of Lands and Mines, 1921*, 15; *Report of the Commissioner of Lands and Mines, 1922*, 3, 18; *Institute of Mines,1920–1921*, 4; *Institute of Mines, 1922–1923*, 4; *British Empire Exhibition, 1924*, 81; Shahabudeen, *Nationalisation of Guyana Bauxite*, 101.
83. David, *The Economic Development of Guyana*, 205–208 and Quamina, *The Mineworkers of Guyana*, 18–19.
84. Ibid. Quamina.
85. *Report of the Commissioner of Lands and Mines, 1913–1914*, 20–25; *Institute of Mines, 1913–1914*, 11; and Quamina, *the Mineworkers of Guyana*, 16.

86. Stockley, *The Geology, Bulletin* 25, 22, 30; D. W. Bishop, *The Bauxite Resources of British Guiana and Their Development, Bulletin* 26, 1–14, Shahabudeen, *Nationalisation of Guyana's Bauxite*, 64–145; Quamina, *The Mineworkers of Guyana*, 16–19; Holloway, *The Aluminum Multinationals*, 39–53.
87. *Report of Commissioner of Lands and Mines, 1918*, 19.
88. United States: Department of State. Letter, Reynolds Metals Company, Richmond, VA to Hon. James F. Byrnes, Secretary of State, Washington, DC, June 20, 1946, and Department of State. Foreign Service Despatch, October 21, 1954. Decimal File (1910–1963) RG 59, 844B.6354/8–1249, 841D.2569/10–2154 (NACP: MD).
89. Ibid. Shahabudeen, *Nationalisation of Guyana's Bauxite*, 97–100 and Holloway, *The Aluminum Multinationals*, 17.
90. Ibid., Shahabudeen, 92–96.
91. Lamur, *The American Takeover*, 1–5, 135–143.
92. Ibid. See Baptiste, "The Exploitation of Caribbean Bauxite," 107–127.
93. Lamur, *The American Takeover*, 1–5, 135–143.d. Lamur.
94. Quamina, *The Mineworkers of Guyana*, 18–19.
95. *Report of the Labour Commission, 1890; British Guiana: Report of the Evidence... the Banana and Fruit Industry Can Best Be Encouraged* (Georgetown, British Guiana, 1892); *British Guiana: Report of the Gold Industry and Crown Lands, 1894–1895* (Georgetown, British Guiana, 1895); *British Guiana: Report of the Balata Committee* (Georgetown, British Guiana, 1912); *Report of the Diamond and Gold Industries, 1928*.
96. *Report of the Commissioner of Lands and Mines, 1917*, 16–19, Appendix 1; *Report of the Commissioner of Lands and Mines, 1918*, Appendix L; *Report of Commissioner of Lands and Mines, 1919*, 18, Table J; *Report of the Commissioner of Lands and Mines, 1920*, Appendix 1; *Report of the Comptroller, 1919*, 7 and Table XIII; and *Report of the Comptroller, 1920*, 6. See Barbara P. Josiah, "The Impact of Gold, Diamonds, and Bauxite Production on the Guyanese Economy," in *Global Development Studies*, 1–2 (1999–2000), 277–296.
97. *Report of the Commissioner of Lands and Mines, 1920*, Appendix 1; *Institute of Mines, 1919–1920*, 1,4; *Report of the Comptroller, 1921*, 5; *Report of the Comptroller, 1922*, 5–7; *Report of the Comptroller, 1923*, 5–7; *Report of the Comptroller, 1924*, 3–5; and *Report of the Comptroller, 1925*, 3–5.
98. *Report of the Commissioner of Lands and Mines, 1930*, 11, 13; *Report of the Commissioner of Lands and Mines, 1934*, 8; *Report of the Commissioner of Lands and Mines, 1938*, 10–12; "$20,000 For Ten Mining Expeditions... Mines Association To Seek Government's Assistance," *The Daily Argosy*, May 20, 1938, 5; "Our Overtaxed Diamond Industry," *The Daily Chronicle*, May 2, 1939, 6; *Report of the Commissioner of Lands and Mines, 1939*, 5–6; *The Daily Argosy*, January 19, 1933, and May 31, 5; June 9, 5; and Lowe, *Historical Analysis of Mining*, 116.
99. "Increased Taxation on Gold... Bauxite... Additional Revenue of $101,000 Expected," *The Daily Chronicle*, November 22, 1939,6; *Report of the Commissioner of Lands and Mines, 1940*, 6–8; *Report of the Commissioner of Lands and Mines,1944*, 7–9; United States: Department of State. Despatch: American Consulate, Georgetown to The Secretary of State, Washington, DC January 6, 1945, "The Diamond Industry in British Guiana." RG 59 Decimal File 844B.6347/-645, (NACP: MD); *Colonial Annual Report, 1946*, 3– 4, 7, 10, 70–71; *Colonial Annual Report, 1948*, 9,17–18,38, 51–52; *British Guiana: Geological Survey Bulletin, 21–27*, 1948–1958; and Veiga, "Artisanal Gold Mining Activities," 3.

100. United States: Department of State. Despatch: American Consul Georgetown to The Department of State, Washington, DC March 3, 1950 "Mineral Trade and Production" and "State of Affairs in Bauxite, Gold, and Diamonds" RG 59 Decimal File 841D 25/3–350 (NACP: MD); *Handbook of Natural Resources*, 23; *Colonial Annual Report, 1948*, 48, 51; *Colonial Annual Report, 1949*, 9, 24, 54, 58–60. Roberta Walker Kilkenny, "And So They've Captured Demerara." A Preliminary Investigation of the United States-Guyana Relations During the Second World War," *Guyana Historical Journal 11* (1990), 68–78. Sydney Dundas, Howell Wood, Winston Valentine, interviews with author, East Coast Demerara, Guyana, 1996.
101. G. M. Stockley, *The Geology of British Guiana and the Development of Its Mineral Resources, Bulletin No. 25* (Georgetown: Geological Survey Department, 1955), 1–20.
102. *British Guiana: Annual Report of the Department of Labour, 1942–1950*, Table I.
103. Quamina, *The Mineworkers of Guyana*, 7.
104. *Colonial Annual Report, 1950*, 3, 23; *Colonial Annual Report, 1954*, Appendix B; *Colonial Annual Report, 1956*, 10–11; and *Report of the Commissioner of Lands and Mines, 1955*, 9. See Bishop, *The Bauxite Resources*, 1–11; *The Economic Development of British Guiana: Report... by the International Bank of Reconstruction and Development* (Baltimore, MD, 1953), 7–8; Quamina, *The Mineworkers of Guyana*, 7; and Grant, *Company Towns*, 4.
105. Stockley, *The Geology of British Guiana*, 2; C. N. Barron, *The Bauxite Resources of British Guiana and Their Development. Bulletin#26* (Georgetown, British Guiana, 1955), 1–14; and *Colonial Annual Report, 1956*, 103–105.
106. *Colonial Annual Report, 1950*, 3, 25, 57–58, Appendix C; *Report of the Commissioner of Lands and Mines, 1956*, 7, Appendix I, Table 8. See Grant, *Company Towns*, 17.
107. *Colonial Annual Report, 1955*, 22, 26; *Report of the Commissioner of Lands and Mines, 1955*, 9; Stockley, *The Geology*, 27; and Stockley, "Mineral Development in the Economy of Guyana." On funds from the Colonial Development Corp. and the US Marshall Plan, United States: Department of State. American Consul, Georgetown. Despatch to the Department of State, Washington, DC Decimal File 841D 25/1–2852, 841D25/1–3151, RG 59, (NACP: MD).
108. *Colonial Annual Report, 1955*, 22, 26; *Report of the Commissioner of Lands and Mines, 1955*, 9; G. M. Stockley, *Mineral Development in the Economy of British Guiana: Report on 1st. Caribbean Geological Conference, Antigua, December 1955* (Georgetown: The Argosy Ltd., 1955), 49–54; and *Report of the Commissioner of Lands and Mines, 1957*, Appendix VIII. 4. Linton "Jack" Adams, interview by author, Den Amstel Village, West Coast Demerara, December 27, 1996.
109. *Report of the Commissioner of Lands and Mines, 1955*, 9, 49 and *Colonial Annual Report, 1955*, 9, 22, 26.
110. *Colonial Annual Report, 1956*, 4 and *Report of the Commissioner of Lands and Mines, 1956*, 7. See Appendix VII.

4 The Perils of Labor in Mining: Migration and Mortality

1. *British Guiana: Register of Deaths, Essequibo* (Districts 3, 4, 16, 17, 1891–1950s); *The Daily Chronicle*, 1890s–1900s; *The Gold Mining Gazette*, 1890s; *Report of the Gold and Forestry Industries: Institute of Mines and Forests of British Guiana, 1908–22* (Report of the Institute of Mines); *British Guiana: Reports of the Department of*

Lands and Mines, 1900s; and *The Daily Argosy*, 1920s–1956, for injuries and deaths. See V. T. Harlow, ed., *The Discovery of the large and beautiful Empire of Guiana by Sir Walter Releigh* (London: The Argonaut Press, 1928, ed.), 1; "Itanami," in *Guiana Sings* (Delaware, Ohio: Informal Music Service, n.d.), 3. Guyana's folk songs in the publication were recorded by Vesta Lowe.

2. *British Guiana: Report of the Crown Surveyor, 1882* (Georgetown: Royal Gazette, 1883), 4–5; *British Guiana: Report of the Commissioner of Mines, 1895–1896*, 2–3; International Monetary Fund. *IMF Staff Country Report, Guyana: Statistical Annex* (Washington, DC: International Monetary Fund), 1997–2010.
3. *Register of Deaths* and *Reports of the Institute of Mines* on origins of miners.
4. Ibid. James Culley, interview by author, Buxton Village, East Coast Demerara, October 24, 1996 and Benjamin Thomas, interview by author, Buxton Village, East Coast Demerara, November 3, 1996.
5. Octavius Barnwell, interview by author, Mocha Village, East Coast Demerara, October 17, 1996.
6. Culley and Thomas, interviews.
7. Ibid.; Joseph Watson, interview by author, Washington, DC, February 5, 1995. See J. M. Reid, *British Guiana Handbook* (Georgetown: The Argosy Printers, 1920), 54–55.
8. Culley, Thomas, Watson, interviews and Howell A. Wood, interview by author, Buxton-Friendship Village, East Coast Demerara, Guyana, October 15, 1996.
9. Ibid.; Watson; Culley and Thomas, and Wood, interviews.
10. *British Guiana: Report of the Labour Commission, 1890* (Georgetown: C. K. Jardine, 1890), 3.
11. Ibid. Howard University (HU): British Parliamentary Papers (BPP) *The West India Royal Commission Report 1897.* Appendix C-Part 11. British Guiana, Section 142, 692. Evidence: Dr. J. M. Rohlehr, MD, CM. on interior deaths and mining disasters in Essequibo County.
12. See *Report of the Labour Commission, 1890* and Response to Circular of July 2, 1890; *Report of the Commission of Mines*, 1884–1902; *Report of the Commissioner of Lands and Mines,* 1903–1950s for views on migration and development. *The Echo,* published by an Afro-Barbadian, often aired opinions on the neglect interior workers.
13. *Register of Deaths, Essequibo, 1891–1895.*
14. *The Echo, The Daily Argosy* and other newspapers for accounts of mishaps. See Alleyne Leechman, *Handbook of British Guiana, 1913* (Georgetown: The Argosy Co. Ltd., n.d.), 203–210.
15. *Gazetteer of Guyana* (Georgetown: Government of Guyana, 1974), 8. *Register of Deaths, Essequibo, 1891–1895.*
16. *Register of Deaths, Essequibo, 1891.* See entries #s 106–110.
17. Ibid., 211–222. Zambia died December 21, 1891.
18. *The Gold Mining Gazette,* October–December 1892, 25.
19. *Register of Deaths, Essequibo, 1893*, Districts 3, 4, 16, 17.
20. *Register of Deaths, Essequibo, 1894*, Districts 3, 4, 16, 17.
21. *Register of Deaths, Essequibo, 1895*, Districts 3, 4, 16, 17. See also *British Guiana: Report of the Registrar General Office*, June 1896.
22. "Dr. A. S. Ozzard," *The Echo*, March 5, 1896, 2.
23. *British Guiana: Report of the Government Agent North West District, 1892–93*, 10; *Report of the Government Agent, 1896–1897*, 9; *British Guiana: Report of the Colonial Civil Engineer*, July 10, 1896, 3–5; *Register of Deaths, Essequibo, 1896–1900; Report of the Commission of Mines, 1896–1901; The Echo*, February 14, 1899, 4.

24. TNA: COOC 111/495. Despatch 208 dated July 20 1897. Enclosure: Dr. J. M. Rohlehr, MD, CM on Government Medical Services and TNA: COOR 111/550. Despatch 107 "Administration of the Colony 'Monster Petition' from Inhabitants Signed by 5,750 Persons on the Neglect of the Recommendations of the Royal Commission" included additional signatures.
25. "Overcrowding at Bartica Hospital," *The Daily Chronicle*, September 21, 1937, 7.
26. *The Gold Mining Gazette*, July 1, 1897, 3 and *The Gold Mining Gazette*, June 30, 1890, 6.
27. *The Mining Gazette*, January 21, 1897 and July 1, 1897, 3.
28. *Register of Deaths, Essequibo, 1899–1900*; and *Report, Commissioner of Mine, 1900–1901*, 5.
29. *Register of Deaths, Essequibo, 1899–1900.*
30. Ibid., and "Miner Bitten by Bushmaster Dies in the Mazaruni," *The Daily Chronicle*, June 10, 1939, 9. See "Boat Accident on the Cuyuni River. Eight Men in Desperate Plight. One Dies on Way at Bartica," *The Daily Argosy*, February 24, 1938, 4; "Pork-knocker Killed, 3 Injured by Falling Tree at Moorabisci Creek, Issano, Mazaruni River," *The Daily Chronicle*, June 3, 1938,1; and "Man Killed In Plunge Down Gold Mine Shaft Accident at Baramani. Others in Hospital," *The Daily Chronicle*, June 10, 1939, 1.
31. Culley and Thomas, interview.
32. *The Gold Mining Gazette*, January 21, 1897, 3–6.
33. *Register of Deaths, Essequibo*, 1900; "Boat Accident in the Mazaruni. Five Lives Lost," *The Daily Argosy*, August 22, 1922, 5; "Frightful Holiday Disaster…No Bodies Found," *The Daily Argosy*, December 24, 1935, 4; J. A. Eytle, a well-known miner of Brickdam, Georgetown owned the vessel; "5 Drowned in Mazaruni Boat Disaster. Boat Sinks at Mora Rapids," *The Daily Chronicle*, May 18, 1937, 1.
34. *The Daily Argosy*, May 25, 1956, 6.
35. *Royal Commission, 1897*, Appendix C. Sec. 162, 682.
36. TNA: COOC 111/539. Despatch dated October 21, 1903. Enclosure: "D. F King, Permission to practice medicine in British Guiana" King's brother drowned March 1902. See Thomas J. Ward, *Black Physicians in the Jim Crow South* (Fayetteville: University of Arkansas Press, 2003) and Margaret Hope Bacon, *But One Race: The Life of Robert Purvis* (Albany: State University of New York Press, 2007) for discrimination against the professionals in the United States.
37. *Report of the Institute of Mines, 1913–1914*, Appendix XIV, *Deaths Occurring in the Interior from 1 January–31 December 1913* for death entry of Alexander Wright. Evelyn A. Wright Toussaint, interview by author, Mocha Village, East Bank Demerara, October 17, 1996.
38. Wright-Toussaint, interview, siblings were Princess, Mildred, and Norah.
39. "Sequel to Mazaruni Boat Accident: Court asked to Presume Man Dead…" *The Daily Argosy*, May 23, 1933, 3 and *Register of Deaths, Essequibo*, 1933, #s 6–12. The others were Solomon Sam, Charles Rutherford, Clement Bobb, John Cummings, George Hart, and Wilfred Legall.
40. Ibid.
41. Norman E. Cameron, *The Evolution of the Negro*, (Westport: Negro University Press, 1970, ed. Reprint 1929, ed.) and Kean Gibson, *The Cycle of Racial Oppression in Guyana* (Lanham: University Press of America, 2003). See Walter Rodney, *A History of the Guyanese Working People, 1881–1905* (Baltimore: Johns Hopkins University Press, 1982); Eusi Kwayana and Tchaiko Kwayana, *Scars of Bondage: A First Study of the Colonial Experience of Africans in Guyana* (Georgetown: Free Press, 2002).

42. Cameron, *The Evolution of the Negro,* 30, on farmers/miners.
43. H. Volpe, "Capitalism and Cheap Labour—Power in South Africa From Segregation to Apartheid," *Economy and Society, 1*; F. Wilson, *Labour in South African Gold Mines, 1911–1969* (Cambridge: Cambridge University Press, 1972), 1–250; A. Jeeves, *Migrant Labour in South Africa's Mining Economy: The Struggle for the Gold Mines' Labour Supply, 1890–1920* (Montreal: McGill-Queens Universities Press, 1985), 1–119; R. M. Packard, *White Plague, Black Labor: Tuberculosis and the Political Economy of Health and Disease in South Africa* (Pietermarizburg: University of Natal Press, 1989), 67–125.
44. Barnwell, interview. See "Caribbean Roundup: Guyana," *Caribbean Life*, December 28, 2004, 3.
45. John Jones, "G. D. F. Herb Doctor Saves Men Hit By Bush Yaws," *The Sunday Chronicle*, February 9, 1979, 9.
46. Barnwell, interview.
47. *Gazetteer of Guyana*, 18.
48. Walter Rodney, *Guyanese Sugar Plantations in the Late Nineteenth Century: A Contemporary Description from the Argosy* (Georgetown: Release Publishers, 1979), 7.
49. Barnwell, interview and *Gazetteer of Guyana*, 57.
50. *Report, Commissioner of Mines*, 1896–1897, 5; *The Echo*, September 19, 1896, 2; and *The Gold Mining Gazette*, September 1896, 24, 3.
51. Royal *Commission, 1897; Report of the Commissioner of Mines, 1896–1897* for registration and licensing. Independent self-financed claim owners included Albert Barrington Gillis, Messrs. Poole, Elias, Fraser, Goring, and A. B. Brown.
52. *The Sunday Chronicle*, March 23, 1930, 4.
53. *The Daily Chronicle*, December 13, 1930, 8.
54. "The Lure of Kurupung.Stranded Diamond Seeker Robbed of His All Hangs Himself in Desperation" *The Daily Chronicle*, February 11, 1922, 2.

5 Aspects of Infrastructure Development: Gold and Diamonds

1. *British Guiana: Report of the Crown Surveyor, 1882* (Georgetown: Royal Gazette Office, 1883), 4–5; H. A. Will, "Colonial Policy and Economic Development in the British West Indies, 1895–1903," *The Economic History Review 23:1* (April 1970), 129–147; Ann Spackman, "The Role of Private Companies in the Politics of Empire: A Case Study of Bauxite and Diamond Companies in Guyana in the Early 1920s," *Social and Economic Studies Vol. 24 #3* (1975), 341–378; Alan Lancaster, "Proposals for Hinterland Settlement and Development of British Guiana, 1884–1900," *History Gazette 59* (1993), 2–24.
2. *British Guiana: Report of the Gold Industry and Crown Lands, 1895* (Georgetown: C. K. Jardine, 1895), 1–5; *British Guiana: Report of the Commissioner of Mines, 1895–96* (Georgetown: C. K. Jardine, 1896), 3–4; and Lancaster, "Proposals for Hinterland Settlement, 8–11, 16–17.
3. TNA: COOC 111/439 British Guiana. Despatches 230–299, May–July 1887; *Report of the Gold Industry, 1895*. Evidence, 1–162; and *British Guiana: Report of the Diamond and Gold Industries, 1928* (Georgetown: The Argosy Co., 1928) 9–34.
4. "Government Accused of Neglecting the Hinterland. Legislator Reveals Facts of Memorandum For Royal Commission. Hon. Jos. Eleazar Address," *The Daily Chronicle* January 28, 1939, 9.

5. Ibid.
6. Ibid.
7. TNA: COOC 111/190 British Guiana. Despatch 98, March–May, 1842. Enclosure: "Administration of the Colony. The African Association of British Guiana"; TNA: COOC 111/553 British Guiana. Despatch 107, 1906. Enclosure: "Administration of the Colony 'Monster Petition' Signed by 5,750 Persons on the Neglect of Recommendations of the Royal Commission, 1897"; TNA: COOC 111/554 British Guiana, 1906. Enclosure: "The People's Association of British Guiana," petitioners now 6,115 persons.
8. Spackman, "The Role of Private Companies in the Politics of Empire," 341–378; Douglas Smith, "The Origins of Porknocking in Guyana," *McGill Studies in Caribbean Anthropology* (1969), 81–92; Alan Lancaster, "Unconquered Wilderness: A Historical Analysis of the Failure to Open Up the Interior of British Guiana, 1838–1919" M. A. Thesis, University of Guyana, 1979; George Lanning and Marti Mueller, *Africa Undermined: A History of Mining Companies and the Underdevelopment of Africa* (New York: Penguin Books, 1979), 19–60.
9. W. W. Kenrick, Gold Commissioner, "Opening Up the Country," *Timehri new series 5* (1891) 237–248; Stafford Combes, "Interior Communications For British Guiana," *Timehri new series 5* (1891), 12–21; H. J. Perkins, *Notes on British Guiana and Its Gold Industry* (London: Sampson Low, Marston, and Company, 1896), 37.
10. I. R. Phimister, "African Labour Conditions and Health in the Southern Rhodesian Mining Industry, 1898–1953," in I. R. Phimister and C. van Onselen, *Studies in the History of African Mine Labour in Colonial Zimbabwe* (Salisbury: Mambo Press, 1978), 102–149. See C. W. Newbury, *The Diamond Rings, Business, Politics and Precious Stones in South Africa, 1867–1947* (Oxford: Clarendon Press, 1989).
11. Will, "Colonial Policy and Economic Development, 129–147; Robert Kubicek, *The Administration of Imperialism: Joseph Chamberlain at the Colonial Office* (Durham: Duke University. Press, 1969), 3–11, 24, 128. See Peter T. Marsh, *Joseph Chamberlain: Enterprise in Politics* (New Haven: Yale Univ. Press, 1994), 332–581.
12. *British Guiana: Report of the Labour Commission, 1890* (Georgetown: C. K. Jardine, 1890), i–xix, 1–66. *Legislative Ordinance #23 of 1890*, for non-Asian migrants. Howard University HU: British Parliamentary Papers (BPP), *Blue Book for 1888*, 56 and 1891, 31; *British Guiana: Report of Evidence... Before the Commission... to Consider... the Banana and Fruit Industry...* (Georgetown: C. K. Jardine, 1892); *Report of the Gold Industry, 1890,* 1–5 on economic consequence of labor shortages.
13. *Report of the Labour Commission, 1890; West India Royal Commission Report, 1897. Appendix C Part II, British Guiana*. See Norman E. Cameron, *The Evolution of the Negro* (Westport: Negro University Press, 1970, ed.), 6–7; Will, "Colonial Policy and Economic Development," 129–147; Walter Rodney, *A History of the Guyanese Working People, 1881–1905* (Baltimore: Johns Hopkins University Press, 1981), 16–18, 32, 46–47; James G. Rose, "The Strikes of 1842 and 1848," in Winston F. McGowan, James G. Rose, and David A. Granger, eds., *Themes in African-Guyanese History* (Georgetown: Free Press, 1998), 158–200; Maurice St. Pierre, *Anatomy of Resistance: AntiColonialism in Guyana, 1823–1966* (London: Macmillan Education Ltd., 1999), 9–34; Kean Gibson, *The Cycle of Racial Oppression in Guyana* (Lanham: University Press of America, 2003), 7–21.
14. *Report of the Gold Industry, 1895*, 1–162 and *Report of the Commissioner of Lands and Mines, 1901.* Ordinance 25, 1902 merged the two departments.
15. J. F. Connolly, "The Riches of the Colony," *Mining Journal*, 1894 and Alvin Thompson, "Gold Mining in British Guiana From Earliest Times to 1939," Unpublished Paper, 1965. See Smith, "The Origins of Porknocking in Guyana," 84, 88.

16. *Report of the Gold Industry, 1895*; Jungen Ostherhamnel, *Colonialism: A Theoretical Overview*, 1995, ed. Translated from the German (Princeton: Markus Weiner, 1997), 60. See Ruth Simms Hamilton, ed., *Routes of Passages: Rethinking the African Diaspora Vol. 1 Part 1* (East Lansing: Michigan State University Press, 2007), 1–33.
17. Will, "Colonial Policy and Economic Development,"137.
18. H. W. Brands, *The Age of Gold: The California Gold Rush and the New American Dream* (New York: Random House/Doubleday Inc., 2002), 334–358, 421, 361–491.
19. Ibid.; H. W. Brands, *The Age of Gold* and *Panamanian Passages/Pasajes Panamenos Exhibition*, Concourse, S. Dillon Ripley Center, Smithsonian Museum, Latino Center, Washington DC, October 14, 2009–May 31, 2010. http://www.latino.si.edu.
20. Sherwood A. J. Lowe, *Historical Analysis of Mining Policy, and Administration in Guyana: 1831–2002* (Georgetown: Pavnik Press, 2003), 100 for cyanide use by 1940s and Raymond E. Dummet, *Eldorado in West Africa: The Gold-Mining Frontier, African Labor, and Colonial Capitalism in the Gold Coast, 1875–1900* (Athens: Ohio University Press), 138, 212, 237, 289, 353 for cyanide use in colonial Ghana.
21. Rodney, *A History of the Guyanese Working People*, 84 on Timinis or Temne, African ethnics, from the "rice belt of West Africa," who grew rice in colonial Guyana.
22. William Russell, "Rice," *Timehri* Vol.5 (1886); A. R. Gilzean, "Rice Cultivation in British Guiana," *Timehri* new series 1(1887); *Royal Commission Report, 1897*, 688–689. Memo dated February 4, 1897, on agricultural industries and government subsidies for the rice industry; Jay Mandle, "Population and Economic Change: The Emergence of the Rice Industry in Guyana, 1895–1915," *Journal of Economic History 30:4* (1970), 158–200; Harry Madramootoo, *The Story of Rice in Guyana* (Georgetown: Guyana Rice Corp., 1973), 1–25; and Rodney, *A History of the Guyanese Working People*, 225, 238.
23. *Report of the Commissioner of Mines, 1895–1896*, 4, on small-scale miners; Michael Moohr, "Patterns of Change in an Export Economy: British Guiana, 1838–1914," Ph.D. Cambridge University, 1971, 260–264 and Moohr, "The Discovery of Gold and the Development of Peasant Industries in Guyana, 1884–1914: A Study in the Political Economy of Change," *Caribbean Studies 15* (1975), 57–72.
24. Arthur de Gobineau, *The Inequality of Human Races, 1853–1855*. Translated by Adrian Collins (New York: G. P. Putnam's Sons, 1915, ed.); Madison Grant, *The Passing of the Great Race; or the Racial Basis of European History* (New York: Charles Scribner, 1916); Jan Nederveen Pieterse, *White on Black: Images of Africa and Blacks in Western Popular Culture* (New Haven: Yale University Press, 1992), 20, 22; Simms Hamilton, *Routes of Passages*, 27; Gregory Michael Dorr, *Segregations' Science: Eugenics and Society in Virginia* (Charlottesville: University of Virginia Press, 2008); and Jonathan Peter Spiro, *Defending the Master Race: Conservation, Eugenics, and the Legacy of Madison Grant* (Burlington: University of Vermont Press, 2009), 2–140 for scientific racism globally and expendable non-Nordics.
25. Deaths recorded in the *Reports of the Institute of Mines*; *Reports of the Commissioner of Lands and Mines*; *Death Registers*; and newspapers gave conflicting information on the number of casualties.
26. HU: BPP British Guiana 28, 1890, 6, 21–22; HU: BPP British Guiana 30, 1890, 317, 328, HU: BPP British Guiana 31, 1891, 16; *Blue Book Annual Report, 1892–1893*, 16–20; and Lancaster, "Unconquered Wilderness," 125–126.
27. Kenrick, Gold Commissioner, "Opening Up the Country," 237–248; Combes, "Interior Communications For British Guiana," 12–21. See *Colonial Annual Reports*

for British Guiana, 1940s–1950s and US Department of State, Foreign Service Despatch, on mining and related developments in British Guiana. RG 59. Decimal Files 741D, 841D, 844 B, 1949–1957 (NACP: MD).

28. HU: BPP British Guiana 30, *1890*, 34; *British Guiana: Report of the Colonial Civil Engineer*, 10 July 1896, 3–5; *Report of the Commissioner of Mines, 1895–1896*, 4, 18; *Report of the Commissioner of Mines, 1896–1897*, 5–6; *Report of the Commissioner of Mines, 1898–1899*, 4–5. See Thompson, "Gold-Mining in British Guiana," 22–23, 27 and Lancaster, "Unconquered Wilderness," 124–128.
29. *British Guiana: Report of the Commissioner of Lands and Mines, 1901–1902*, 9; *Report of the Commissioner of Lands and Mines, 1902–1903*, 14; *Report of the Commissioner of Lands and Mines, 1910–1911*, 6; *Report of the Commissioner of Lands and Mines, 1920*, 13; *British Guiana: Report of the Director for Public Works, 1920*, 8–10; *British Guiana: Report of the Commissioner of the North West District, 1920*, 3–4; *Report of the Institute of Mines, 1920/21*, 5–6; and *British Guiana Combined Court, First Special Session, 1921* (Combined Court, C. S. O. 2812/21).
30. *Report of the Commissioner of Lands and Mines, 1922*, 13–14 and *British Guiana: Report of the British Guiana Commission, April 1927* (London: H. M. Stationery Office, 1927), 32–39.
31. *Report of the Director of Public Works, 1933*, 3; *Report of the Director of Public Works, 1934*, 34; *The Daily Argosy*, January 19 and 25, 1933; and November 2 and 8, 1933; *Report of the Department of Lands and Mines, 1937*, 13; and *Report of the Department of Lands and Mines, 1939*, 4–5. Lowe, *Historical Analysis of Mining Policy*, 87–88 on Lancaster, "Unconquered Wilderness," 124–128.
32. Lowe, *Historical Analysis of Mining Policy*, 101–106; Will, "Colonial Policy and Economic Development," 129–147.
33. *Colonial Annual Reports for British Guiana, 1946–1956*, on infrastructure development plans; "The Mining Industry: Better Arrangements For Bringing Men Down," *The Daily Argosy* January 5, 1936, 8; "Hinterland Travel Difficulties: Hearing of Case Adjourned," *The Daily Argosy*, December 4, 1937, 7, 9. See, Lowe, *Historical Analysis of Mining Policy*, 101–106.
34. *Colonial Annual Reports for British Guiana, 1946–1956* and Lowe, *Historical Analysis of Mining Policy*, 101–106.
35. *The Economic Development of British Guiana. Report of a Mission Organized by the International Bank for Reconstruction and Development* (Baltimore: The Johns Hopkins University Press, 1953), 283–290; Lowe, *Historical Analysis of Mining Policy*, 84–110; US Department of State from the American Consul, Georgetown, Memo "The U.S. Marshall Plan". Decimal file 841D 25/1–2852, 841D 25/1–3151, R G 59, (NACP:MD).
36. Will, "Colonial Policy and Economic Development," 133–134.
37. *Report of the Commissioner of Mines, 1896–1897*, 5; *British Guiana Combined Court, First Special Session, 30/1924. Report of the Committee to Consider... Providing Inland Communication Between Georgetown and Potaro Landing and Between Wismar and Mallali, C. S. O. 2812/24* (Georgetown: The Argosy Co. Ltd., 1924); *Colonial Annual Report, 1946*, 72; and *50th Anniversary THD—Fifty Years of Service Transport and Harbours Dept. January 1931–December 1980* (Georgetown: Guyana Printers, 1980), 13–14.
38. TNA: COOC 111/593 British Guiana. Despatch dated January 5, 1914 Enclosure: "Proposal for Interior Railway"; TNA: COOR 111/610 British Guiana. Despatch 20 dated January 13, 1917 Enclosure: "Railway from Georgetown to Brazilian Border."

See "The Colony on the Verge of Crisis: Hope in Foreign Capital," *The Daily Argosy* August 20, 1922, 4; and Brands, *The Age of Gold: The California Gold Rush*, 334–358, 421, 361–491.

39. *Report of the Committee on Providing Inland Communication Between Georgetown and Potaro Landing and Between Wismar and Mallali*, 2–7; Appendix 1, *Report of Evidence*, 2–22, and Appendix 111, i–vii; and Lowe, *Historical Analysis of Mining Policy*, 86–87.
40. *Report of the Commissioner for Lands and Mines, 1930*, 4; "Hinterland Railway Development to be urged on Royal Commission. Development Association to be Formed and Public Meeting at Buxton," *The Daily Argosy* September 25, 1938, 9; *Colonial Annual Report for 1946*, 72; *Colonial Annual Report for 1956, 8*; and *The Economic Development of British Guiana, 264–274.*
41. HU: BPP British Guiana 28, 1890, 6, 21–22; HU: BPP British Guiana, 1892–1893, 4, 16–20*; Report of the Colonial Civil Engineer, 1892–1893*, 6; and Lancaster, "Unconquered Wilderness," 125–126.
42. *British Guiana: Report of the Colonial Steamer Services, 1920*, 6; *Report of the Commission of Lands and Mines, 1901–1902*, 9; and *Report of the Commissioner of Lands and Mines, 1902–1903*, 14.
43. HU: BPP British Guiana, 1892–1893, 4; *Report of the Colonial Civil Engineer, 1892–1893*, 6; and Perkins, *Notes on British Guiana and Its Gold Industry*, 31–35.
44. James Culley, interview by author, Buxton Village East Coast Demerara, October 24, 1996 and Lionel Arthur, interview by author, Mocha Village, East Bank Demerara, November 17, 1996. See Perkins, Notes *on British Guiana and Its Gold Industry,* 93; *Report of the Colonial Steamer Service,* 1920, 6; *50th Anniversary THD*, 13–25; *Colonial Annual Report, 1946*, 71–73; and *Colonial Annual Report, 1950*, 94–95.
45. *Report of the Commissioner of Lands and Mines, 1902–1903*, 14.
46. Ibid., and *Report of the Commissioner of Lands and Mines, 1916*, 17–18.
47. *Report of the Commissioner of Lands and Mines, 1902–1903*, 14 and *Report of the Commissioner of Lands and Mines, 1916*, 17–18.
48. *Report of the Colonial Steamer Service, 1920*, 6; *50th Anniversary THD*, 13–25; "Mineral Resources of British Guiana" (Georgetown: *The Daily Argosy*), 40–41; and *The Economic Development of British Guiana*, 56–57, 259–262.
49. *Report of the Colonial Steamer Service*, 1920, 6; *50th Anniversary THD*, 13–25; "Mineral Resources of British Guiana" (Georgetown: *The Daily Argosy*), 40–41; and *The Economic Development*, 56–57, 259–262.
50. *Report of the Gold Industry*, 4; Connolly, "The Riches of the Colony;" Perkins, *Notes on British Guiana and Its Gold Industry,*, 93; and Thompson, "Gold-Mining in British Guiana," 22–23, 27.
51. *Report of the Colonial Civil Engineer, 1904–1905*, 11; *Report of the Commissioner of Mines, 1920 Amendments to the Upper River Boat Regulations, 1912* of 15 September 1920, 15; *Commission Report on the Diamond and Gold Industry, 1928*, on river transportation, 19–20; *Report of the Commissioner of Lands and Mines, 1920*, 15; *British Guiana: Rules, Regulations, By-Laws, and Orders in Council, 1924*; and *The Economic Development of British Guiana*, 259–262.
52. United States: Department of State, American Consul, British Guiana Foreign Service Despatch 66 dated October 6, 1957. Enclosure: "The Political, Economic, and Social Progress in British Guiana." RG 59 741D.00/10–1957 (NACP: MD). See *Colonial Annual Report, 1952*, 7, 12.
53. *Air Navigation Ordinance 1925-Proclamation Annulling 1922 Air Navigation (Colonies and Protectorates) Order-in-Council and 1927 Navigation Ordinance*

(Colonies, Protectorates, and Mandated Territories, Orders, 1927. British Guiana: Rules, Regulations, By-Laws, and Orders in Council 1929 (Georgetown: The Argosy Company Ltd, 1930) and H. S. Burrowes, *Fifty Years of Flying in British Guiana* (Georgetown: The Guiana Graphic Ltd., 1963), 1–7.

54. *British Guiana: Report of the Post Master General, 1892–1893*, 29; *Report of the Post Master General, 1896–1897*, 5; and Burrowes, *Fifty Years of Flying*, 8–21.
55. "Deputy Boundary Commissioner Dies in the Jungle... Futile Efforts to Rush Him to the City," *The Daily Chronicle*, October 10, 1934, 4; "Air Service To Brazilian Boundary Urged. City Businessmen Ask Government. Boundary Commission Tragedy Cited," *The Daily Chronicle*, October 28, 1934, 4; "Art Williams Granted Flying License," *The Daily Chronicle*, November 10, 1934, 4; "British Guiana Needs $50,000,000 To Develop Hinterland. Aircraft Should Precede Roads," *The Daily Argosy*, April 27, 1937, 6.
56. *Colonial Annual Report, 1946*, 7–18, 38, 70–77; *Colonial Annual Report, 1947*, 41; *Colonial Annual Report, 1948*, 84; *Colonial Annual Report 1949*, 12, 96–98; *Colonial Annual Report, 1951*, 11,13; *Colonial Annual Report, 1954*, 153. *Air Navigation-Colonial Air Navigation (General) (Amendment) Regulations 1953 #18/1953 British Guiana Regulations* (Georgetown: The Argosy Ltd, 1953), B73–B86; and Burrowes, *Fifty Years of Flying*, 21–43.
57. *Report of the Commissioner of Lands and Mines, 1953*, 8; *Colonial Annual Report, 1956*; and *The Economic Development of British Guiana*, 294–296.
58. *Colonial Annual Report, 1954*, 153 and Burrowes, *Fifty Years of Flying*, 21–43.
59. *Report of the Post Master General, 1896–1897*, 5; *50th Anniversary THD*, 13–14.
60. *Colonial Annual Report, 1954*, 154–155.
61. "Improving the Amenities of Bush Life For Mazaruni Diamond Miners," *The Daily Chronicle*, August 24, 1930, 4; *Colonial Annual Report, 1946*, 68 and 1947, 41; *The Economic Development*, 297–298.
62. Smith, "The Origins of Porknocking in Guyana," 81–92.
63. Ibid.
64. *Report of the Commissioner of Lands and Mines, 1918*, 80 on mining companies by 1894 and *Report of the Gold Industry, 1895*, evidence, Im Thurn, Government Agent, NWD, 57–63.
65. *West India Royal Commission Report, 1897*, Appendix C Part 11 British Guiana, Section 142, 690, evidence, Dr. J. M. Rohlehr, MD.
66. *Report of the Commissioner for Essequibo/Pomeroon, 1903–1904*, 4; *Report of the Commissioner of Lands and Mines, 1906–1907*, 13, 17, *Report of the Commissioner/Essequibo/Pomeroon, 1906–1907*, 3–4. See *Report of the Commissioner of Lands and Mines, 1934*, 10.
67. *Great Britain: West India Royal Commission, 1938–39*. (London: H. M. Stationery Office, 1940), 9.
68. *Great Britain: Report on the British Guiana and British Honduras Settlement Commission*, 194; *Colonial Annual Report, 1948*, 3–4, 48; and *Colonial Annual Report, 1949*, 9, 24, 54, 59.
69. *Colonial Annual Report, 1950*, 57.
70. *British Guiana: Report of the Department of Labour, 1942–1950*, 4; *Colonial Annual Report, 1952*, 27; *Report of the Commissioner of Lands and Mines, 1953*, 10; and "Frank" Francis H. Proctor, interview with author, Upper Tulse Hill, London, June 21, 2010.
71. Dr. Rohlehr, Evidence, *West Indian Royal Commission, 1897* and Dr. A. S. Ozzard, *The Echo*, February 4, 1899, 6.

72. Phimister, "African Labour Conditions and Health," 102–149.
73. Marcel Fredericks, John Lennon, Paul Mundy, and Janet Fredericks, eds., *Society and Health in Guyana: The Sociology of Health Care In a Developing Nation* (Durham: Carolina Academic Press, 1986), 1–117.
74. Lowe, *Historical Analysis of Mining Policy*, 24–25 quoting Minutes of the Combined Court 7 Dec. 1886.
75. TNA; COOC 111/434 British Guiana, 1886. Enclosure: "Reorganizing of Medical Service. 87/1886 A Bill To Establish A Government Medical Service and To Enforce the Registration of Practitioners in Medicine and Surgery" (*Medical Ordinance #5/1886* required British credentials); K. O. Laurence, "The Development of Medical Services in British Guiana and Trinidad, 1841–1873," *Jamaica Historical Revue IV* (1964), 60–67; and Chris Curtis, ed., *Demerara Doctor, An Early Success Against Malaria:... Physician George Giglioli, 1897–1975* (London: Smith Gordon & Co. Ltd., 2006) , 60–67, 130–132.
76. *Report of the Commissioner of Mines, 1895–1896*, 5 and *West India Royal Commission, 1897,* Dr. Rohlehr and several African-Guyanese miners gave evidence on the situation.
77. *The Echo*, March 5, 1896, 2; *British Guiana: Report of the Surgeon General, 1897–1898*, 4; and *Report of the Commissioner of Mines, 1898–1899*, 5.
78. *The Echo*, March 5, 1896,2 and Curtis, *Demerara Doctor*, 54–73, 130–132.
79. H. I. Perkins, "Seven Months Up the Puruni River," *Timehri* 2 (1885), 35 and *Report of the Commissioner of Mines, 1894; Report of the Diamond and Gold Industries, 1928*, 22–25, 36; Lowe, *Historical Analysis of Mining Policy*, 79–80.
80. TNA; COOC 111/455 British Guiana, 1890. Despatch 40 dated February 13, 1890. Enclosure: "Registering of Dr. J. M. Rohlehr." See "Death Report" *The Daily Argosy*, February 23, 1922, 5; *The Daily Chronicle*, 23 February 23, 1922, 2; *Report of the Diamond and Gold Commission, 1928*, 22–25 on interior health needs; Harold Lutchman, "Middle Class Colonial Politics: A Study of Guyana with Special Reference to the Period, 1920–31," PhD., Manchester University, 1970, 62–63; Arthur J. and Elma Seymour, *Dictionary of Guyanese Biography* (Georgetown: By the authors, 1984–86), 76.
81. TNA: COOC 111/606 British Guiana, 1916. Despatch dated July 15, 1916. Enclosure: "Medical Service Recommendation. Appointment of Dr. L. R. Sharples."
82. "Legislative Council. Electives and Public Health Appointment," *The Daily Chronicle,* November 29, 1939, 7; and Lutchman, "Middle Class Colonial Politics," 62–63.
83. "Legislative Council. Electives and Public Health Appointment." European, Dr. P. A. T. Sneath, appointed without the credentials, *The Daily Argosy,* January 29, 1938, 6.
84. *The Echo*, March 5, 1896, 2.
85. *Mineral Resources of British Guiana* (Georgetown: The Daily Chronicle Printers, 1935), 74–77.
86. *Report of the Diamond and Gold Industries, 1928*, 22–25, 36. See *British Guiana: The British Commission on the Financial Situation of British Guiana. Report* (London: H. M. Stationery Office, 1931), 36; and *Mineral Resources of British Guiana*, 77.
87. *Report of the Diamond and Gold Industries, 1928*, 22–25.
88. *Colonial Annual Report, 1946*, 49, 51–52.
89. *Colonial Annual Report, 1946*, 49, 51–52; *Colonial Annual Report, 1947*, 55; *Colonial Annual Report, 1949*, 16–17; *The Economic Development of British Guiana*, 9–10. See Fredericks, *Society and Health in Guyana*; Seecharran, *Sweetening "Bitter Sugar;"* and Curtis, *Demerara Doctor.*

90. *Colonial Annual Report, 1948*, 50; *Colonial Annual Report, 1949*, 16–17; *Colonial Annual Report, 1953*; *The Economic Development of British Guiana*, 9–1; and Fredericks, *Society and Health in Guyana*.
91. *Colonial Annual Report, 1948*, 50; *Colonial Annual Report, 1949*, 16–17; *Colonial Annual Report, 1953; Report of the Surgeon General; Report of the Registrar General; Report of the Commissioner of Lands and Mines; The Economic Development of British Guiana*, 9–10; Fredericks, *Society and Health*; Curtis, *Demerara Doctor*, 45–259; and Selwyn R. Cudjoe, *Caribbean Visionary: A. R. F. Webber and the Making of the Guyanese Nation* (Jackson: University of Mississippi Press, 2009), 118–121.
92. *Report of the Surgeon General; Report of the Registrar General; Report of the Commissioner of Lands and Mines; The Economic Development of British Guiana*, 9–10; Fredericks, *Society and Health*; Curtis, *Demerara Doctor*, 45–259.
93. *Colonial Annual Report, 1952*, 7, 12 and US Foreign Service Despatch 66, 1957.
94. US Foreign Service Despatch 66, 1957.
95. *Report of the Gold Industry, 1895*, 1, 4–5, 118. and *Report of the Diamond and Gold Industries, 1928*, 13–15, 2–24. Aged miners interviewed within earshot of their families; they offered no such information.
96. *Report of the Gold Industry, 1895*, 1, 4–5, 118 and *Report of the Commissioner of Lands and Mines, 1900–1901*, 7.
97. *Report of the Diamond and Gold Industries, 1928*, 13–15, 2–24.
98. Ibid.
99. Ibid. Dr. Rohlehr, evidence, West India Royal Commission, 1897.
100. *Report of the Commissioner of Lands and Mines, 1900–1901*, 7.
101. *Colonial Annual Report, 1946*, 50; *Colonial Annual Report, 1947*, 49; *Colonial Annual Report, 1948*, 3–4, 48; and *Colonial Annual Report, 1949*, 9, 24, 54, 59.
102. *The Manchester Guardian*, May 30, 1952 in US Department of State R G 59. 841D.00 Colonial/British Guiana1952 RG 59 (NACP: MD).
103. *Colonial Annual Report, 1946*, 3, 71; *Colonial Annual Report 1949*, 13. See G. M. Stockley, *The Geology of British Guiana and the Development of its Mineral Resources, Bulletin #25*, 1955.
104. Critical views of development plans in local newspapers of the 1940s–1950s. See TNA: COOC 111/821/6. West Indian Dept., British Guiana, 60596/A *1950*, "Colonial Development Corporation Projects, Press Criticism," and *Colonial Annual Report, 1956*, 1–4, 171–177.

6 Another Approach: Organizing Bauxite Production

1. TNA: COOC 111/598 British Guiana, 1914. Enclosures: "Application for Bauxite Lands," July 21, 27; October 7; November 25, 28; December 11, 16, 1914; *The Daily Argosy*, September 17, 1922, 16, "Sharp Rise in Bauxite Exports," *The Daily Chronicle*, June 13, 1936, 1; "The Bauxite Industry at Mackenzie," in "Commercial Review;" *The Daily Chronicle*, January 11, 1937, 4; "Mackenzie City On The Improve," *The Daily Chronicle*, August 7, 1937, 8; "Demerara Bauxite Company to Increase Output," *The Daily Chronicle*, March 6, 1938, 1; *Royal Commission Inquiry, 1939*, Orde-Browne on British Guiana, 4; and "Undesirable Conditions in Mining Industry. Men Stranded, Unpaid, Unfed (Gold &Diamonds);" and "Happy Labouring Force at Mackenzie (Bauxite)" *The Daily Chronicle*, August 26, 1939, 8.

2. "HappyLabouring Force at Mackenzie(Bauxite)". See Chris Curtis, ed., *Demerara Doctor. An Early Success Against Malaria: The Autobiography of a Self-Taught Physician, George Giglioli, 1897–1975* (London: Smith-Gordon, 2006), 45–108.
3. Cedric Grant, *Company Towns in the Caribbean: A Preliminary Analysis of Christianburg-Wismar-Mackenzie* (Georgetown: Government Printing Office, 1971); 1–79; M. Shahabudeen, *Nationalisation of Guyana's Bauxite: The Case of Alcan* (Georgetown: Ministry of Information and Culture, 1981), 64–145; Odida T. Quamina, *The Mineworkers of Guyana: The Making of a Working Class* (Atlantic Highlands, NJ: Zed Books, 1989), 13–16.
4. Quamina, *The Mineworkers of Guyana*,13–16.
5. Ibid.
6. Quamina, *The Mineworkers of Guyana*, 17. See Ann Spackman, "The Role of Private Companies in the Politics of Empire: A Case Study of Bauxite and Diamond Companies in Guyana in the Early 1920s," *Social and Economic Studies* 23:3 (1975), 341–378 and Carlo Lamur, *The American Takeover: Industrial Emergence and Alcoa's Expansion in Guyana and Suriname, 1914–1921* (Dordrecht, Holland: Floris Publications, 1985), 39, 53–73, 135–143.
7. Shahabudeen, *Nationalisation of Guyana's Bauxite*, 121–122; Odida T. Quamina, "The Social Organization of Plantation Mackenzie: An Account of Life in the Guyana Mining Enterprise" (Geneva: U. N. Research Institute for Social Development, 1981), 14, 17; and *The Mineworkers of Guyana*, 19–23, 37.
8. Maurice St Pierre, "Industrial Unrest in Mackenzie, Guyana," in *McGill Series in Caribbean Anthropology 5* (1969), 66–80 and Quamina, *The Mineworkers of Guyana*, 21–23, 29–42.
9. TNA: CO 941/171/-24/958, Colonial Office List, British Guiana, 63. Grant, *Company Towns in the Caribbean*, 1–7; Maurice St. Pierre, "The Sociology of Decolonization, The Case of Guyana Bauxite Ltd.," *New World Quarterly* 5:4 (1972), 50–62; and Shahabudeen, *Nationalization of Guyana's Bauxite* , 99–145.
10. Grant, *Company Towns in the Caribbean*, 1–7; St. Pierre, "The Sociology of Decolonization," 50–62; and Shahabudeen, *Nationalization*, 99–145.
11. Albert M. Craig, William A. Graham, Donald Kagan, Steven Ozment, and Frank M. Turner, *The Heritage of World Civilizations* (Prentice Hall: Upper Saddle River, NJ; Prentice Hall, 2009, 9th. ed.), 731–757.
12. Craig, Graham, Kagan, Ozment, and Turner, *The Heritage of World Civilizations* 766–776. See Gerald Horne, *Cold War in a Hot Zone: The United States Confronts Labor and Independence Struggles in the British West Indies* (Philadelphia: Temple University Press, 2007), 1–28, 157–211; and Daniela Spenser, "Standing Conventional Cold War History on Its Head," in Gilbert M. Joseph and Daniela Spenser, eds., *In From the Cold: Latin America's New Encounter With the Cold War* (Durham: Duke University Press, 2008), 381–395. Spenser noted the Hoover Institute Press Guide (Palo Alto, 2001) for Record Group 89, Russian State Archives Microfilm Collection of the Archives of the Communist Party and Soviet State, Contemporary History ((RGANT) as a new source for Cold War History).
13. G. M. Stockley, *The Geology of British Guiana and the Development of its Mineral Resources* (Georgetown: Geological Survey Department, 1955), 28; and Steven Holloway, *Aluminum Multinationals and the Bauxite Cartel* (New York: St. Martin's Press, 1988), 9–10.
14. *West India Royal Commission, 1939.* Orde-Browne on British Guiana, 4; "Undesirable Conditions in Mining Industry. Men Stranded, Unpaid, Unfed (Gold &Diamonds)."

15. *Reports of the Institute of Mines and Forests; Reports of the Commissioner of Land and Mines; British Guiana: Census of the Population, 1931, 1946; Report of the Local Government, 1955*, 17. See Maurice St. Pierre, "Race, the Political Factor and the Nationalization of the Demerara Bauxite Company, Guyana," *Social and Economic Studies* 24 (1975), 482; Grant, *Company Towns in the Caribbean*, 8; Quamina, *The Mineworkers of Guyana*, 20–22; and Curtis, *Demerara Doctor*, 48.
16. "Sharp Rise in Bauxite Exports," 1; "Mackenzie City On The Improve," *The Daily Chronicle,* August 7, 1937, 8; Quamina, "The Social Organization," 24, St. Pierre, "The Sociology of Decolonization," 51–54.
17. Grant, *Company Towns*, 11–14; St. Pierre, "The Sociology of Decolonization, 51–54 and "Race, the Political Factor," 482–484; Quamina, *The Mineworkers of Guyana*, 19–25.
18. Grant, *Company Towns*, 11–14; St. Pierre, "The Sociology of Decolonization, 51–54 and "Race, the Political Factor," 482–484; Quamina, *The Mineworkers of Guyana*, 19–25.
19. Quamina, *The Mineworkers of Guyana*, 19–25.
20. Shahabudeen, *Nationalisation of Guyana's Bauxite*, 121–122; Odida T. Quamina, "The Social Organization of Plantation Mackenzie: An Account of Life in the Guyana Mining Enterprise" (Geneva: U. N. Research Institute for Social Development, 1981), 14, 17; and *The Mineworkers of Guyana*, 19–23, 37.
21. Grant, *Company Towns*, 11. See St. Pierre, "Race, the Political Factor," 484–489.
22. *The Mackenzie Miner,* 1:13 (August 1948), 1; and *The Mackenzie Miner* 4:15 (July 1951), 4.
23. Curtis, *Demerara Doctor*, 47–48.
24. *Census of the Population, 1921, 1946; The Daily Argosy*, September 17, 1922, 16; "Sharp Rise in Bauxite Exports," *The Daily Chronicle*, June 13, 1936, 1; "Demerara Bauxite Company to Increase Output," *The Daily Chronicle*, March 6, 1938, 1; St. Pierre, "Race, the Political Factor," 482; and Quamina, *The Mineworkers of Guyana,* 21–24.
25. Quamina, *The Mineworkers of Guyana*, 21–24.
26. Quamina, *The Mineworkers of Guyana*, 26 and Curtis, *Demerara Doctor*, 45–108.
27. St. Pierre, "Race, the Political Factor," 481–489 and Eunice Williams, interview by author, Brooklyn, NY, February 28, 2006.
28. Ibid.
29. Darrell Gray, interview by author, Queenstown Village, Essequibo County, Guyana, January 3, 1997.
30. George E. Robert Adams, interview by author, Victoria Village, East Coast Demerara, November 23, 1996.
31. Eunice Williams, interview.
32. "Mr. J. A. M. Proctor Honored By U.S. Engineering Institute," *The Daily Chronicle,* January 13, 1936, 4 and "Mr. J. A. M. Proctor's Success. Associate of American Electrical Institute. Foreman Electrician at Mackenzie," *The Daily Argosy*, January 14, 1936, 8.
33. "Frank" Francis Hector Proctor, interview by author, Upper Tulse Hill, London, 21 June 21, 2010, offspring—Noel, Godfrey, Patsy, Melvin, Ingrid, Michael, and Royston.
34. Harold "Paris" Thompson, interview by author, Clapton, London, May 30, 2010. Bertram, Jacqueline Dube, Ingrid David, Hazel and Cheryl Matthews—offspring.
35. *British Guiana: Report of the Commissioner of Lands and Mines, 1924–1925*, 14–15.

36. Curtis, *Demerara Doctor*, 40–108, 116–167, 194–260.
37. Ibid.
38. Ibid.
39. Ibid.
40. Curtis, *Demerara Doctor*, 49–108.
41. Ibid.
42. *Report of the Commissioner of Lands and Mines, 1924–1925*, 14–15.
43. Ibid.
44. A. E. Lord-Isaacs, "Bauxite Company Spends $10,000 A Year To Prevent River Contamination," *The Daily Chronicle*, April 16, 1939, 18.
45. Curtis, *Demerara Doctor*, 60–67.
46. Ibid.
47. *Report of the Commissioner of Lands and Mines*, 14–15; and Curtis, *Demerara Doctor*, 49–108.
48. *Report of the Commissioner of Lands and Mines*, 14–15.
49. Curtis, *Demerara Doctor*, 40–80, 116–167, 194–260. See *Colonial Annual Reports*, 1940s–1950s; "Public Health Problems in British Guiana. Rockefeller Foundation Experts Delighted With Progress," *The Daily Chronicle*, September 10, 1939, 11; Marcel Fredericks, John Lennon, Paul Mundy, and Janet Fredericks, eds., *Society and Health in Guyana: the Sociology of Health Care in a Developing Nation* (Durham: Carolina Academic Press, 1986); Clem Seecharan, *Sweetening "Bitter Sugar": Jock Campbell, The Booker Reformer in British Guiana, 1934–1966* (Kingston: Ian Randle Publishers, 2004), 56–59, 76, 157, 176, 223, 363–409; and Selwyn R. Cudjoe, *Caribbean Visionary: A. R. F. Webber and the Making of the Guyanese Nation* (Jackson: University of Mississippi Press, 2009), 118–121.
50. *Report of the Commissioner of Lands and Mines*, 14–15.
51. Ibid.
52. *British Guiana: Colonial Annual Report, 1950*, 3–5, 24–25, 28–29; *The Economic Development of British Guiana: Report of a Mission Organized by the International Bank for Reconstruction and Development* (Baltimore: The Johns Hopkins University Press, 1953), 10; and Seecharran, *Sweetening "Bitter Sugar,"* 76.
53. Grant, *Company Towns*, 11–12; St. Pierre, "The Sociology of Decolonization," 51–53; Quamina, "The Social Organization," 2, 13, 24 and *The Mineworkers of Guyana*, 22, 32–34, 40–41, 92–94.
54. Quamina, "The Social Organization," 4, 24 and *The Mineworkers of Guyana*, 29–32.
55. Quamina, "The Social Organization," 21–25 and *The Mineworkers of Guyana*, 39–42.
56. "Grand Steamer Excursion To Mackenzie City, Sunday July 30, 1922," *The Daily Argosy*, July 19, 1922, 1; and "You are Invited to Wismar August Monday," *The Daily Chronicle*, June 19, 1939, 7.
57. Quamina, "The Social Organization," 21–25 and *The Mineworkers of Guyana*, 39–42.
58. Quamina, "The Social Organization," 21–25 and *The Mineworkers of Guyana*, 39–42.; and *The Mackenzie Miner*, 4:1 (July 1951), 4.
59. St Pierre, "Race, the Political Factor, 482; and Curtis, *Demerara Doctor*, 45–48.
60. St Pierre, "Race, the Political Factor, 482; Curtis, *Demerara Doctor*, 47, 53; Quamina, "The Social Organization," and *The Mineworkers of Guyana*. See Leslie P. Cummings, *Geography of Guyana*, (London: Collins Press, 1975, ed.), 21.
61. Quamina, "The Social Organization," and *The Mineworkers of Guyana*.

62. Cummings, *Geography of Guyana*, 29.
63. Quamina, *The Mineworkers of Guyana*, 35–37; St. Pierre, "The Sociology of Decolonization," 52; and Ashton Chase, *A History of Trade Unionism in Guyana, 1900 to 1964* (Georgetown: New Guyana Co. Ltd., 1964), 126–132.

7 Evolving Relations: Mining and Trade Unionism

1. TNA: COOC 111/553. Despatch 107, 1906. Enclosure: "Administration of the Colony 'Monster Petition' from the Inhabitants. Signed by 5,750 People on the Neglect of the Recommendations of the Royal Commission, 1897" and Kimani Nehusi, "The Causes of the Protest of 1905," in Winston F. McGowan, James G. Rose, and David A. Granger, eds., *Themes in African-Guyanese History* (Georgetown: Free Press, 1998), 251–276.
2. Ashton Chase, *A History of Trade Unionism in Guyana, 1900 to 1964* (Georgetown: New Guyana Company Ltd., 1964), 34–38, 59, 124–132, 262–264; Alan Lancaster, "Proposals for Hinterland Settlement and Development of British Guyana, 1884–1890," in *History Gazette 59* (1993), 11–15; and Maurice St. Pierre, *Anatomy of Resistance: AntiColonialism in Guyana, 1823–1966* (London: Macmillan Education Ltd., 1999), 46–50.
3. Douglas Smith, "The Origins of Porknocking in Guyana," *McGill Studies in Caribbean Anthropology* (1969), 83, n.14. Quoting H. I. Perkins, "Seven Months in the Puruni River," in *Timehri 2* (1888), 335; Sherwood A. J. Lowe, *Historical Analysis of Mining Policy, Law & Administration in Guyana: 1831–2002* (Georgetown: Pavnik Press, 2003), 60. See *British Guiana: Report of the Commissioner of Mines, 1894*, 5.
4. Albert M. Craig, William A. Graham, Donald Kagan, Steven Ozment, and Frank M. Turner,,*The Heritage of World Civilizations* (Upper Saddle River, NJ: Prentice Hall, 9th ed., 2009), 743–747.
5. O. Nigel Bolland, *The Politics of Labour in the British Caribbean: The Social Origins of Authoritarianism and Democracy in the Labour Movement* (Kingston: Ian Randle Publishers, 2001), 658–662 and Boland, *Colonialism and Resistance in Belize: Essays in Historical Sociology* (Kingston: University of the West Indies Press, 2003), 8, 163–170, 172–192; and Gerald Horne, *Cold War in A Hot Zone: The United States Confronts Labor and Independence Struggles in the British West Indies* (Philadelphia: Temple University Press, 2007) , 1–28, 157–211. See Barbara P. Josiah, "Organizing Within the Diaspora: Claude H. A. Denbow, Howard University, and the League of Coloured Peoples," *The Journal of African American History*, 95:2 (Spring 2010), 229–247.
6. Bolland, *The Politics of Labour*, ix–xiii, 1–19, 80–81, 117–119, 181–191.
7. Norman E. Cameron, *The Evolution of the Negro* (Westport: Negro University Press, 1970. Reprint. 1929 ed.), 6–10, 18, 20–32. See St. Pierre, *Anatomy of Resistance*, 4–5, 17–18, 46–57, 68–70 and Barbara P. Josiah, "Creating Worlds: A Study of Mutuality and Financing among African Guyaneses 1800s–1950s," in *The Journal of Caribbean History*, 38:1 (2004), 106–127.
8. *British Guiana: Report of the Labour Commission, 1890* (Georgetown: C. K. Jardine, 1890); *British Guiana: Report of the Gold Industry and Crown Lands Commission, 1895* (Georgetown: C. K. Jardine, 1895); and *British Guiana: 1928 Report of the Diamond and Gold Industries* (Georgetown: The Argosy Company Ltd., 1928).
9. Cameron, *The Evolution of the Negro*, a comprehensive study on the initiatives of African Guyanese in all aspects of the country's development. See Walter Rodney, *A History of the Guyanese Working People, 1881–1905* (Baltimore: The Johns Hopkins University Press, 1982, ed.), 58, 97–99, 101–102, 144, 170; and Barbara

P. Josiah, "A. B. Brown, L. F. S. 'Forbes' Burnham, Herbert N. Critchlow, Claude H. A. Denbow, Winifred I. V. Gaskin, and Jane Phillips Gay," in Colin Palmer, ed., *Encyclopedia of African-American Culture and History: The Black Experience in the Americas, 2nd. Ed.* (Farmington Hills, MI: Thompson Gale/ Macmillan, 2006, ed). See Magnus Morner, *Adventurers and Proletariats: The Story of Migrants in Latin America* (Pittsburg: University of Pittsburg Press, 1985), 67–85, 100–129 and Walter Nugent, *Crossings: The Great TransAtlantic Migrations* (Bloomington: Indiana University Press, 1992), 112–135.

10. "B. G. Labour Union. Third Anniversary Celebrations," *The Daily Chronicle,* January 17, 1922, 8; Hazel M. Woolford, "The Origins of the Labour Movement," in Winston F. McGowan, James G. Rose, and David A. Granger, eds., *Themes in African-Guyanese History* (Georgetown: Free Press, 1998), 277–295; St. Pierre, *Anatomy of Resistance,* viii–11, 16–34, 60–73; Josiah, "Organizing Within the Diaspora," 229–247 and Josiah, "From Shaw University to 'the Booker T. Washington of British Guiana:' Hon. E. F. Fredericks and Contributions of HBCUs to the African Diaspora," Paper presented at the *Association for the Study of African American Life and History, 95th Annual Meeting,* Raleigh, NC, September 29–October 3, 2010.
11. Chase, *A History of Trade Unionism,* 34–38, 59, 124–132, 262–264; Vere T. Daly, *The Making of Guyana* (London: Macmillan Education Ltd., 1974), 190–195; Maurice St. Pierre, "Race, the Political Factor and the Nationalization of the Demerara Bauxite Company, Guyana," *Social and Economic Studies* 24:4 (1975), 481–503; and Odida T. Quamina, *The Mineworkers of Guyana: The Making of a Working Class* (Atlantic Highlands, NJ: Zed Books, 1987), 9, 21–38, 43–48, 54–98.
12. *Report of the Labour Commission, 1890,* 25, evidence, Rev. D. J. Reynolds, on excess workers. See, Quamina, *The Mineworkers of Guyana,* 60–61, 99 and Chase, *A History of Trade Unionism,* 132.
13. *British Guiana: Colonial Report, 1947,* 6, 19; US State Department, the American Consulate, British Guiana. Enclosure: "One Day Strike at Demba" May 7, 1950. RG 59 File 841D.062/Box 4814 841D-42 (NACP: MD); Chase, *A History of Trade Unionism,* 124–133, 187; Cedric Grant, *Company Towns in the Caribbean: A Preliminary Analysis of Christianburg-Wismar-*Mackenzie (Georgetown: Government Printers, 1971), 14, 20; Quamina, *The Mineworkers of Guyana,* 54–90; and St. Pierre, *Anatomy of Resistance,* 65.
14. Chase, A History of Trade Unionism, 124–133, 187; Grant, *Company Towns,* 14, 20; Quamina, *The Mineworkers of Guyana,* 54–90; and St. Pierre, *Anatomy of Resistance,* 65.
15. *Report of the Gold Industry, 1895,* 1, 45, 118, 133 and *Ordinance #8 of 1894; British Guiana: Report of the Commissioner of Mines, 1895–1896,* 5; *British Guiana: Report of the Commissioner of Lands and Mines, 1900–1901,* 7; *Report of the Diamond and Gold Industries, 1928,* 25. See Frederick A. Johnston, *Class, Race, and Gold: A Study of Class Relations and Racial Discrimination in South Africa* (London: Routledge, 1976), 1–90, 168–200; Grantley Walrond, "Small Gold Mines: The Production and Declaration Problem-Another Dilemma for a Small Developing Country," in *Natural Resources Forum,* 10:4 (1986); *Diamonds, Gold, and South Africa,* www.geology.ucdavis.edu/cowen, March 2000.
16. Alan Carter, interview by author, Stabroek Market, Georgetown, December 14, 1996.
17. *Report of the Institute of Mines,* June–August, 1890.
18. *Report of the Gold Industry, 1895,* 9–10; Chase, *A History of Trade Unionism,* 34–35; and Lowe, *Historical Analysis of Mining Policy,* 63–64, 79.
19. *The Gold Mining Gazette,* April 1–June 24, 1892.

20. *The Gold Mining Gazette,* June–August, 1890, April–June, 1892; *The Echo*, August 23, 1894, August 7, 1897, 1–2, February 4, 1899; *Report of the Commissioner of Mines, 1898–1899,* 4 on Section 13, *Mining Regulations 15, 1896*. See Chase, *A History of Trade Unionism*, 34–35 and Lowe, *Historical Analysis of Mining Policy*, 54–56.
21. *Report of the Commissioner of Mines, 1897–1898; The Gold Mining Gazette*, July 1, 1897, 2–3; and June 30, 1899; *The Echo*, February 4, 1899; *Report of the Institute of Mines, 1919–1920*, 6.
22. *Report of the Institute of Mines, 1913–1914*, 7.
23. *The Gold Mining Gazette,* July 1, 1897, 2–3; *The Echo*, February 4, 1899; *Report of the Institute of Mines, 1919–1920*, 6.
24. *Report of the Institute of Mines, 1913–1914*, 7; and *Report of the Institute of Mines, 1915–1916*, 1. See Chase, *A History of Trade Unionism*, 34–35; Smith, "Origins of Porknocking," 84 and Lowe, *Historical Analysis of Mining Policy*, 54–56.
25. Ibid., Chase, *A History of Trade Unionism*, 34–35.
26. Ibid.
27. *Report of the Institute of Mines, August 1894–June 1896; Report of the Institute of Mines, 1913–1914, Report of the Institute of Mines 1914–1915*. See H. J. Perkins, *Notes on British Guiana and Its Gold Industry* (London: Sampson Low, Marston, and Co., 1896), 23.
28. *Report of the Institute of Mines, 1914–1915*; and *Report of the Institute of Mines 1920–1921*.
29. TNA: COOC 111/613. Despatch 25 dated September 6, 1917. Enclosure: Memo by the Commissioner of Lands and Mines.
30. James G. Rose, "The Strikes of 1842 and 1848," in Winston F. McGowan, James G. Rose, and David A. Granger, eds., *Themes in African-Guyanese History* (Georgetown: Free Press, 1998), 158–200; "B. G. Labour Union. Third Anniversary Celebrations;" Woolford, "The Origins of the Labour Movement," 277–295; St. Pierre, *Anatomy of Resistance*, viii–11, 16–34, 60–73; Josiah, "Organizing Within the Diaspora," 229–247 and Josiah, "From Shaw University to 'the Booker T. Washington of British Guiana:' Hon. E. F. Fredericks."
31. TNA: COOC 111/553. Despatch 107, 1906. Enclosure: "Administration of the Colony 'Monster Petition' from the Inhabitants; Chase, *A History of Trade Unionism*, 48–51; Rodney, *A History of the Guyanese Working People*, 190–216; Neheusi, "The Causes of the Protest of 1905," 251–276; and Barbara P. Josiah, "Hubert Nathaniel Critchlow," in Colin Palmer, ed., *Encyclopedia of African-American Culture and History: The Black Experience in the Americas, 2nd. Ed.* (Farmington Hills, MI: Thompson Gale/ Macmillan, 2006, ed.), 557–558.
32. "Employer Charged With Breach of Contract. Edward Harvey Before the Court. Indebtedness to Labourers Alleged," *The Daily Argosy*, May 4, 1939, 1; and "Fine House and Car While His Labourers Starve. Defaulting Gold Claim Holder," *The Daily Argosy*, May 4, 1939, 7.
33. "The Depression of Labourers in the Goldfields," *The Echo*, February 4, 1899, 1.
34. Ibid.
35. Ibid.
36. Perkins, *Notes on British Guiana*, 12 and *Report of the Commissioner of Mines, 1898–1899*, 4.
37. Lowe, *Historical Analysis of Mining Policy*, 54, for quote on absence of trade unions by Edward Braddon in the *British Mining Journal*, May 28, 1904.
38. TNA: COOC 111/553. Despatch 107, 1906. Enclosure: "Administration of the Colony 'Monster Petition' from the Inhabitants and Neheusi, "The Causes of the Protest of 1905," 251–276.

39. "Fellow-Men. William Alexander Brown. 'Pigeon Island' Brown," *The Daily Argosy*, February 27, 1937. Supplement 10–11.
40. "The Depression of Labourers in the Goldfields."
41. *The Gold Mining Gazette*, July 10, 1890, 3; *Report of the Gold Industry, 1895*, 10.
42. Amended Mining Laws 1931 applied. Major G. St. Orde-Brown, *Report: Labour Conditions in the West Indies* (London: H. M. Stationery Office, 1939), 170–171.
43. *Report of the Commissioner of Mines, 1894*, 5; *Mining Ordinance, 1894; Report of the Gold Industry, 1895*, 4, 72–73; *Report of the Institute of Mines, 1913–1914*, 7 and *1919–1920*, 6; *Report of the Diamond and Gold Industries, 1928*, 25. See Johnston, *Class, Race, and Gold*, 4; and *Lowe, Historical Analysis of Mining Policy*, 60–61.
44. H. I. Perkins, "The Future Prosperity of the Colony," *The Mining Gazette*, March 14, 1895, 6; "The New Mining Regulations" *The Daily Argosy*, September 5, 1896; *The Mining Laws of the British Empire and Foreign Countries. Imperial Institute* (London: H. M. Stationary Office, 1927); *Report of the Diamond and Gold Industries, 1928*, Appendix iv, 44. See Johnston, *Class, Race and Gold*; Michael Hunt, *Mining Laws in Western Australia, 3rd ed.* (NSW: The Freedom Press, 2001); and Martin Chanock, "South Africa, 1841–1924: Race, Contract, and Coercion," in Douglas Hay and Paul Craven, eds., *Masters, Servants and Magistrates in Britain and the Empire, 1562–1955* (Chapel Hill: University of North Carolina Press, 2004), 338–364.
45. *Report of the Gold Industry, 1895*, 4–5, 118, 133; *Report of the Commissioner of Mines, 1895–1896*, 5.
46. *Report of the Commissioner of Mines, 1894*. See *Ordinance #8, 1894. The New Mining Regulations.*
47. *Report of the Diamond and Gold Industries, 1928*. Appendix IV, 44 refers to South Africa and searching of native laborers in Chapter 8 of the *Precious Stones Ordinance, 1903* (Transvaal); Part XIII of the *Precious Stones Ordinance, 1904 (Orange Free State), Precious Stones (Alluvial) Amendment Act, 1909*, and in the *Native Labour Regulation Act, 1911* and Volume II Parts I and III of the *Mining Laws of the British Empire*. See *Precious Stones Ordinance 1927*, 102, 88 and 108. See Smith, "The Origins of Porknocking in Guyana," 83, n.14; and; Lowe, *Historical Analysis of Mining Policy*, 60.
48. Perkins, "Seven Months Up the Puruni River," 335; *Report of the Diamond and Gold Industries, 1928*, 22–25, 36; Lowe, *Historical Analysis of Mining Policy*, 79.
49. *Report of the Commissioner of Mines, 1894*. See *Ordinance #8, 1894. The New Mining Regulations.*
50. H. A. Will, "Colonial Policy and Economic Development in the British West Indies, 1895–1903," *Economic History Review* 23:1 (April 1970), 129–147.
51. *Report of the Labour Commission, 1890*, i–vii. See Arthur de Gobineau, *The Inequality of Human Races, 1853–1855*. Translated by Adrian Collins (New York: G. P. Putnam Sons, 1915, ed.); Jan Nederveen Pieterse, *White on Black: Images of Africa and Blacks in Western Popular Culture* (New Haven: Yale University Press, 1992); and Ruth Simms Hamilton, ed., *Routes of Passage: Rethinking the Africa Diaspora Vol.1 Part 1* (East Lansing: Michigan State University Press, 2007), 1–32.
52. *Report of the Gold Industry, 1895*, 5.
53. "Mazaruni Claim-Raiders Jailed," *The Daily Chronicle*, June 1, 1935, 4; *The Daily Chronicle*, October 25, 1930, 4.
54. *Report of the Commissioner of Mines, 1898–1899*, 4; *Report of the Commissioner of Mines, 1900–1901*, 7; (Dept. of Lands and Mines merged/established 1902) *Colonial Annual Report, 1948*, 50; US Department of State, American Consulate, Georgetown, British Guiana. Enclosures: "The Diamond Industry in British Guiana," dated

January 6,1945, 844B .6347/-645; "Mineral Trade and Production," dated March 3, 1950, 841D.25/3–350; "British Guiana Industrial Diamonds" dated June 20, 1951, 841D2535/6–2051; and "British Guiana Industrial Diamonds" dated November 27, 1951, 841D.2535/12–751, RG 59 (NACP: MD).

55. TNA: COOC 111/633, British Guiana, 1952. Colonies 58/14/45 dated January 30, 1952 one CO18498/82/91/51 from the governor to the secretary of state for the colonies and *Colonial Report, 1956*, 1.
56. Ibid.
57. Ibid.
58. TNA: COOC 111/553. Enclosure: "The People's Association of British Guiana." See Chase, *A History of Trade Unionism*, 13–27; and Woolford, "The Origins of the Labour Movement," 277–295; and St. Pierre, *Anatomy of Resistance*, 46–50.
59. COOC 111/553 Enclosures dated December 7, 1906 and December 26, 1906.
60. TNA: COOC 111/547 British Guiana, 1905. Despatch 304 "Riots, Strikes of Labourers at Peter's Mines," (B. G. Consolidated Goldfields Ltd.). See Rodney, *A History of the Guyanese Working People*, 194.
61. Chase, A History of Trade Unionism, 38.
62. Ibid., 17.
63. Chase, *A History of Trade Unionism*, 13; Woolford, "The Origins of the Labour Movement," 277–295.
64. "The Labour Commission. Evidence of Mr. H. Critchlow. Labour Union's Attitude Explained," *The Daily Chronicle*, February 11, 1936, 6; February 20, 1936, 8; *Colonial Annual Report, 1951*, 25, 58; *Colonial Annual Report, 1954*, 36; Chase, *A History of Trade Unionism*, 13–27; Vere T. Daly, *A Short History of the Guyanese People* (London, Macmillan Education, Ltd., 1975), 287–288; Quamina, *The Mineworkers of Guyana:*, 54–55; St. Pierre, *Anatomy of Resistance*, 44–50; BGMA officials, J. G. Stephens, Buxton; M. A. Goring, Victoria; J. Mingo, Belladrum and Paradise; J. Briggs, Den Amstel; W. Bourne and W. Brown, Essequibo; E. Pile, Pouderoyen; W. D. David, Plaisance; J. Mansfield, J. A. E. Patterson, J. Lucie Griffith, Georgetown.
65. *Report on the Labour Conditions in the West Indies*, Appendix iv, 183; *Colonial Annual Report, 1946*, 20–21 for Trade Unions (Amendment) Ordinance passed by the Legislative Council 1946; and *Colonial Annual Report, 1952*, 32–34 on Industrial Relations. See Chase, *A History of Trade Unionism*, Appendix, 273–277.
66. *Colonial Annual Report, 1947*, 18–19; *Colonial Annual Report, 1950*, 28; *Colonial Annual Report, 1955*, 28–30; *British Guiana: Annual Report of the Department of Labour, 1956*, Table X(a) List of Trade Unions; and St. Pierre, *Anatomy of Resistance*, 51–54.
67. CO 950/30 *West India Royal Commission, Ninth Session 29/9/38* Memo of Evidence: Witnesses Dr. Harold Moody and Mr. Blackman, League of Coloured Peoples—Includes International African Service Bureau and Negro Welfare Association (London: National Archives).
68. Josiah, "Organizing Within the Diaspora." See Woolford, "The Origins of the Labour Movement," 277–295; Franklin W. Knight, *The Caribbean: The Genesis of Fragmented Nationalism* (New York: Oxford University Press, 1990), 177–181; and Horne, *A Cold War in a Hot Zone*, 1–28, 157–211.
69. Maurice St. Pierre, "Industrial Unrest in Mackenzie, Guyana," *McGill Studies in Caribbean Anthropology*, (1969), 65–80 and St. Pierre, "The Sociology of Decolonization: The Case of Guyana Bauxite," *New World Quarterly* 5:4 (1972), 50–62; Grant, *Company Towns*, 1–6; Mohammed. Shahabudeen, *Nationalisation of*

Guyana's Bauxite: The Case of Alcan (Georgetown: Ministry of Information, 1981), 121–139; and Quamina, *The Mineworkers of Guyana*, 16–18, 99–100.

70. Odida T. Quamina, "The Social Organization of Plantation Mackenzie: An Account of Life in the Guyana Mining Enterprise," (Geneva: U. N. Research Institute for Social Development, 1981), 14, 17 and Quamina, *The Mineworkers of Guyana*, 19–23, 37; and Shahabudeen, *Nationalisation of Guyana's Bauxite*, 121–122.
71. Quamina, "The Social Organization of Plantation Mackenzie," 14, 17 and Quamina, *The Mineworkers of Guyana*, 19–23, 37; and Shahabudeen, *Nationalisation of Guyana's Bauxite*, 121–122.
72. *Report of the Diamond and Gold Industries, 1928*, 36 on regulations in South Africa; Daly, *A Short History of the Guyanese People*; Chase, *A History of Trade Unionism*, 34–37, 48–263; and Lowe, *Historical Analysis of Mining Policy*, 60. See Graham, et al., *The Heritage of World Civilizations*, 646; and Thomas E. Skidmore and Peter H. Smith, *Modern Latin America* (New York: Oxford University Press, 2001), 6–10, 44.
73. *Report of the Diamond and Gold Industries, 1928*, 36.
74. Interviews with former workers and residents, Darrell Gray, Eunice Williams, Harold "Paris" Thompson, Frank Proctor et al. who enjoyed life in the bauxite belt.
75. United States: Department of State, Foreign Service Despatch 66, 1957, American Consulate. Enclosure: "Paper on the Political, Economic, and Social Progress of British Guiana," 741D.00/10–1557, RG 59 (NACP:MD); "Demerara Bauxite Company To Increase Output," *The Daily Chronicle*, March 6, 1938, 1; Quamina, "The Social Organization," 24 and *The Mineworkers of Guyana*, 40.
76. *Report of the Commissioner of Lands and Mines, 1925*; Chase, *A History of Trade Unionism*, 28–33; Daly, *A Short History of the Guyanese*, 286–287; and Chris Curtis, *Demerara Doctor, An Early Success Against Malaria: The Autobiography of a Self-Taught Physician George Giglioli, 1897–1975* (London: Smith Gordon and Co. Ltd., 2006), 45–108.
77. Josiah, "Organizing Within the Diaspora." Guyana, LCP established in 1937. See Chase, *A History of Trade Unionism*, 124–133, 187, 262; Grant, *Company Towns*, 11–20; Shahabudeen, *Nationalisation of Guyana's Bauxite*, 121–124; Quamina, *The Mineworkers of Guyana*, 1–2,6, 14–26, 35–91; St. Pierre, "Industrial Unrest," 66–78, "Race the Political Factor," 481–493, and *Anatomy of Resistance*, 63–65.
78. Chase, *A History of Trade Unionism*, 124–133, 187, 262; Grant, *Company Towns*, 11–20; Shahabudeen, *Nationalisation of Guyana's Bauxite*, 121–124; Quamina, *The Mineworkers of Guyana*, 1–2, 6, 14–26, 35–91; St. Pierre, "Industrial Unrest," 66–78; "Race the Political Factor," 481–493; *Anatomy of Resistance*, 63–65.
79. St. Pierre, "The Sociology of Decolonization, 52–53.
80. Chase, *A History of Trade Unionism*, 124–133, 187, 262; Grant, *Company Towns*, 11–20; Shahabudeen, *Nationalisation of Guyana's Bauxite*, 121–124; Quamina, *The Mineworkers of Guyana*, 1–2, 6, 14–26, 35–91; St. Pierre, "Industrial Unrest," 66–78; "Race the Political Factor," 481–493; *Anatomy of Resistance*, 63–65.
81. Chase, *A History of Trade Unionism*, 126–127; Grant, *Company Towns*, 15–20; St Pierre, "Industrial Unrest," 66–78; 51–55; 482–489; Shahabudeen, *Nationalisation of Guyana's Bauxite*, 121–124; and Quamina, *The Mineworkers of Guyana*, 6, 35–38, 55.
82. TNA: COOC 111/795/6 British Guiana, 1948 "Labour Situation Bauxite Company, 1946–48," and U. S. Department of State. Correspondence from Reynolds Metals Company, Richmond, VA to Hon. James F. Byrnes, secretary of state dated June 20, 1946 and Foreign Service Despatch dated October 26, 1954. Decimal File (1910–1963) RG 59, 844B.6354/8–1249, 841D.2569/10–2154 (NACP: MD) See Chase,

A History of Trade Unionism, 126–127 and Quamina, *The Mineworkers of Guyana*, 56–57.

83. TNA: COOC 111/795/7 British Guiana, 1949, "Labour Situation in Bauxite Company, 1949." See *Colonial Annual Report, 1947*, 16–19 and Review Section 6, 18; *Colonial Annual Report, 1948*, 51.
84. *Colonial Annual Report, 1953*, 28–30, 46; Chase, *A History of Trade Unionism*, 126–128; St. Pierre, "Race the Political Factor," 492–493 and *Anatomy of Resistance*, 63–65; and Quamina, *The Mineworkers of Guyana*, 56–57.
85. *Colonial Annual Report, 1954*, 36; Chase, *A History of Trade Unionism*, 129–133, 187, 262; Daly, *A Short History of the Guyanese*, 295; Quamina, "The Social Organization," 26 and *The Mineworkers of Guyana*, 58–63, 88–91.
86. United States: Department of State. American Consul British Guiana. Despatch 66, 1957 Enclosure: "Paper on the Political, Economic, and Social Progress;" and Chase, *A History of Trade Unionism*, 132.
87. United States: Department of State. Despatch 66, 1957 Enclosure: "Paper on the Political, Economic, and Social Progress."
88. *Colonial Annual Report, 1949*, 20, 25; and *Colonial Annual Report, 1956*, 25–31.
89. *Colonial Annual Report, 1956*, 34–38.
90. *Colonial Annual Report, 1955*, 21, 25–26, 29–33; Chase, *A History of Trade Unionism*, 262; and Grant, *Company Towns*, 16–17.
91. *British Guiana: Annual Report of the Dept of Labour, 1956*, 13–16. See Table X(a) List of Trade Unions-31/12/56 and *Colonial Annual Report, 1956*, 34–38.
92. Quamina, *The Mineworkers of Guyana*, xii–xiii, 1–2, 5–6, 7–12, 69–110.

8 Internal Migration and Village Dynamics: Families and Communities Coping

1. *Royal Gazette: British Guiana, Jan. 19, 1869* on *Ordinance I of 1864* to restrict movement of labor; *British Guiana: Report of the Labour Commission, 1890* (Georgetown: C. K. Jardine, 1890), 54; Norman E. Cameron, *The Evolution of the Negro* (Westport: Negro Universities Press, 1990. Reprint 1929, ed.), 20, 27, 95; *Who Is Who in British Guiana* (Georgetown: *The Daily Chronicle*, 1940), 29–30, and Barbara P. Josiah, "The Other Migration: African Diaspora Sailors from Latin America and the Caribbean in the United States Civil War Navy," Paper presented at the *Association for the Study of World African Diaspora Conference*, Rio de Janeiro, Brazil, October 5–7, 2005. See *United States Naval Records Collection: Muster Rolls of Ships. Records of the Bureau of Navy Personnel*, RG 24 (National Archives, Washington, DC).
2. H. J. Perkins, *Notes on British Guiana and Its Gold Industry* (London: Sampson, Low, Marston & Co., 1896), 11–12, 36–39; Cedric Grant, *Company Towns in the Caribbean: A Preliminary Analysis of Christianburg-Wismar-MacKenzie* (Georgetown: Guyana Government Printers, 1971), 6–19; Odida T. Quamina, *The Mineworkers of Guyana: The Making of a Working Class* (Atlantic Highlands, NJ: Zed Books, 1987), 18–45.
3. Darrell Gray, interview by author, Queenstown, Essequibo, January 3, 1997; Frank H. Proctor, interview by author, Upper Tulse Hill, London, England, June 21, 2010; and other former gold and diamond miners interviewed in Washington, DC, US, Guyana, and London, England, 1995–2010.
4. Allan Young, *The Approaches to Local Self-Government in British Guiana* (London: Longmans, Green, and Co., Ltd, 1958), 143–161; Carl B. Greenidge, *Empowering A*

Peasantry in a Caribbean Context: The Case Of Land Settlement Schemes in Guyana, 1865–1985 (Kingston: University of the West Indies Press, 2001), 34–81; and Kean Gibson, *The Cycle of Racial Oppression in Guyana* (Lanham: University Press of America, 2003), 9–21. See Cameron, *The Evolution of the Negro*, 6–14, 18–21; Walter Rodney, *A History of the Guyanese Working People, 1881–1905* (Baltimore: The Johns Hopkins University Press, 1982), 31–89; James G. Rose, "African-Guyanese Business," in *Emancipation: African Guyanese Achievement* 2 (1994); Barbara P. Josiah, "Craftspersons in Guyana: Identity and Knowledge Transfer from Slavery to Present," in Emory J. Tolbert, ed., *Perspectives on the African Diaspora Volume One: To 1800* (Boston: Houghton Mifflin, 1998), 323–349 and "Creating Worlds: Aspects of Mutuality and Financing among African Guyanese, 1800s–1950s, *The Journal Of Caribbean History*, 38:1 (2004), 106–127; and David Granger, ed., *Emancipation: The African-Guyanese Magazine*, for the community's initiatives.

5. See Cameron, *The Evolution of the Negro*, 36–37 and Young, *The Approaches to Local Self-Government*, 162–188 on drainage and irrigation. Alan Adamson "Monoculture and Village Decay in British Guiana, 1854–1872," *The Journal of Social History* (1970), 386–405; Rodney, *A History of the Guyanese Working People*; 1–18, 76–84, 101–108; Brian Moore, *Race, Power, and Social Segmentation in a Colonial Society: Guyana After Slavery, 1838–1891* (New York: Gordon and Breach Science Publishers, 1987), 145–146 and Moore, "African Guyanese Political Disempowerment during the Nineteenth Century," in Winston F. McGowan, James G. Rose, and David A. Granger, eds., *Themes in African Guyanese History*, (Georgetown: Free Press, 1998), 235–250; and Kimani Nehusi, "The Causes of the Protest of 1905,"in Winston F. McGowan, James G. Rose, and David A. Granger, eds., *Themes in African-Guyanese History* (Georgetown: Free Press, 1998), 251–276, capture disempowerment.
6. Cameron, *The Evolution of the Negro*, 3–95. See Barbara P. Josiah After Emancipation: Aspects of Village Life in Guyana, 1869–1911," *The Journal of African American (Negro) History* (Winter 1997), 105–121.
7. Cameron, *The Evolution of the Negro*, 43. See A. R. F. Webber, *Centenary History and Handbook of British Guiana* (Georgetown: The Argosy Co. Ltd., 1931), 235.
8. *British Guiana: Report of the Inspector of Villages, 1889*, 7 compiled from *British Guiana: Report of the Comptroller of Customs and Excise.*
9. *Report of the Labour Commission, 1890*, 3, 1–5, 15–17 on West Coast Berbice and Mahaicony farmers who worked as miners and took earnings back to villages; Cameron *The Evolution of the Negro*, 25–30, 95.
10. *British Guiana: Report of the Local Government Board, 1911–1912*, 22; Cameron, *The Evolution of the Negro*, 59; *British Guiana: Report of the District Commissioner, East Coast Demerara, 1934*, on sugar cane, coconuts, plantains, rice, breadfruit, and livestock. See *Community Development Programme Golden Grove-Nabaclis* (Georgetown: Ministry of Local Government and Cooperative Development, 1956), 2.
11. *British Guiana: Report of the Department of Lands, 1901; British Guiana: Report of the Department of Mines, 1901*, and Cameron, *The Evolution of the Negro*, 30–32, 44–56.
12. TNA: CO 1031/1691, 1954–1956. *Report by F. A. Brown on Land Settlement in British Guiana, Appendix II*, West Indian Department File #170/67/01. Enclosure: "Extract From the Small Farmers Committee Report" of 1931.
13. *The Daily Argosy*, May 6, 1934, 12. See Rodney, *A History of the Guyanese Working People*, 68, note 32 on "Petition of Mocha Residents Forwarded by the British Anti-Slavery Society, 4 Nov., 1887" on effort by capitalist owners of Plantation Diamond to seize the

villagers' land; and Joseph Pollydore, "A Brief History of Mocha-Arcadia and Its People, 1894–1994," *Emancipation: African Guyanese Achievement* 3 (1995), 40–43.

14. Special Section, *The Echo*, January 22, 1897, on mining workers and coping mechanisms to deal with the adverse effects of indentured immigration on their labor and wages; Pollydore, "A Brief History of Mocha-Arcadia," 40–43; and James G. Rose, "The Strikes of 1842 and 1848" in Winston F. McGowan, James G. Rose, and David A. Granger, *Themes in African-Guyanese History* (Georgetown: Free Press,1998), 158–200.
15. *The Mining Gazette, Reports of the Institute of Mines and Forests, Reports of the Commissioner of Lands and Mines, Death Registers*, and newspapers identified persons from Mocha-Arcadia as mining workers. See Rodney, *A History of the Guyanese Working People*, 60–89 and Gibson, *The Cycle of Racial Oppression*, 1–21, for travails of villagers.
16. Joseph Watson, interview by author, Washington, DC, February 5, 1995,Winston Valentine, interview by author, Golden Grove Nabaclis Village, East Coast Demerara, October 26, 1996, and Benjamin Thomas, interview by author, Buxton Village, East Coast Demerara, November 3, 1996.
17. Rex Bosson and Benson Varon, *The Mining Industry and Developing Countries* (Oxford: Oxford University Press, 1977), 52–63.
18. Michael D. Bordo, "Gold Standard: The Concise Encyclopedia of Economics. Library of Economics and Liberty www.econeis.org/Library/Enc/Gold Standard. html, 2009, 1–6, http://www.econlib.org/library/Enc/GoldStandard.html and *Gold Standard*, 1–11; http://en.wikipedia.org/wiki/Gold_standard.
19. Michael P. Todaro, *Internal Migration in Developing Countries: A Review of Theory, Evidence, Methodology, and Research Priorities* (Geneva: I. L. O., 1976), 1–14, 26–27, 75–82. See, R. Jolly, ed., "Rural Urban Migration: Dimensions, Causes, Issues, and Policies in Education in Africa," in *Africa: Research and Action* (Nairobi: East Africa Publishing House, 1970); Donald Larson and Yair Mundlak, "On the Intersectoral Migration of Agricultural Labor," *Economic Development and Cultural Change* 45.2 (1997); and George Reid Andrews, *Blackness in the White Nation: A History of Afro-Uruguay* (Chapel Hill: The University of North Carolina Press, 2010), 96.
20. Aderanti Adepoju, "Migration and Socio-Economic Change: Policy and Research," in Jorge Balan, ed., *Why People Move: Comparative Perspectives on the Dynamics of Internal Migration* (Paris: UNESCO Press, 1981), 317–336. See Magnus Morner, *Adventurers and Proletariats: The Story of Migrants to Latin America* (Pittsburg: The University of Pittsburg Press, 1985), 67–85, 100–129; Walter Nugent, *Crossings: The Great Transatlantic Migrations, 1870–1914* (Bloomington: Indiana University Press, 1992), 112–135; and Lok Siu, "Migration Stories: Serial Migration and the Production of Home and Identity in Transnationalsm," in Andrew R. Wilson, ed., *The Chinese in the Caribbean* (Princeton: Markus Weiner Publishers, 2004), 159–190.
21. See *International Monetary Fund. Staff Country Reports. Guyana: Statistical Annex*, July 1995 #95/68, 39, 41; November 1996 #96/123, 45–46; February, 1998 #98/13, 10, 35, 37; June 1999 #99/52, 62; January 2001 #01/01, 33; and March 2010 #10/293, 4, 6 (International Monetary Fund Publications: Washington, DC).
22. Cameron, *The Evolution of the Negro*, 20, 53–63.
23. *The Echo*, January 18, 1896, 4 and Cameron, *The Evolution of the Negro*, 14–30. See Adepoju, "Migration and Socio-Economic Change," 317–336.
24. *The Echo*, January 18, 1896, 4 and Special Section, *The Echo*, January 22, 1897, on mining workers and coping mechanisms to deal with the adverse effects of indentured

immigration on their labor and wages; *The Echo*, July 6, 1897, on pursuit of education by African Guyanese was unwise; and *The Echo*, August 20, 1898, 1 on remarks by Corry in the presence of the governor at a ceremony to commemorate Queen Victoria at Den Amstel Village., WCD. TNA: COOC 111/607, British Guiana 1917. Enclosure: James McFarlane Curry retired April 1, 1917, @65 years old; Magistrate's Clerk for 35 years.

25. *Howard University Bulletin. Annual Catalogue, 1869*–1956; *Who Is Who in British Guiana, 1938–1940* (Georgetown: The Daily Chronicle, 1948); and Arthur and Elma Seymour, *Dictionary of Guyanese Biography*, (Georgetown: By the authors, 1984), identifies foreign trained—Samuel Herbert, J. M Rohlehr, Fred T. Wills, Robert S. Miller, Neville J. Abbensetts, J. N & C. C. Nicholson, Hyacinth Davis, Frank Williams, A. T. Peters, J. N. Fox, A. M. Cunningham, C. H. A. Denbow, Enid Denbow. Dr. Rohlehr also practiced at Bourda Hospital and its dispensary in the city and Victoria Dispensary, East Coast Demerara.
26. Edith Campbell Brown, *The Life Story of Andrew Benjamin Brown* (Georgetown: British Guiana Lithographic Co., 1947); *Who Is Who in British Guiana*; Seymour and Seymour, *Dictionary of Guyanese Biography*; Vere T. Daly, *A Short History of the Guyanese People* (London: Macmillan Education Ltd., 1975), 267–275. See, Barbara P. Josiah, "Brown, Andrew Benjamin," in Colin Palmer, ed., *Encyclopedia of African-American Culture and History: The Black Experience in the Americas* (Farmington Hills, MI: Thomson Gale & Macmillan Reference USA, 2006 ed.), 338–339.
27. Norman E. Cameron, *The History of Queen's College* (Georgetown: F. A. Persick Ltd., 1951); *The Evolution of the Negro*, 66–87; and *The History of Education in Guyana*, 1808–1957, Appendix and Table. *The Tribune*, May 25, 1930, 6, the school at UNIA's Liberty Hall organized by Paul Peters prepared students for the London Matriculation, Cambridge Senior, and Junior Examinations. See also *1st. Anniversary Publication, The County High School Alumni, 1956* (MD, 1997); Deon F. Abrams, "Education Tradition: Buxton Village" *Emancipation: African-Guyanese Achievement* 2:13 (2004–2005), 57–59.
28. *Howard University Bulletin. Annual Catalogue, 1869–1950*; Volda Williams, "Dr. Claude Hicks Augustus Denbow, 1911–1979," in *Emancipation: African-Guyanese Achievement* 2 (1994), 20–22; Barbara P. Josiah, "Organizing Within the Diaspora: Claude H. A. Denbow, Howard University, and The League of Coloured Peoples," *The Journal f African American History* 95:2 (Spring 2010), 229–247; Josiah, "From Shaw University to 'the Booker T. Washington of British Guiana:' Hon. E. F. Fredericks and Contributions of HBCUs to the African Diaspora," Paper presented at the *95th Annual Meeting, Association for the Study of African American Life and History Annual Conference*, Raleigh, NC, September 29,–October 3, 2010.
29. *The Economic Development of British Guiana. Report by the International Bank for Reconstruction and Development* (Baltimore: Johns Hopkins University Press, 1953), 23–40, 27–76, 351–355; and *British Guiana: Annual Report of the Labour Department, 1956*, 12.
30. Quamina, *The Mineworkers of Guyana*, 18–45.
31. United States: Department of State. American Consulate, British Guiana. Foreign Service Despatch 66 dated October 15, 1957. Enclosure: "Political, Economic and Social Progress in British Guiana" RG 59. 741D.00/10–1557 (NACP: MD), *Annual Report of the Labour Dept, 1956*, 13, and Quamina, *The Mineworkers of Guyana*, 40–41.
32. United States: Department of State. Despatch 66: "Political, Economic and Social Progress in British Guiana," and *Annual Report of the Labour Dept, 1956*, 13.

33. Darrell Gray, interview with author, Queenstown Village, Essequibo, January 3, 1997 (Gray died 1999); Edward "Guitar" Levans, interview with author, Vergenoegen Village, East Bank Essequibo, October 30, 1996. Levans, my cousin, a former headmaster, born at Den Amstel Village, WCD, son of a gold and diamond miner; his mother Salome assisted him to attend Central High School, Georgetown. His sibling Ronald, a gold and diamond seeker, drowned and was buried at Sorrow Hill, Bartica. Cambridge Academy first located on Hadfield Street; relocated to 226–227 South Road, Bourda; another branch was on Charlotte Street; a private high school in Georgetown which educated thousands of rural and urban students whose parents included miners. Robert Pinkerton was a long term principal. Dulcina Ross Armstrong, a founding member of the LCP provided such accommodations in Regent Street, Georgetown.
34. Joan O. LaRoc, interview by author, September 22, 2007, Beltsville, MD. LaRoc is Daniels' cousin.
35. Lenox Griffith, interview (telephone) by author, (Florida), July 1, 2006. Many students from Linden, the bauxite mining area attended Cambridge Academy including the Williams 'sisters and Maureen Fraser.
36. Josiah, "After Emancipation: Aspects of Village Life in Guyana, 103–121," on Guyana's Births and Marriage Registers.
37. Rowena Mingo, interview by author, Brooklyn, NY, November 16, 1997; Philomena Murray, interview by author, Zorg-en-Vlygt Village, Essequibo, November 19, 1996(died 2006); Betty Sheila Simon, interview by author, Riverdale, MD, USA, August 4–6, 2001 (born Uitvlugt, Guyana, died 2003). See Pollydore, *"A Brief History of Mocha-Arcadia*, 40–43; Gershom O. Browne, *A History of Bagotville Village* (Georgetown: The Free Press, 1999); and *The New Victoria: A Souvenir Magazine 170th Year. Redevelopment Through Revival of Our Traditions & Culture1* (September. 2009).
38. Cameron, *The Evolution of the Negro*, 20, 26, 30, 53–63.
39. *A Summary of Research on Internal Migration in Latin America: Bolivia, Brazil, and Peru* (Geneva: Secretariat for Migration, World Council of Churches, 1969), 6–11.
40. Cameron, *The Evolution of the Negro*; 29–30, 53–63 and Young, *The Approaches to Local Self-Government*, 33, 42, 53 60–66,147–53.
41. TNA: COOC, British Guiana, 1840. Enclosure: Papers Relative to the Affairs of British Guiana, dated May 15, 1840; TNA: CO 114/46–48, 1890. Sessional Papers, British Guiana, September 30, 1890. Appendix V. List of Persons to whom transports of Lots in Winkle Village are to be passed, 38–39; *Inducements to the Colored People of the United States to Emigrate to British Guiana* (Boston: Kidder and Wright, 1840), 24; and William N. Arno, *History of Victoria Village, East Coast Demerara* (By the author, ND) 7, 19.
42. Biplab Dasgupta, "Rural-Urban Migration and Rural Development: Causes and Consequences," in Jorge Balan, ed., *Why People Move: Comparative Perspectives on the Dynamics of Internal Migration* (Paris: UNESCO Press, 1981), 43–58.
43. *Daily Argosy*, January 8, 1933 and July 18, 1934, on the cassava and the coconut oil industry. Warder-Matthews family, interview by author, Stabroek Market, Georgetown, November 29, 1996 (basket makers); Norma Bryan, interview by author, Queenstown Village, Essequibo, November 24, 1996 (straw hats); Pearl Allen, interview by author, Richmond Village, Essequibo, November 23, 1996, (cassareep); Desiree Burnett, interview by author, Kingstown, Georgetown, October 19, 1996, (straw hats); Edna Adams, interview by author, Golden Grove-Nabaclis

Village, October 26 1996 (cassava bread); Paulette Cato (Kingston), interview with author, Victoria Village, E. C. D., November 18, 1996 (coconut oil); Franklin and Henley "Piggy" Assanah (fishermen) interview by author, Victoria Village, E. C. D. October 26, 1996. See Cameron, *The Evolution of the Negro*, 22–23, 42–63, 95; and Desmond F. Saul, *Family History of Patrick and Mildred Saul* (Victoria, East Coast Demerara: By the author, 1997), 9–10, 12–14.

44. Elsa M. Chaney, "Scenarios of Hunger in the Caribbean: Migration, Decline of Smallholder Agriculture and the Feminization of Farming," Working Paper #18, Michigan State University, March 1983, 9–10; and "Women Who Go, and Women Who Stay Behind," in *Migration Today* 10:3–4 *(1982)*, 8–13.
45. Marcello M. Veiga, "Artisanal Gold Mining Activities in Suriname," *United Nations Industrial Development Organization Report, 1997*, December 1–19; .Marieke Heemskerk, "Gender and Gold Mining: The Case of the Maroons of Suriname," Working Paper #269, *Women and International Development*, International Center (East Lansing: Michigan State University, 2000), 1–33; Fergus Mackay, "Mining in Suriname: Multinationals, the State and the Maroon Community of Nieuw Koffiekamp," in Lyuba Zarsky, ed., *Human Rights and the Environment Conflicts and Norms in a Globalizing World* (London: Earthscan Publications Ltd., 2002), 57–78; and "Small-Scale Gold Mining in Suriname," http://www.heemskerk.sr.org/GoldMining/GoldMining.html.
46. Interview, Mingo.
47. Ibid. See Judith A. Carney, *Black Rice: The African Origins of Rice Cultivation in the Americas* (Cambridge: Harvard University Press, 2001) and Edda Fields-Black, *Rice Farmers in West Africa and the African Diaspora* (Bloomington: Indiana University Press, 2008) on African-descent women in rice farming.
48. Interview, Mingo.
49. Information supplied by former miners and Evelyn Wright, interview by author, Mocha Village East Bank Demerara, October 17, 1996, her father, a diamond seeker drowned in the interior.
50. Interview, Watson, James Culley, interview by author, Buxton Village, E. C. D. October 24, 1996, Richard Dodson, interview by author, Blankenburg Village, W. C. D, December 29, 1996, and Sydney Dundas, interview by author, Anns Grove Village, E. C. D. December 31, 1996.
51. Lionel Arthur, interview with author, Mocha Village, East Bank Demerara, November 17, 1996.
52. Ibid.
53. Ibid.
54. Ibid., and Walter Bagot, interview by author, Den Amstel Village, W. C. D. December 29, 1996.
55. Lennox "Mboya" Wood, "Mammy Fiffee: A Tradition of Bush Medicine," in *Caribbean Daylight*, May 6, 1995, 8, 17 and *County High School, 1956, Alumni 1st Anniversary Publication, MD* (1997), 18–22; Josiah, "Craftspersons in Guyana," 323–349. See Cameron, *The Evolution of the Negro*, 102–103; K. O. Laurence, "The Development of Medical Services in British Guiana, 1841–1873," *Jamaican Historical Review IV* (1964), 60–67; and Marcel Fredericks John Lennon, Paul Mundy, and Janet Fredericks, eds., *Society and Health in Guyana: The Sociology of Health Care in a Developing Nation* (Durham: Carolina Academic Press, 1986); Deborah A. Lachman-White, C. Dennis Adams, and Ulric O. D. Trotz, *A Guide To Medicinal Plants of Coastal Guyana* (London: Commonwealth Science Secretariat, 1987), iv–v, 1–304; and A. J. Mc R. Cameron "Ethno-Medicine…Long Ago," *Stabroek News*, June 3, 1989, 11.

56. Interview, Watson.
57. *The Daily Chronicle*, December 13, 1930, 8; *The Daily Argosy*, March 9, 1933, 8; August 16, 1933; and November 23, 1933.
58. *The Daily Argosy*, January 5, 6, 9, 1934.
59. *The Daily Argosy*, January 12, 1934, 1.
60. *The Daily Argosy* January 10, 20, 30 and February 10, 1934. Josephus Headley was my fraternal grandfather.
61. Information on Charles Egerton Denbow I, County sergeant major provided by Dr. Charles E. Denbow II, late professor of Medicine and Cardiologist, University of the West Indies Hospital, Mona, Jamaica, September, 2007, from family papers, "Life of Charles Egerton Denbow."
62. Alan Carter, interview by author, Stabroek Market, Georgetown, December 31, 1996.
63. Pollydore Miggins, interview by author, Den Amstel Village, W. C. D. December 20, 1996, and David Hatton, interview by author, Edmonton, London November 24, 1997 and Queens, NY, November 20, 2008.
64. *The Tribune*, April 15, 1934, 1.

9 Knowledge Transfer and Cooperativism: Agriculture and Mining Eras

1. TNA: COOC 111/171, British Guiana. Despatch 48 dated April 9, 1840. Enclosure: "Affairs of British Guiana. Statement of Transports of Land Executed in Favour of Purchasers (formerly Apprenticed Labourers)... From August 1, 1838 to April 8, 1840"; James Rodway, *The Story of Georgetown* (Georgetown: Guyana Heritage Society, 1997. Reprint 1920, ed.), 202; Norman E. Cameron, *The Evolution of the Negro* (Westport: Negro Universities Press, 1970. Reprint 1929, ed.), 2–10; Barbara P. Josiah, "After Emancipation: Aspects of Village Life in Guyana, 1869–1911," *The Journal of African American (Negro) History 82 (1997)*, 105–121; Josiah, "Craftspersons in Guyana: Identity and Knowledge Transfer From Slavery To Present," in Emory J. Tolbert, *Perspectives on the African Diaspora. Volume One: To 1800* (Boston: Houghton Mifflin, 1998), 323–349; and Josiah, "Creating Worlds: A Study of Mutuality and Financing among African Guyanese, 1800s–1950s," *The Journal of Caribbean History* 38:1 (2004), 106–127.
2. Walter Rodney, *A History of the Guyanese Working People, 1881–1905* (Baltimore: The Johns Hopkins University Press, 1981), 75 and James G. Rose, " The Strikes of 1842 and 1848," in Winston F. McGowan, James G, Rose, and David A. Granger, eds., *Themes in African-Guyanese History* (Georgetown: Free Press, 1998), 158–200. See *Royal Gazette*, January 19, 1869, on legislation restricting the number of locals leaving the colony.
3. *British Guiana: Report of the Labor Commission, 1890* (Georgetown: C. K. Jardine, 1890); i–xix, 1–166; *British Guiana: Report of the Commissioner of Mines, 1895–96*, 3; Douglas Smith, "The Origins of Porknocking in Guyana," *Mc Gill Studies in Caribbean Anthropology* 5 (1969), 84; Rodney, *A History of the Guyanese Working People*, 169–170; and Kimani Neheusi, "The Causes of the Protest of 1905," in Winston F. McGowan, James G. Rose, and David A. Granger, eds., *Themes in African Guyanese History* (Georgetown: Free Press, 1998), 251–276. In the early 1900s, Secretary Hargreaves, Institute of Mines, suggested prosecution for those who facilitated locals migrating externally.
4. Toyin Falola and A. Adebayo, *Culture, Politics, and Money among the Yoruba* (New Brunswick: Transaction Pub., 2000), ix–127 and Josiah, "Creating Worlds: A Study of Mutuality," 106–127.

5. Shirley Ardener, "The Comparative Study of Rotating Credit Associations," *Journal of the Royal Anthropological Society* 94 (1964), 201–229; Aubrey W. Bonnett, *Institutional Adaptation of West Indian Immigrants to America: An Analysis of Rotating Credit Associations* (Washington, DC: University Press of America, 1981); Shirley Ardener and Sandra Burman, eds., *Money-Go-Rounds: The Importance of Rotating Savings and Credit Associations for Women* (Washington, DC: Berg Publishers, 1995), 1–2; and Falola and Adebayo, *Culture, Politics, and Money among the Yoruba*, ix–127.
6. Ardener, "The Comparative Study of Rotating Credit Associations," and Josiah, "Creating Worlds: A Study of Mutuality," 106–111.
7. TNA: COOC 111/110 British Guiana, 1830. Enclosure: "Account of the Berbice Savings Bank," dated March 12–September 2, 1830; *Ordinance #13, 1828*, referred to by J. Graham Cruickshank, "The Beginnings of Our Villages," *Timehri* 7:3 *(1921)*, 64–76, 69 n.1; Donna Danns, *The Development and Demise of the Savings Bank System in Guyana* (Georgetown: Bank of Guyana, 1986), 7–9, 18.
8. Danns, *The Development and Demise of the Savings Bank System.*
9. TNA: COOC 111/161, British Guiana. Enclosure: "Conditions of the Hill Coolies. Summary Return of Labourers, Domestic Servants... Emigrated to British Guiana Since August. 1834. Official Returns, Special Justices, Sept. 1838"; TNA: COOC 111/183, British Guiana, 1841, Enclosure: "Returns of Immigrants from Rio de Janeiro and Captured Africans." See J. Graham Cruickshank, "Among the Aku (Yoruba) in Canal # 1," *Timehri* 3:4 (1917), 70–81; Cruickshank, "African Immigrants After Freedom," *Timehri* 3:6 (1919), 74–85; Monica Schuler, "Kru Migration to British and French Guiana, 1841–1857," in Paul Lovejoy, ed., *Africans in Bondage: Studies in Slavery and the Slave Trade* (Madison: University of Wisconsin Press, 1986), 155–202; and David Northrup, *Indentured Labor in the Age of Imperialism, 1834–1922* (New York: Cambridge University Press, 1995), 2, 4–10, 16–28, 45–51.
10. Rodney, *A History of the Guyanese Working People*, 34–35, 86; James G. Rose, "The Repatriation Controversy and the Beginning of an East Indian Village System," *Guyana Historical Journal, University of Guyana* 1 (1989), 51–67; Carl B. Greenidge, *Empowering A Peasantry in a Caribbean Context: The Case of Land Settlement Schemes in Guyana, 1865–1985* (Kingston: University of the West Indies Press, 2001), 24–47; and Kean Gibson, *The Cycle of Racial Oppression in Guyana* (Lanham: University Press of America, 2003), 12–13.
11. Cruickshank, "Among the Aku (Yoruba) in Canal # 1" and "African Immigrants After Freedom," Josiah, "After Emancipation," 105–121; "Craftspersons in Guyana," 323–349; and "Creating Worlds: A Study of Mutuality," 111.
12. M. J. and F. S. Herskovits, *Trinidad Village* (New York: Alfred Knopf, 1947), 76–78, 257–265 and A. F. and D. Wells, *Friendly Societies in the West Indies: Report on a Survey* (London: H. M. Stationery Office, 1953), 11–12.
13. Henry Coleridge, *Six Months in the West Indies in 1825* (Westport: Negro Universities Press, 1970 ed.), 98–102, Josiah, "Providing for the Future: The World of the African American Depositors of Washington, DC's Freedmen's Savings Bank, 1865–1874," *The Journal of African American History* 89:1 (2004), 1–16.
14. TNA: COOC 111/440 British Guiana, 1887. Enclosure 392: "Report of Committee-Post Office Savings Banks, July 1887;" Section 36, *Savings Bank Ordinance, #6, 1889* dated September 11, 1889, cited in Danns, *The Development and Demise of the Savings Bank System,* 14; Howard University (HU): British Parliamentary Papers (BPP), 1910–1911, Section JJ 1; and "Birth, Growth, and Development of Banking in British Guiana. How Post Office and Other Banks Were Formed," *The Daily Chronicle,* December 29, 1936, 6.

15. TNA: COOC 111/190, British Guiana, 1842. Despatch 498 Enclosure: "The African Association of British Guiana;" Cameron, *The Evolution of the Negro*, 11; Alan Young, *The Approaches to Local Self-Government in British Guiana* (London: Longman and Green, 1958), 152–161; Rodney, *A History of the Guyanese Working People*, 108–109, 128–143; and Rose, "The Strikes of 1842 and 1848," 158–200.
16. *The Laws of British Guiana Vol. 1, 1774–1865* (Oxford: Oxford University Press, 1895) Ordinance #4, 1851 dated February 17, 1851 and Ordinance #1, 1852. See Cameron, *The Evolution of the Negro*, 13–31; Young, *The Approaches to Local Self-Government*, 47–64.
17. Rodney, *A History of the Guyanese Working People*, 108–109, 128–143, Brian Moore, "The Social and Economic Subordination of the Guyanese Creoles," in Winston F. McGowan, James G. Rose, and David A. Granger, eds., *Themes in African Guyanese History* (Georgetown: Free Press, 1998), 141–157; Eusi Kwayana, *Buxton-Friendship in Print & Memory* (Georgetown: Red Thread Press, 1999), 58–71.
18. Vere T. Daly, *A Short History of the Guyanese People* (London: Macmillan Education Ltd., 1975), 210–219; Rodney, *A History of the Guyanese Working People*, 108–109, 128–143; and Moore, "The Social and Economic Subordination," 141–157.
19. Young, *The Approaches to Local Self-Government*, 47–64; Reuben Kartick, "The Village Policy of Francis Hicks, 1862–1868," *History Gazette* 32 (1991), 1–8; and Kwayana, *Buxton-Friendship in Print & Memory*, 64–68.
20. Cameron, *The Evolution of the Negro*, 16–20; Young, *The Approaches to Local Self-Government*, 90–142; Brian Moore, "African-Guyanese Political Disempowerment during the Nineteenth Century," 235–250 and Kimani Neheusi, "The Causes of the Protest of 1905," in Winston F. McGowan, James G. Rose, and David A. Granger, eds., *Themes in African Guyanese History* (Georgetown: Free Press, 1998), 276–295.
21. Cameron, *The Evolution of the Negro*, 16–20; Young and *The Approaches to Local Self-Government*, 90–142.
22. Cameron, *The Evolution of the Negro*, 16–20; Young and *The Approaches to Local Self-Government*, 90–142.
23. Cameron, *The Evolution of the Negro*, 20–30; *Who Is Who In British Guiana* (Georgetown: *The Daily Chronicle*, 1940), 26; and Rodney, *A History of the Guyanese Working Peoples*, 58.
24. Cameron, *The Evolution of the Negro*, 27, 30, 95 on tax payments by gold and diamond miners; Lionel Arthur, interview with author, Mocha Village, E. B. D. November 17, 1996 and Winston Valentine, interview with author Golden Grove/Nabaclis Village, E. C. D. October 26, 1996. See Aderanti Adepoju, "Migration and Socioeconomic Change: Policy and Research," in Jorge Balan, ed., *Why People Move: Comparative Perspectives on the Dynamics of Internal Migration* (Paris: The Unesco Press, 1981), 317–336.
25. *British Guiana: Report of the Inspector of Villages, 1888*, 3; Cameron, *The Evolution of the Negro*, 20–23, 30.
26. *Report of the Inspector of Villages, 1889*, 7.
27. Ibid. *Report of the Inspector of Villages, 1890*, 3–7, Danielstown villagers overpaid rates by $14.76; *Report of the Inspector of Villages. Office of the Central Board of Health and Villages, 1892–1893*, 3–7; *Report of the Inspector of Villages, 1895–1896*, 3–6. See Cameron, *The Evolution of the Negro*, 27, 30, 95 on miners' tax payments ; and Moore, "The Political Disempowerment," 235–250.
28. Cameron, *The Evolution of the Negro*, 20.
29. *Report of the Inspector of Villages, 1892–1893*, 7 and Cameron, *The Evolution of the Negro*, 20.

30. Cameron, *The Evolution of the Negro*, 20. See *The Laws of British Guiana, 1803–1921* (London: Waterlow and Sons Ltd., 1923).
31. *British Guiana: Report of the Post Office, 1896–1897*, 151.
32. *Report of the Post Office Savings Banks*, 1900–1956; BPP *1910–1911*, JJ-2; "Birth, Growth, and Development of Banking in British Guiana," 6; Danns, *The Development and Demise of the Savings Bank System*, 4–36; and Josiah, "Creating Worlds: A Study of Mutuality," 111–114. See Table: Selected Post OfficeBank Branches, 113.
33. Danns, *The Development and Demise of the Savings Bank System*, 4–36, Table 7 and Josiah, "Creating Worlds: A Study of Mutuality," 111–114.
34. Lionel Arthur, Darrell Gray, Sydney Dundas, and Howell Wood, interviews by author, Guyana, 1996–1997.
35. Danns, *The Development and Demise of the Savings Bank System*, 4–39.
36. *British Guiana: Report of the Postmaster General, 1932* (Georgetown: The Argosy, 1933), 134.
37. "Post Office Savings Bank as Prosperity Barometer…Better Facilities for Miners," *The Daily Chronicle*,April 5, 1934, 4.
38. *British Guiana: Colonial Annual Report, 1953* (London: H. M. Stationery Office, 1954), 20–24.
39. Danns, *The Development and Demise of the Savings Bank System*, 11.
40. *Report of the Inspector of Villages, 1895–1896*, 3–6.
41. *Report of the Inspector of Villages, 1895–1896*, 3–6; *Report of the Inspector of Villages, 1897–1898; 1898–1899; and 1899–1900; Reports of the Central Board of Health, and Reports of the Local Government Board, 1900–1921* duplicate information.
42. *Report of the Inspector of Villages, 1901–1902* and *1902–1903*; *Report of the Inspector of Villages, 1902–1903*, Table 1. See Cameron, *The Evolution of the Negro*, 28–30.
43. TNA: COOC 111/607 British Guiana, 1916. Enclosure: "James McFarlane Corry, Magistrate's Clerk, West Coast Judicial District." Corry retired after thirty-five years in the Colonial Public Service at sixty-five years old. See Cameron, *The Evolution of the Negro*, 28–32.
44. *British Guiana: Report of the Commissioner of Lands and Mines, 1905–1906*, 11–13 and *1906–1907*, 9, 12.
45. *Report of the Commissioner of Lands and Mines, 1906–1907*, 9; *1907–1908*, 7, 10; *1908–1909*, 8–9. Appendix B and E; 1911–1912, 12–13; and *British Guiana: Report of the North Western District, 1907–1908*, 6. See Cameron, *The Evolution of the Negro*, 30.
46. *Report of the Inspector of Villages, 1911–1912*. Cameron, *The Evolution of the Negro*, 29–30.
47. *British Guiana: Report of the Banker's Committee of the Local Government Board and on the Cooperative Credit Banks Established in the Colony for the Year 1920* (Georgetown: The Argosy, 1921), 1–2; Cameron, *The Evolution of the Negro*, 8–32; and James G. Rose, "Africans in Guyanese History," in *Emancipation: African Guyanese Achievement* 2 (1994), 19–20.
48. *Report of the Banker's Committee*, 3–5. *Ordinance #9, 1914*, established the Cooperative Credit Banks.
49. *Report of the Banker's Committee*, 1.
50. *Report of the Banker's Committee*, 3–5; Cameron, *The Evolution of the Negro*, 28–32, 95; and "Anns Grove" in *Emancipation: The African-Guyanese Magazine* #7 (1999–2000), 32.

51. *British Guiana: Report of the Banker's Committee of the Local Government Board and on the Cooperative Credit Banks Established in the Colony for the Year 1930* (Georgetown: The Argosy, 1931), 1–10.
52. *Report of the Banker's Committee*, 3–7; and Cameron, *The Evolution of the Negro*, 30–32. See "The Cooperative Credit Societies of British Guiana," *The Daily Chronicle*, September 5, 1934, 4; and "Cooperative Credit Banks in British Guiana. Extending their Usefulness," *The Daily Chronicle*, September 5, 1934, 4.
53. *Report of the Inspector of Villages*, 1913–1916 and 1917–1956; "West Demerara Chairmen's Union," *The Daily Chronicle*, February 2, 1937, 2, 8. See Cameron, *The Evolution of the Negro*, 30.
54. *The Argosy*, January 15, 1956, 7 and February 10, 1956, 8.
55. *British Guiana*: *Report of the Labor Commission, 1890*, (Georgetown: C. K. Jardine, 1890), 16–17 identified farmers who also worked as miners and took their earnings back to the villages. See Cameron, *The Evolution of the Negro*, 25–30, 95.

10 African Continuities, Jewels, and Economic Linkages to Mining

1. Sir Walter Raleigh, *The Discovery of The Large, Rich, and Beautiful Empire of Guiana* (London: Hakluyt, 1845. Reprint, 1596 ed.), 1 and C. A. Harris and J. A. J. Villiers, *Storm Van's Gravesande: The Rise of British Guiana* (Nendeln, Liechtenstein: Kraus Reprint Ltd., 1967. Reprint, 1911 ed.), 61–62.
2. James E. Newton and Ronald L. Lewis, *The Other Slaves: Mechanics, Artisans, and Craftsmen* (Boston: G. K. Hall & Co., 1978), ix–xv, 171–245; and Beatrice Bridglall and Edmund W. Gordon, "The Nurturance of African American Scientific Talent," *The Journal of African American History* 89: 4 (Fall 2004), 331–347, note 6 quoting Lisa Jardine, *Ingenious Pursuits: Building the Scientific Revolution* (New York, 1996).
3. S. Allen Counter and David L. Evans, *I Sought My Brother: An African American Reunion* (Cambridge: Massachusetts Institute of Technology Press, 1981), 24–272; Jean Libby, "Technological and Cultural Transfer of African Iron making Into the Americas and the Relationship to Slave Resistance," 1993, http://www.afrigeneas.com/slavedata/Paper-LSU-1492–1992.html, 1–22; Judith Carney, *Black Rice: The African Origins of Rice Cultivation in the Americas* (Cambridge: Harvard University Press, 2001), 1–177; Claudius Fergus, "African Secret Societies: Their Manifestations and Functions in West Indian Culture," Paper presented, *The Thirty-Third Annual Conference of the Association of Caribbean Historians*, April 2–6 2001, University of the West Indies, St. Augustine, Trinidad and Tobago; Gwendolyn Midlo Hall, *Slavery and African Ethnicities in the Americas: Restoring the Links* (Chapel Hill: University of North Carolina Press, 2005), xiii–172.
4. TNA: CO 295/323, British Guiana (T) 71/391–446, 1817–1834, *Colonial Slave Registry 10,* henceforth, *Colonial Slave Registry 10*; TNA: CO111/183 Vol.7, 1841 "Return of Immigrants Rio de Janeiro and Captured Africans" Magistrate C. H. Strutt, Berbice; Guy Grannum, *Tracing Your West Indian Ancestors: Sources in the Public Record Office* (London: Public Record Office, 1990), 4–5, 23, 42, 88.
5. O. Nigel Bolland, *The Politics of Labour in the British Caribbean: The Social Origins of Authoritarianism and Democracy in the Labour Movement* (Kingston: Ian Randle Publishers, 2001), 658–665; Walter Rodney, *A History of the Guyanese Working People, 1881–1905* (Baltimore: The Johns Hopkins University Press, 1981), 1–30, 120–173; and Kean Gibson, *The Cycle of Racial Oppression in Guyana* (Lanham, MD: University Press of America, 2003), 1–21.

6. Joseph Harris, *The African Diaspora Map I and II*, (Atlanta: Clarke University Cartographic Services, 1990); Joseph E. Inikori, "The Origin of the African Diaspora: The Slave Trade From Africa," *Tarikh*, 5:4 (1978), 1–19; and James Walvin, *Atlas of Slavery* (Harlow, UK: Pearson Education Ltd., 2006), 23–120.
7. Maureen Warner-Lewis, *Guinea's Other Suns: The African Dynamic in Trinidad Culture* (Dover, MA: The Majority Press, 1991), xv–192; Michael A. Gomez, *Exchanging Our Country Marks: The Transformation of African Identities in the Colonial and Antebellum South* (Chapel Hill: The University of North Carolina Press, 1998), 1–290; Jacqueline L. Tobin and Raymond G. Dobard, *Hidden in Plain View: A Secret Story of Quilts and the Underground Railroad* (New York: Doubleday, 1999), 1–192; Toyin Falola and Akarmu Adebayo, *Culture, Politics, and Money among the Yoruba* (New Brunswick: Transaction Publishers, 2000), ix–127; Sheila S. Walker, ed., *African Roots/American Cultures: Africa in the Creation of the Americas* (Lanham, MD: Rowman & Littlefield Publishers, Inc., 2001), 1–358; and Midlo Hall, *Slavery and African Ethnicities in the Americas*; xiii–172.
8. *Colonial Slave Registry 10*; Grannum, *Tracing Your West Indian Ancestors*), 4–5, 23, 42, 88. See David Eltis, "The Traffic in Slaves Between the British West Indian Colonies, 1807–1833," in *Economic History Review Vol. 2#25* (1972), 55–64; Paul Fransworth, "From Past to the Present: An Exploration of the Formation of African-Bahamian Identity During Enslavement," in Jay B. Haviser, ed., *African Sites: Archaeology in the Caribbean* (Princeton: Markus Wiener Publishers & Kingston, Jamaica: Ian Randle Publishers, 1999), 94–130; Barbara P. Josiah, "Craftspersons in Guyana: Identity and Knowledge Transfer From Slavery to Present," in Emory J. Tolbert, *Perspectives on the African Diaspora. Volume One: To 1800* (Boston: Houghton Mifflin, 1997), 323–349; Josiah, "Tattoos, Body Marks, and Changing Identities: The Impact of Culture and Occupation on African American Sailors in the United States Civil War Navy, " paper presented at the *Annual Conference, The Association for the Study of African American Life and History*, Los Angeles, CA, October 1997; and Josiah, "Caribbean Migration to Guyana: Identity, Patterns, and Contributions From Slavery to W. W. I," in Emory J. Tolbert, ed., *Perspectives on the African Diaspora. Volume II: Since 1800* (Boston: Houghton Mifflin, 1999), 285–304.
9. Jaclyn J. Grier and Laurie Mercier, eds., *Mining Women: Gender in the Development of a Global Industry, 1670–2005* (New York: Palgrave Macmillan, 2006), 15–16 and Libby, "Technological and Cultural Transfer of African Ironmaking Into the Americas."
10. Sherwood A. J. Lowe, *Historical Analysis of Mining Policy, Law & Administration in Guyana: 1831–2002* (Georgetown: Pavnik Press, 2003), 17.
11. Raleigh, *The Discovery of The Large, Rich, and Beautiful Empire of Guiana*, 1; Harris and Villiers, *Storm Van's Gravesande: The Rise of British Guiana*, 61–62. See, William Sharp, *Forsaken But For Gold: An Economic Study of Slavery and Mining In the Colombian Choco, 1680–1834* (Chapel Hill: University of North Carolina Press, 1970); Ivan Van Sertima, *They Came Before Columbus The African Presence in Ancient America* (New York: Random House, 2003. Reprint, 1976, ed.) and Aline Helg, *Liberty and Equality in Caribbean Colombia, 1770–1835* (Chapel Hill: University of North Carolina Press, 2004), 44–46, 64, 67, 72, 140, 171;); and Kari Lydersen, "Who Went With Columbus? Dental Studies Give Clues." *The Washington Post*, May 18, 2009, A6.
12. *Colonial Slave Registry 10*. See entries # 120 and 123 dated July 17, 1817, Moses owned by John Kirton.

13. William Bosman, *A New and Accurate Description of the Coast of Guinea: Divided Into the Gold, the Slave, and the Ivory Coasts*. Translated from the Dutch (London: Frank Cass & Co. Ltd., 1967. Reprint, 1704 ed.), 119–128.
14. Ibid., Bosman, *A New and Accurate Description of the Coast of Guinea*. See Norman E. Cameron, *The Evolution of the Negro* (Reprint 1929, ed. Westport: Negro University Press, 1970), 146–170 and Patrick McNaughton, *Mande Blacksmith: Knowledge, Power, and Art in West Africa* (Bloomington: Indiana University Press, 1988), 22–39, 101–145.
15. Jerome S. Handler, "The Middle Passage and the Material Culture of Captive Africans," *Slavery and Abolition* 30:1 (March 2009), 1–26.
16. Grier and Mercier, *Mining Women*, 15–16.
17. Ibid.
18. Grier and Mercier, *Mining Women*, 15–16 and T. E. Anin, *Gold in Ghana* (London: Selwyn Publishers Ltd., 2nd ed., 1990), 1–16, 63, 69, 75.
19. Ibid. Richard W. Boyle, *Gold: A History and Genesis of Deposits* (New York: Van Nostrand Reinhold, 1987) 5–6; Anin, *Gold in Ghana*, 1–8, 63, 69, 75; and Raymond E. Dummett, *Eldorado in West Africa: The Gold-Mining Frontier, African Labor, and Colonial Capitalism in the Gold Coast, 1875–1900* (Athens: Ohio University Press, 1998), 1–4, 26–27, 65–66, 70, 266, 291–293.
20. Timothy F. Garrard, *Akan Weights and the Gold Trade* (London: Longman Group, 1980), xi–xiv, 1–36, 99–331; Herman Schadt, *Goldsmith's Art*. Translated from the German by Ann Potter Schadt. (Stuttgart: Arnoldsche, 1996), 7–13; Grier and Mercier, *Mining Women*, 15–16, 40–57; and Godehard Lenzen, *The History of Diamond Production and the Diamond Trade*. Translated from the German by F. Bradley. (Hendon: Barrie and Jenkin Ltd., 1970); and Peter Greenhalgt, *West African Diamonds, 1919–1983: An Economic History* (Manchester: Manchester University Press, 1985), 28–29.
21. *Colonial Slave Registry 10* occupations include goldsmiths, blacksmiths, and engineers. See Candice Goucher, "African-Caribbean Metal Technology: Forging Cultural Survivals in the Atlantic World," in Jay Haviser, ed., *African Sites Archeology in the Caribbean* (Kingston: Ian Randle Publishers and Princeton: Markus Wiener Publishers, 1999), 143–196.
22. Libby, "Technological and Cultural Transfer of African Ironmaking Into the Americas." and Hall, *Slavery and African Ethnicities*. 66–70.
23. Raleigh, *The Discovery of the Large, Rich, and Beautiful*, 1 and Harris and Villiers, *Storm Van's Gravesande*, 61–62.
24. Harold G. Levine, "African Explorers of the New World," *Crisis* (1962), 321–332; Jay B. Haviser, "Introduction, " in Jay B. Haviser, ed., *African Sites: Archaeology in the Caribbean* (Princeton: Makcus Wiener Publishers & Kingston, Jamaica: Ian Randle Publishers, 1999),1–6; George Reid Andrews, *Afro-Latin America* (Oxford: Oxford University Press, 2004) 70; Helg, *Liberty and Equality in Caribbean Colombia*; Hall, *Slavery and African Ethnicities*, 19–20, 66–70, 105. See Peter Wood, *Black Majority* (New York: Knopf, 1974); Daniel C. Littlefield, *Rice and Slaves* (Baton Rouge: Louisiana State University Press, 1981); Carney, *Black Rice: The African Origins of Rice Cultivation in the America;* and Carney, "With Grains in Her Hair: Rice in Colonial Brazil," in *Slavery and Abolition 25: 1* (2004), 1–27 and Carney, "Rice and Memory in the Age of Enslavement: Atlantic Passages to Suriname" in *Slavery and Abolition 26:3* (2005), 325–347; and Edda Fields, *Rice Farmers in West Africa and the African Diaspora*. (Bloomington: Indiana University Press, 2008).
25. Hall, *Slavery and African Ethnicities*, 67–68, 113–114, 116.

26. D. Graham Burnett, *Masters of All They Surveyed; Exploration, Geography, and A British Eldorado* (Chicago: University of Chicago Press, 2000), 13, 26,198 and *Lowe, Historical Analysis of Mining Policy*, 6–110.
27. Bridglall and Gordon, "The Nurturance of African American Scientific Talent," 331–347. See *Colonial Slave Registry, 10* ; Dummett, *Eldorado in West Africa*, 66,70; and Newton and Lewis, *The Other Slaves: Mechanics, Artisans, and Craftsmen*, ix–xv, 171–245.
28. Colonial Slave Registry, Slave Registers for 1832, Demerary and Essequibo, #312; Libby, "Technological and Cultural Transfer of African Ironmaking Into the Americas"; McNaughton, *Mande Blacksmith*, 22–39, 101–145; Haviser, "African-Caribbean Material Culture" in *African Sites: Archaeology in the Caribbean*, 143–275; and Bridglall and Gordon, "The Nurturance of African American," 331–332.
29. *Report on the Census, British Guiana, 1841–1931* and *The West Indian Census, 1946, British Guiana.*
30. TNA: COOC 111/183 British Guiana, 1841. See Monica Schuler, "Kru Migration to British and French Guiana, 1841–1857 in Paul Lovejoy, ed., *Africans in Bondage: Studies in Slavery and the Slave Trade* (Madison: University of Wisconsin Press, 1986), 155–202 and Barbara P. Josiah, "After Emancipation: Aspects of Village Life in Guyana, 1869–1911," *The Journal of African American (Negro) History 82* (1997), 105–121.
31. *Official Gazette, British Guiana, 1869*; James Rodway, *History of British Guiana from the year 1668*. 3 volumes, (Georgetown: J. Thompson, 1891–1894), 217–219, *Report of the Labour Commission, 1890*, 27; Cameron, *The Evolution of the Negro*, 27, 95; *Who Is Who In British Guiana*, (Georgetown: The Daily Chronicle Ltd., 1940), 138. Gamiel Herry, interview with author (telephone), Atlanta Georgia, October 2, 2005, his uncle left Grenada to mine gold in Colombia.
32. TNA: COOC 295/323, Trinidad and Tobago, 1887, and United States Naval Records Collection of the Office of Navy Records and Library, Record Group 45 and Bureau of Naval Personnel, RG 25 (Washington, DC: National Archives).
33. Cameron, *The Evolution of the Negro*, 45 and Bridgelall and Gordon, "The Nurturance of African American," 331–332.
34. *Who Is Who In British Guiana*, #s106, 26, 603; Joseph and "Piggy" Assanah, interview by author, Victoria Village, Guyana, October 26, 1996; and Cameron, *The Evolution of the Negro*, 89.
35. *British Guiana: Report of the Gold Industry and Crown Lands Commission, 1895* (Georgetown: C. K. Jardine Printers, 1895), 1–5 and Cameron, *The Evolution of the Negro*, 146–170.
36. J. B. Harrison, *British Guiana and Its Resources* (London: The West India Committee, 1907), 28; H. J. Perkins, *Notes on British Guiana and Its Gold Industry* (London: Sampson Low, Marston and Co., 1896), 10–11, Cameron, *The Evolution of the Negro*, 61–62 .
37. Harrison, *British Guiana and Its Resources*, 28.
38. *British Guiana: Report of the Diamond and Gold Commission, 1928* (Georgetown: The Argosy Ltd., 1928), 30. See Clem Seecharan, *"Tigers in the Stars:" The Anatomy of Indian Achievement in British Guiana, 1919–1929* (London: Macmillan Education Ltd., 1997), 2–16 for occupational categories of East Indians.
39. K. O. Laurence, *A Question of Labour: Indentured Immigration Into Trinidad and British Guiana, 1875–1917* (New York: St. Martins Press, 1994), 417–418 and *Preinvestment Study on the Enhancement of the Jewelery Industry in Guyana* (Dublin: Kilkenny Development Partners, 1994).

40. Carney, *Black Rice,* 4–11, quoting Sydney W. Mintz introduction to the 1990 edition of Melville Herskovits' *The Myth of the Negro Past.* See, Libby, "Technological and Cultural Transfer of African Ironmaking Into the Americas"; and Bridglall and Gordon, "The Nurturance of African American Scientific."
41. Cameron, *The Evolution of the Negro,* 44–56, 146–186; and Lowe, *Historical Analysis of Mining Policy,* 17. See Eric de Kolb, *Ashanti Goldweights* (New York: Gallery d'Hautbarr, 1968), 1–16; Ingrid Kuntzsch, *A History of Jewels and Jewellery* (New York: St. Martin's Press, 1981), 18–21; Hugh Tait, ed., *Jewelery, 7,000 Years: An International History and Illustrated Survey from the Collections of the British Museum* (New York: Harry N. Abrams, Inc., 1987), 11–20, 177–182, Translated from the German by Sheila Marnie; Deb Kumar Banerjee, *The Goldsmiths. A Study of An Occupational Group in Calcutta* (Calcutta: Calcutta University Press, 1983), 1–21; M. S. Shukla, *Panorama of Gems and Jewellery in Indian Historical Setting* (Varanasi, India: Kala Prakashan, 2003). See Harry P. Whitlock, *The Story of Gems* (New York: Lee Furman, 1936), 67–100; Michael Weinstein, *The World of Jewel Stones* (New York: Sheridan House, Inc., 1958), 3–46; Angela Fisher, *Africa Adorned* (New York: Abrams, 1984); John Sampson White, *The Smithsonian Treasury: Minerals and Gems* (Washington, DC: Smithsonian Institution Press, 1991), 7, 58; Victoria Finlay, *Jewels: A Secret History* (New York: Random House, 2006), 339–44, 218, 222–223, 236, 319–369; Blake Gopniz, "Taking a Shine To Jewelry", *The Washington Post,* 6 November 6, 2008, M3, on *Bedazzled: 5,000 Years of Jewelry.* Exhibition at the Walters Art Museum, Baltimore, MD; *Ghana Investment Profile: Fine and Custom Jewelery.* www.gipc.org.gh; and "New Delhi. Craftspersons get a Global Platform" *The Hindi,* January 30, 2011, 1. The World Crafts Council Global Jewellery Summit, 'Abhushan' included Africans, Asians, Latin Americans, and others.
42. *The Census Report, British Guiana, 1891.*
43. *The Census Reports, 1891* and 1911, Sections vii & xi–xv information on the number of gold workers, watchmakers, jewelers of African descent. In the *Census Report, 1931,* Section V 12–13, 28–29, 122–123.
44. *The Census Report, 1931,* Section V 12–13, 28–29, 122–123.
45. *Colonial Slave Registry, 10.* Cameron, *The Evolution of the Negro,* 146–186 and Goucher, African-Caribbean Metal Technology," 143–196.
46. *British Guiana: Death Registers, 1890s–1950s; Reports of the Institute of Mines and Forests of British Guiana, 1890s–1920s*; and contemporary newspapers. See Chapter 4.
47. See Carney, *Black Rice* and Bridglall and Gordon, "The Nurturance of African American."
48. Walter Bagot, interview by author, Den Amstel Village, West Coast Demerara, December 29, 1996.
49. Clifford Hector, interview by author, Hadfield Street, Georgetown, December 20, 1996, and Maurice July, interview by author, Peter's Hall Village, East Bank Demerara, December 22, 1996.
50. Hector, interview by author and July, interview by author. Other era jewelery businesses were S. M. Daniels, 90 Regent St., E. M. Reid, 19 Lombard St., and S. P. Moore 14 Camp St.
51. Hector, interview by author.
52. *Report of the Labour Commission, 1890,* evidence, T. J. Wakefield, 15–17; *Report of the Gold Industry, 1895,* 1–5, 93, and evidence, Hugh Sproston, 3. *Who Is Who in British Guiana,* 104–105; W. Henry Skerrett, "Jewelers Dedicated to Their Work,"

in *Stabroek News*, January 25, 1994, 9; and Skerrett, "Gaskin and Jackson, Silver Anniversary 25 Years," in *The Sunday Chronicle*, May 1991, 10.

53. Skerrett, "Jewelers Dedicated to Their Work," 9; and Skerrett, "Gaskin and Jackson, Silver Anniversary, 10.
54. Skerrett, "Gaskin and Jackson, Silver Anniversary," 10.
55. *Report of the Gold Industry, 1895*, 1–5; W. A. Swain, "The Gold and Diamond Mining Industry in Guyana," Unpublished Master's Thesis, University of London, 1980, 70–73; interviews, Hector and July; and "The Cultural Modification of Teeth," http://www.uic.edu/classes/osci/osci590/6–1. See United States: Department of State, American Consulate, British Guiana. Foreign Service Despatch "Mineral Trade and Production Annual Review" dated March 3, 1950. RG 841D.25/3–350 (NACP: MD).
56. *Report of the Labor Commission Report, 1890,* evidence, George Garnett, 1; *British Guiana: Report of the Postmaster General, 1896–1897,* 151; "Death of Mr. Edward Nassy. Noted B. G. Boat-Builder," *The Daily Chronicle*, June 20, 1935, 4. Nassy died June 19, 1935.
57. "Frightful Holiday Disaster... No Bodies Found," *The Daily Argosy*, December 24, 1935, 4.
58. See E. Daniel Potts and Annette Potts, "The Negro and the Australian Gold Rushes, 1852–1857," *Pacific Historical Review xxxvii:4* (November 1968), 381–399; Ian Duffield, "Martin Beck and Afro-Blacks in Colonial Australia," *Journal of Australian Studies* (1985), 3–20; Martha Putney, *Black Sailors: Afro-American Seamen and Whalemen Prior to the Civil War* (Westport: Greenwood Press, 1987); Michael Cohen and Michael Platz, *Black Men of the Sea* (New York: Dodd Mead and Co., 1978); George Baker Farr, *Black Odyssey: The Seafaring Traditions of Afro-Americans* (New York: Peter Lang Publishers, 1989) ; Rudolph M. Lapp, *Blacks in Gold Rush California* (New Haven: Yale University Press); for colonizers and emigrationists, Teresa Meade and Gregory Alonso Pirio, "In Search of the Afro-American 'Eldorado': Attempts by North American Blacks to Enter Brazil in the 1920s," *Luso-Brazilian Review XXXV: 1* (1988), 85–110; and Ramla M. Bandele, *Black Star: African American Activism in the International Political Economy* (Chicago: University of Illinois Press, 2008).
59. "Diamonds in Surinam. Samples Received Locally Between Corentyne and Nikerie," *The Daily Argosy* October 25, 1922, 4.
60. Ibid.
61. Ibid. See Guyana's Death Registers, 1890s–1950s for deceased African Americans and Chapters 1 and 4.
62. *Report of the Labor Commission Report, 1890*, evidence, George Garnett, 14 and *Report of the Postmaster General, 1896–1897,* 151.
63. *Report of the Labor Commission Report, 1890,* 15 and Michael Moohr, "The Discovery of Gold and the Development of Peasant Industries in Guyana, 1884–1914: A Study in the Political Economy of Change," in *Caribbean Studies* 15: 2 (1975), 57–72.
64. *The Echo*, January 18, 1896, 4. See Douglas Smith, "The Origins of Porknocking in Guyana" *McGill Studies in Caribbean Anthropology* (1969), 81–92.
65. J. Graham Cruickshank, "The Beginnings of Our Villages," *Timehri* 7:3 (1921), 69.
66. Kimani Neheusi, "The Causes of the Protests of 1905," in Winston F. McGowan, James G. Rose, and David A. Granger, eds., *Themes in African Guyanese History* (Georgetown: Free Press, 1998), 276–295.
67. James Rose, "African Guyanese Business," *Emancipation: The African Guyanese Achievement* 2 (1994), 35–37. See unpublished family papers, "Life of Charles

Egerton Denbow." Dr. Charles Egerton Denbow II, interview by author, Bowie, MD, September 28, 2007.

68. Cameron, *The Evolution of the Negro*, 20–32.
69. Edith Campbell-Brown, *The Life Story of Andrew Benjamin Brown* (Georgetown: British Guiana Lithographic, 1947), 13–14; and Barbara P. Josiah, "Andrew Benjamin Brown," in Colin A. Palmer, ed., *Encyclopedia of African-American Culture and History: The Black Experience in the Americas, 2nd ed.* (Farmington Hills, MI: Thomson Gale & Macmillan Reference, 2006 ed.), 338–339. On investments of Brown, Maclean Ogle, and other African Guyanese such as J. G. Stephen, Edward Levans, and Josephus Headley (my fraternal grandfather) See *The Daily Argosy*, January 3, 1956, 6; February 1, 1956, 1; and *The Daily Argosy*, 25 May 1956.
70. "The Negro and His Money. Example of Mr. J. G. Stephen", *The Daily Chronicle*, January. 1935, 6 and *The Daily Argosy*, January 3, 1956, 6.
71. *Who Is Who In British Guiana*, #s26, 106, 385, 502. Edward "Guitar" Levans, interview with author, Vergenoegen Village, East Bank Essequibo, October 30, 1996.
72. Dr. Walter B. Hope, interview by author, Howard University, Washington, DC, July 14, 2006, Aaron Hope's son. Alexander B. Gillis was his maternal grandfather.
73. Howell Wood, interview by author, Buxton-Friendship Village, East Coast Demerara, 15 0ctober 15, 1996.
74. *British Guiana: Report of the Department of Lands and Mines, 1911–1912*. Appendix B & E; 1913–1914, 12 and *Report of the Institute of Mines, 1913–1914*, 9.
75. *The Daily Argosy*, January 21, 1933, quoting *The Sunday Despatch*, January 1, 1933.
76. Ibid.
77. Ibid. Brown's remarkable discovery of gold is legendary in Guyana to this day.
78. Ibid.
79. *Report of the Department of Lands and Mines, 1937* referring to earlier spectacular gold finds.
80. "Fellow-Men. William Alexander Brown. "Pigeon Island Brown," *The Daily Argosy*, February 27, 1937. Supplement 10–11 and "Pigeon Island" Brown In Court. Charged With Disciples With Disorderly Conduct," *The Daily Chronicle*, March 12, 1937, 7.
81. Denis Angoy, "The King of Diamonds," *Guyana Chronicle*, May 7, 1967, and Rickey Singh, "Eldorado . . . More Than Just A Legend," *Sunday Graphic*, May 21, 1967, 8.
82. Ibid.
83. Angoy, "The King of Diamonds."
84. Ibid.
85. "Frederick Mahaica Eulogised Before Village Home Burial," *Stabroek New*, December 24, 1991, 9.
86. Interview with the grandson, Alfred Ince, by Calvin Niles, a family member, St. Phillips, Barbados, January 20, 2011.
87. Robert W. Boyle, *Gold: History and Genesis of Deposits* (New York: Van Nostrand Reinhold, 1987), 1; Rex Bosson and Benson Varon, *The Mining Industry and Developing Countries* (Oxford: Oxford University Press, 1977), 3–52; Perry Nance, *A Prospectors' Guide to Gold in Australia* (Wellington: A. H. Reed and A. W. Reed, Ltd., 1982), 12–13; Arthur B. Cummins and Ivan A. Given, eds., *SME Mining Engineering Handbook* (New York: Society of Mining Engineers . . . Inc., 1993), 9.2–9.8, 6–13.
88. Boyle, *Gold: History and Genesis*, v.
89. Mohammed Shahabudeen, *Nationalisation of Guyana's Bauxite: The Case of Alcan* (Georgetown: Ministry of Information, 1981), 124–126.
90. Ibid., and Odida T. Quamina, *The Mineworkers of Guyana: The Making of A Working Class* (Atlantic Highlands, NJ: Zed Books, 1987), 7, 99.

91. Shahabudeen, *Nationalisation of Guyana's Bauxite*, 124–126.
92. Ibid.
93. Funeral Service For the Late James Edmund Orna, March 14, 1914–April 7, 2007, held at St. Aidan's Anglican Church, Wismar, Linden, Wednesday, April 18, 2007, @2.30 P.M.
94. Ibid. Keith Easton, interview by author, Lanham, MD, December 8, 2007.
95. Ransford Palmer, *Pilgrims From the Sun: West Indian Migration to America* (New York: Twayne Publishers, 1995) and Vivienne Francis, *With Hope in Their Eyes: The Compelling Story of the Windrush Generation* (London: Nia, 1998). See International Monetary Fund. Staff Country Reports. Guyana: Statistical Annex, 10.293 (Washington, DC: International Monetary Fund Publications, 2010), 4, 6.

Selected Bibliography

Archives

College Park, MD. National Archives at College Park.
Georgetown. Guyana National Archives.
National Archives and Records Administration, Washington, DC.
Richmond, United Kingdom. National Archives of the UK.

Libraries

Caribbean Research Centre, University of Guyana.
Howard University Libraries, Washington, DC.
Library of Congress, Washington, DC.
London School of Economics, London.

Government Documents

British Guiana: Report of the Crown Surveyor, 1882.
British Guiana: Report of the Labor Commission, 1890.
British Guiana: Report of the Titles and Land Commissioners Claims to Land, Berbice and Essequibo, 1893.
British Guiana: Report of the Gold Industry and Crown Land Commission, 1895.
British Guiana: Report of the Commissioner of Mines, 1890–1903.
British Guiana: Report of the Lands and Mines Department, 1904–1957.
British Guiana: Mining Industries: Gold, Diamonds, Bauxite, Ironstone, and other Mineral Products: Commissioner of Mines. The Permanent Exhibition Committee, 1918.
British Guiana: Report of the British Guiana Commission, 1927. London: H. M. Stationery Office, 1928.
British Guiana: Report of the Diamond and Gold Commission, 1928.
Colonial Office: Report of the British Guiana and British Honduras Settlement Commission. London: H. M. Stationery Office, 1948.
Gazetteer of Guyana. Georgetown: Guyana Government, 1974.
Register of Births, Demerara and Essequibo, 1869, 1898, 1901, 1911.
Register of Index of Births, Demerara and Essequibo, 1897.
Register of Deaths, Demerara and Essequibo, 1890–1956.
West Indian Royal Commission, 1897. Report and Recommendations.
West Indian Royal Commission, 1938–1939. Recommendations. London: H. M. Stationery Office, 1940.

Guyana National Archives

Report of the Decennial Census of the Population of British Guiana, 1861–1911.

Howard University Libraries (Founders)

British Parliamentary Papers, Colonies General, 1890–1899. Dublin: Irish University Press, 1900.

British Parliamentary Papers, Colonies West Indies, 1877–1898. London: Harrison and Sons, 1899.

Library of Congress

British Guiana: Administrative Reports, 1888–1957.

British Guiana: Colonial Annual Reports, 1946–1957.

British Guiana: Report on the Blue Books for Statistics, 1895–1957.

British Guiana: Report of the Banker's Committee... the Cooperative Credit Banks, 1920, 1930.

The Gold Mining Gazette.

Report of the Institute of Mines and Forests of British Guiana, 1892–1920s.

Report of the Decennial Census of the Population of British Guiana, 1891–1931.

The West Indian Census, 1946.

National Archives at College Park and Washington, DC

Records of the United States Department of State, RG 59.

United States Naval Records Collection of the Office of Navy Records and Library, RG 45.

Bureau of Naval Personnel, RG 25.

National Archives, UK (Public Record Office)

CO 111, Guyana (British Guiana) Original Correspondence.

CO 114, Guyana (British Guiana) Sessional Papers.

CO 115, Guyana (British Guiana) Government Gazettes.

CO 116, Guyana (British Guiana) Miscellanea.

CO 345, Guyana (British Guiana) Registers of Correspondence, 1850–1951.

CO 941 Guyana (British Guiana) Colonial Office Lists.

CO 1031Guyana (British Guiana) Original Correspondence from 1951.

T/71, Records of the Treasury, Colonial Slave Registry. British Guiana.

Newspapers

The Daily Argosy, Georgetown, Guyana.

Caribbean Daylight, Brooklyn, NY, USA.

Caribbean Life, Brooklyn, NY, USA.

The Echo, Plaisance, Guyana.

The Glasgow Herald, Glasgow, Scotland, UK.

The Daily Chronicle, Georgetown, Guyana.

Guyana Graphic, Georgetown, Guyana.

New York Amsterdam News, NY, USA.
The New York Times, NY, USA.
The People, Georgetown, Guyana.
The Sentinel, Georgetown, Guyana.
The Sunday Despatch, London, England.
Sunday Stabroek News, Georgetown, Guyana.
The Tribune, Plaisance, Guyana.
The Washington Post, Washington, DC USA.

THESIS

Josiah, Barbara P. "Providing for the Future: The World of the Black Depositors of the Washington, DC Branch of the Freedmen's Savings Bank, 1865–1874." Unpublished Masters Thesis, Howard University, 1992.

Lancaster, Alan. "An Unconquered Wilderness: A Historical Analysis of the Failure to Open Up the Interior of British Guiana, 1838–1919." Unpublished Masters Thesis, University of Guyana, 1979.

Swain, W. A. "The Gold and Diamond Mining Industry in Guyana.' Unpublished Masters Thesis, University of London, 1980.

DISSERTATIONS

Josiah, Barbara P. "Internal Migrants: African Guyanese and the Development of the Gold and Diamond Industries, 1890–1956." Unpublished PhD. Diss., Howard University, 1998.

Moohr, Michael. "Patterns of Change in an Export Economy, British Guiana, 1830–1914." Unpublished PhD. Diss., Cambridge University, 1971.

ARTICLES AND BOOKS

Adams, Kathleen, "Work Opportunity and Household Organization Among the Barima Caribs of Guyana."*Anthropos, Vol. 74* (1979).

Adamson, Alan H. "Monoculture and Village Decay in British Guiana, 1854–1872." *The Journal of Social History* (1970).

———. *Sugar Without Slaves: The Political Economy of British Guiana, 1838–1914*. New Haven: Yale University Press, 1972.

Adepoju, Aderanti. "Migration and Socio-Economic Change: Policy and Research," in Jorge Balan, ed. *Why People Move: Comparative Perspectives on the Dynamics of Internal Migration*. Paris: The Unesco Press, 1981.

Amin, Samir. "Growth is Not Development." *Development Forum United Nations Center for Economic and Social Information, 1:3* (April 1973).

Angoy, Dennis. "The King of Diamonds." *Guyana Chronicle*, May 7, 1967.

Anin, T. E. *Gold in Ghana*. London: Selwyn Publishers Ltd., Second Ed., 1990.

Ardener, Shirley, "A Comparative Study of Rotating Credit Associations." *Journal of the Royal Anthropological Institute Journal, 94* (1964).

Ardener and Sandra Burman, eds., *Money-Go-Rounds: The Importance of Rotating Credit Associations for Women* (Washington, DC: Berg Publishers, 1995.

Arno, William, *A History of Victoria Village, East Coast Demerara*. Georgetown: British Guiana Lithographic Company, Ltd., 1964.

Banerjee, Deb Kumar. *The Goldsmiths: A Study of an Occupational Group in Calcutta.* Calcutta: Calcutta University Press, 1983.

Baptise, Fitzroy Andre. "The Exploitation of Caribbean Bauxite and Petroleum, 1914–1945," *Social and Economic Studies 37:1&2* (1988).

Blackburn, Robin. *The Overthrow of Colonial Slavery, 1776–1848*. London: Versco, 1988.

Bolland, O. Nigel. *The Politics of Labour in the British Caribbean: The Social Origins of Authoritarianism and Democracy in the Labour Movement.* Kingston: Ian Randle Publishers, 2001.

———. *Colonialism and Resistance in Belize: Essays in Historical Sociology.*

Belize: Cubola Books & Kingston: University of the West Indies Press, 2003, ed. Bonnett, Aubrey W. *Institutional Adaptation of West Indian Immigrants to America: An Analysis of Rotating Credit Associations.* Washington, DC: University Press of America, 1981.

Bordo, Michael D. Gold Standard: The Concise Encyclopedia of Economics. Library of Economics and Liberty www.econeis.org/Library/Enc/Gold Standard.html, 2009.

Bosson, Rex and Benson Varon. *The Mining Industry and Developing Countries.* Oxford: Oxford University Press, 1977.

Boyle, Robert W. *Gold: History and Genesis of Deposits.* New York: Van Nostrand Reinhold, 1987.

Bracewell, S. "The Search for Minerals in British Guiana," *Timehri* 29 (1950).

Brands, H. W. *The Age of Gold: The California Gold Rush and the New American Dream*. New York: Random House, Inc., 2002.

Braveboy-Wagner, Jacqueline A. *The Venezuela-Guyana Border Dispute: British Colonial Legacy in Latin America.* Boulder: Westview Press, 1984.

Bridglall, Beatrice L. and Edmund W. Gordon. "The Nurturance of African American Scientific Talent," *The Journal of African American History Vol. 89:4* (Fall 2004).

Brooks, George E. *The Kru Mariner of the Nineteenth Century.* Newark, DE: Liberian Studies Association, 1972.

Buck, E. C. "Proposed Railway Development of the Hinterland of British Guiana," *Timehri* (1919).

Buckner, Timothy R. "The Slave Trade Apex in the Eighteenth Century." Toyin Falola and Kevin D. Roberts, eds. *The Atlantic World, 1450–2000.* Bloomington: Indiana University Press, 2008.

Burnett, D. Graham. *Masters of All They Surveyed: Exploration, Geography, and A British El Dorado.* Chicago: University of Chicago Press, 2000.

Burns, E. Bradford. *The Poverty of Progress: Latin America in the Nineteenth Century.* Berkeley: University of California Press, 1980.

Burrowes, H. S. *Fifty Years of Flying in British Guyana. Georgetown:* The Guiana Graphic Ltd., 1963.

Buxton Pride. Vol.3 #1 (March 1993).

Buxton Pride. Special Reunion Edition. 153 Years Anniversary Pub. (Aug. 1993).

Cameron, Norman E. *The Evolution of the Negro.* Reprint, 1929–1934. Westport: Negro University Press, 1970.

———. *The History of Education in British Guyana: 150 Years of Education in Guyana.* Georgetown: Autoprint Guyana Ltd., 1968.

Campbell-Brown, Edith M. *The Life Story of Andrew Benjamin Brown.* Georgetown: British Guiana Lithographic, 1947.

Canterbury, Dennis C. *Guyana's Gold Industry: Evolution, Structure, Impacts & Non-Wage Benefits.* Turkeyen: University Of Guyana, 1998.

Carberry, Edward. *Inducements to the Colored People of the United States to Emigrate to British Guiana.* Boston: Kidder and Wright, 1840.

Carney Judith A. *Black Rice: The African Origins of Rice Cultivation in the Americas.* Cambridge: Harvard University Press, 2001.

———. "With Grains in Her Hair: Rice in Colonial Brazil," *Slavery and Abolition 25:1* (April 2004).

———. "Rice and Memory in the Age of Enslavement: Atlantic Passages to Suriname" *Slavery and Abolition 26:3* (December 2005).

Carrington, Selwyn H. H. "The United States and Canada: The Struggle for the British West Indian Trade," *Social and Economic Studies 37:3–4* (1988).

Chaney, Elsa M. "Women Who Go, and Women Who Stay Behind," *Migration Today 10:3–4* (1982).

Chase, Ashton. *A History of Trade Unionism in Guyana, 1900 to 1964.* Georgetown: New Guyana Company Ltd., 1964.

CIMBUX, Inc. The Committee For the Improvement of Buxton, Inc., Presents "Log Gaff" A Play by Keith Easton and Kwesi Oginga, December 18, 2004.

Combes, Stafford. "Interior Communications for British Guiana," *Timehri new series 5* (1891).

Connolly, H. J. C. *Economic Geological Survey.* London: Crown Agents for the Colonies, 1925.

Counter, S. Allen and David L. Evans, *I Sought My Brother: An African American Reunion*. Cambridge: Massachusetts Institute of Technology Press, 1981.

Cowell, Alan. "A New Generation of DeBeers: Diamonds and It Seems the Oppenheimers Are Forever," *New York Times*, (April 25, 2003).

Craig, Albert M., William A. Graham, Donald Kagan, Steven Ozment, and Frank M. Turner, eds. *The Heritage of World Civilization.*9th ed. Upper Saddle River, New Jersey: Prentice Hall, 2009.

Craton, Michael. "Continuity Not Change: The Incidence of Unrest Among Ex-Slaves in the British West Indies, 1838–1876," in Hilary Beckles and Verene Shepherd, eds.*Caribbean Freedom: Economy and Society From Emancipation to the Present.* Kingston, Jamaica: Ian Randle Publishers and Princeton: Markus Wiener Publishers, 1993.

Cruickshank J. Graham. "Among the Aku (Yoruba) in Canal #, West Bank Demerara," *Timehri 3:4*) (1917).

———. "African Immigrants After Freedom," *Timehri 3:6* (1919).

———. "The Beginnings of Our Villages," *Timehri 7:3* (1921).

Cudjoe, Selwyn R. *Caribbean Visionary: A. R. F. Webber and the Making of the Guyanese Nation.* Jackson: University of Mississippi Press, 2009.

Cummings, Leslie P. *Geography of Guyana.* London: Collins Clear Type Press, 1975.

Cummins, Arthur B and Ivan A. Given. *SME Mining Engineering Handbook.* New York: Society of Mining Engineers…Inc., 1973.

Curtis, Chris, ed. *Demerara Doctor, An Early Success Against Malaria: The Autobiography of a Self-Taught Physician, George Giglioli, 1897–1975.* London: Smith Gordon & Co. Ltd., 2006.

Daly, Vere T. *The Making of Guyana.* London: Macmillan Education Ltd, 1974.

———. *A Short History of the Guyanese People.* London: Macmillan Education Ltd, 1975.

Danns, Donna F. *The Development and Demise of the Savings Bank System in Guyana.* Georgetown: Bank of Guyana, 1986.

Dasgupta, Biplab. "Rural-Urban Migration and Rural Development: Causes and Consequences," in Jorge Balan, ed. *Why People Move: Comparative Perspectives on the Dynamics of Internal Migration.* Paris: The UNESCO Press, 1981.

David, Wilfred. *The Economic Development of Guyana, 1953–1964.* Oxford: Oxford University Press, 1984.

D'Souza, Dinesh, "Two Cheers for Colonialism." *The Chronicle of Higher Education*, May 2002.

Dorr, Gregory Michael. *Segregation's Science: Eugenics and Society in Virginia*. Charlottesville: The University of Virginia Press, 2008.

Duignan, Peter and L. H. Gann, eds. *Colonialism in Africa, 1870–1960. Volume 4* The Economics of Colonialism. (Cambridge: Cambridge University Press, 1975.

Dumett, Raymond E. *El Dorado in West Africa: The Gold Mining Frontier, African Labor, and Colonial Capitalism in the Gold Coast, 1875–1900*. Athens: University of Ohio Press, 1998.

Dunn, W. A. "The Gold Industry of British Guiana," *Timehri 3:2* (1912).

Elder, J. D. *African Survivals in Trinidad and Tobago*. London: Karia Press, 1988.

Elkins, Caroline. *Imperial Reckoning: The Untold Story of Britain's Gulag in Kenya*. New York: Henry Holt & Co., 2005.

Eltis, David. "The Traffic in Slaves Between the British West Indian Colonies, 1807–1833," *Economic History Review 2:25* (1972).

Eltis, David and David Robinson. *Extending the Frontiers: Essays on the New Transatlantic Slave Trade Database*. New Haven: Yale University Press, 2008.

Eudell, Demetrius L. *The Political Languages of Emancipation in the British Caribbean and the U.S. South*. Chapel Hill: University of North Carolina Press, 2002.

Falola, Toyin and Akarmu Adebayo. *Culture, Politics, and Money among the Yoruba*. New Brunswick: Transaction Publishers, 2000.

Farley, Rawle, "The Rise of the Village Settlements of British Guiana," *Caribbean Quarterly 3* (1953).

———. "The Shadow and the Substance," *Caribbean Quarterly 4:2* (1955).

Fergus, Claudius. "African Secret Societies: Their Manifestations and Functions in West Indian Slave Culture." Paper presented at the *33rd Annual Conference of the Association of Caribbean*, University of the West Indies St. Augustine, Trinidad and Tobago. April 2–6 2001.

Ferguson, Niall. *Empire: The Rise and Demise of the British World Order and the Lessons For Global Power*. New York: Basic Books, 2003.

50th. Anniversary, THD. Fifty Years of Service, Transport and Harbours Department, January 1931–December 1980. Georgetown: Guyana National Printers Ltd., 1980.

Fields-Black, Edda. *Rice Farmers in West Africa and the African Diaspora*. Bloomington: Indiana University Press, 2008.

1st. Anniversary Publication. County High School Alumni, 1956. MD 1997.

Fisher, Angela. *Africa Adorned*. New York: Harry N. Abrams, 1984.

Fletcher, L. P. "The Decline of Friendly Societies in Grenada: Some Economic Aspects," *Caribbean Studies 12:2* (July 1972).

———. "The Decline of Friendly Societies in Barbados," *Caribbean Studies 15:4* (1976).

———. "The Friendly Societies in St. Lucia and St. Vincent," *Caribbean Studies. 18:3–4* (Oct. 1979).

———. "Politics, Public Policy, and Friendly Societies in Trinidad and Tobago," *Social and Economic Studies 39:3* (1990).

Francis, Vivienne. *With Hope in Their Eyes: The Compelling Story of the Windrush Generation*. London: Nia, 1998.

Francis, William and John Mullen, eds. H*andbook of British Guiana, 1922*. Georgetown: The Argosy Co. Ltd., 1922.

Fredericks, Marcel, Paul Mundy, and Janet Fredericks, eds. *Society and Health in Guyana: The Sociology of Health Care In a Developing Nation*. Durham: Carolina Academic Press, 1986.

Frost, Diane. *Work and Community Among West African Migrant Workers Since the Nineteenth Century.* Liverpool: University of Liverpool Press, 1999.

Garrard, Timothy F. *Akan Weights and the Gold Trade.* London: Longman Group, 1980.

Gibson, Kean. *The Cycle of Racial Oppression in Guyana.* Lanham: University Press of America, 2003.

Gilsean, A. R. "Rice Cultivation in British Guiana," *Timehri new series 1* (1887).

Girvan, Norman. "Rethinking Development: Out Loud," in *Rethinking Development.* Judith Wedderburn, ed. Kingston: Institute of Social and Economic Research, (1991).

Gobineau, Arthur de. *The Inequality of Human Races, 1853–1855.* New York: G. P. Putnam's Sons, 1915, ed. Godehard, Lenzen. *The History of Diamond Production and the Diamond Trade.* Translated from the German by F. Bradley. Hendon: Barrie and Jenkin, 1970.

Gold Standard. http://en.wikipedia.org/wiki/Gold_Standard, 4/2/2009.

Gomez, Michael A. *Exchanging Our Country Marks: The Transformation of African Identities in the Colonial and Antebellum South.* Chapel Hill: U. N. C. Press, 1998.

———. *Reversing Sail: A History of the African Diaspora.* Cambridge: Cambridge University Press, 2005.

Gonzales, Michael, *Plantation Agriculture and Social Control in Northern Peru, 1875–1933.* Austin: University of Texas Press, 1985.

Goslinga, Cornelius C. H. *The Dutch in the Caribbean and in the Guianas, 1680–1791* Dover: Van Gorcum and Co., 1985.

Graham, Ronald. *The Aluminum Industry and the Third World.* London: Zed Books,1982.

Grant, Cedric. *Company Towns in the Caribbean: A Preliminary Analysis of Christianburg-Wismar-Mackenzie.* Georgetown: Government Printers, 1971.

Grant, Madison. *The Passing of the Great Race; or the Racial Basis of European History.* New York: Charles Scribner, 1916.

Granum, Guy. *Tracing Your West Indian Ancestors:Sources in the Public Record Office.* London: Public Record Office, 1995.

Greaves, Thomas and William Culver, eds. *Miners and Mining in the Americas.* Manchester: Manchester Press, 1985.

Green, William A. *British Slave Emancipation, the Sugar Colonies, and the Great Experiment, 1830–1865.* Oxford: Oxford University Press, 1976.

Greenidge, Carl B. *Empowering A Peasantry in a Caribbean Context: The Case of Land Settlement Schemes in Guyana, 1865–1985.* Mona, Kingston: The University of the West Indies Press, 2001.

Grier, Jaclyn J. and Laurie Mercier, eds. *Mining Women: Gender in the Development of a Global Industry, 1670–2005.* New York: Palgrave Macmillan, 2006.

Hall, Douglas. "The Flight From the Estates Reconsidered: The British West Indies, 1838–1842," in Hilary Beckles and Verene Shepherd, eds. *Caribbean Freedom: Economy and Society From Emancipation to the Present.* Ian Randle Publishers: Kingston, Jamaica and Princeton: Markus Wiener Publishers, 1993.

Handbook of Natural Resources of British Guiana. Georgetown: The Daily Chronicle Printers and Publishers, 1946.

Handler, Jerome S. "The Middle Passage and the Material Culture of Captive Africans," in *Slavery and Abolition 30:1* (March 2009).

Harlow, V. T., Ed. *The Discovery Of The Large And Beautiful Empire Of Guiana By Sir Walter Raleigh.* London: The Argonaut Press, 1928, ed.

Harris, C. A. and J. A. J Villiers. *Storm Van's Gravesande: The Rise of British Guiana.* Reprint Hakluyt Society, 1911, ed. Nende/Lechtenstein: Kraus Reprint Ltd., 1967.

Harrison, J. B. *British Guiana and Its Resources.* London: The West India Committee, 1907.

Haviser, Jay B., Ed. *African Sites: Archaeology in the Caribbean*. Kingston, Jamaica: Ian Randle Publishers and Princeton: Markus Wiener Publishers, 1999.

Hay, Douglas and Paul Craven, eds. *Masters, Servants, and Magistrates in Britain and the Empire, 1562–1955*. Chapel Hill: University of North Carolina Press, 2004.

Heemskerk, Marieke. "Gender and Gold Mining: The Case of the Maroons of Suriname." *Women and International Development*. East Lansing: Michigan State Univ., 2000.

Helg, Aline. *Liberty and Equality in Caribbean Colombia, 1770–1835*. Chapel Hill: University of North Carolina Press, 2004.

Herskovits, M. J. and F. S. *Trinidad Village*. New York: A. Knopf, 1947.

Higman, B. W. Slave *Populations of the British Caribbean, 1807–1834*. Baltimore: The Johns Hopkins University Press, 1984.

Hochschild, Adam. *King Leopold's Ghost: A Story of Greed, Terror, and Heroism in Colonial Africa*. Boston: Houghton Mifflin, 2006, ed.

Holloway, Steven. *Aluminum Multinationals and the Bauxite Cartel*. New York: St. Martin's Press, 1988.

Holloway, Thomas. *Immigrants on the Land: Coffee and Society in Sao Paulo, 1886–1934*. Chapel Hill: University of North Carolina Press, 1980.

Holt, Thomas C. *The Problem of Freedom: Race, Labor, and Politics in Jamaica and Britain, 1832–1938*. Baltimore: The Johns Hopkins University Press, 1992.

Hope Bacon, Margaret. *But One Race: The Life of Robert Purvis*. Albany: State University of New York Press, 2007.

Horne, Gerald. *Cold War in a Hot Zone: The United States Confronts Labor and Independence Struggles in the British West Indies*. Philadelphia: Temple University Press, 2007.

Hunter, Hope. "Gold in British Guiana, Its History and Prospects," *Timehri 6* (December 1892).

Hyam, Ronald. "The Colonial Office Mind," *The Journal of Imperial and Commonwealth History viii: 1* (1979).

Ibekwe, Walter. "African Cultural Survivals in the New World: The Examples of Brazil and Cuba," *Tarikh 5:4* (1978).

Imperial Institute: The Mining Laws of the British Empire and Foreign Countries, Vol.ix, British Guiana. London: H. M. Stationery Office, 1927.

Inikori, Joseph E. "The Origin of the African Diaspora: The Slave Trade from Africa," *Tarikh 5:4* (1978).

———. *Africans and the Industrial Revolution in England: A Study in International Trade and Economic Development*. Cambridge: Cambridge University Press, 2002.

Johnson, Robert. *British Imperialism*. New York; Palgrave Macmillan, 2003.

Jones, John. "Bush Yaws," *Sunday Chronicle*, February 18, 1979.

Josiah, Barbara P. "After Emancipation: Aspects of Village Life in Guyana, 1869–1911," *The Journal of Negro (African American) History 82* (1997).

———. "Tattoos, Body Marks, and Changing Identities: The Impact of Culture and Occupation on African American Sailors in the United States Civil War Navy," Paper presented at *Annual Conference, The Association for the Study of African American Life and History*, Los Angeles, CA. Oct. 1997.

———. "Craftspersons in Guyana: Identity and Knowledge Transfer From Slavery To Present," in Emory J. Tolbert, ed., *Perspectives on the African Diaspora. Volume One: To 1800*. Boston: Houghton Mifflin, 1998.

———. "Caribbean Migration to Guyana: Identity, Patterns, and Contributions From Slavery to WWI." Emory J. Tolbert, ed. *Perspectives on the African Diaspora. Volume Two: Since 1800*. Boston: Houghton Mifflin, 1999.

———. "The Impact of Gold, Diamond, and Bauxite Production on the Guyanese Economy, 1890–1956." *Global Development Studies. International Development Options*, (2000).

———. "Creating Worlds: A Study of Mutuality and Financing among African Guyanese, 1800s–1950s," *The Journal of Caribbean History 38:1* (2004).

———. "The Other Migration: African Diaspora Migrants from Latin America and the Caribbean in the United States Civil War Navy." Paper presented at the *Association for the Study of World African Diaspora Conference*, Rio de Janeiro, Brazil, 5–7 October 5–7, 2005.

———. "Claude H. A. Denbow." Colin Palmer, ed. *Encyclopedia of African-American Culture and History: The Black Experience in the Americas.* 2nd. Ed. Farmington Hills, MI: Macmillan Reference (USA) Gale, 2006, ed.

———. "Organizing Within Diaspora: Dr. Claude H. A. Denbow, Howard University, and the League of Coloured Peoples." *The Journal of African American History* 95:2 (Spring 2010).

———. "From Shaw University to 'the Booker T. Washington of British Guiana' Hon. E. F. Fredericks and Contributions of HBCUs to the African Diaspora." Paper presented at the *95th Annual Meeting, Association for the Study of African American Life and History Annual Conference*, Raleigh, NC October 3, 2010.

Kartick, Reuben. "The Village Policy of Francis Hicks, 1862–1868," *History Gazette 32* (1991).

Kelley, Robin D. G. *Freedom Dreams: The Black Radical Imagination.* Boston: Beacon Press, 2002.

Kenrick, W. W. "Opening Up the Country," *Timehri new series 5* (1891).

Knight, Franklin W. *The Caribbean: The Genesis of a Fragmented Nationalism.* New York: Oxford University Press, 1990.

Kubicek, Robert. *The Administration of Imperialism: Joseph Chamberlain at the Colonial Office.* Durham: Duke University Press, 1969.

———. *Economic Imperialism in Theory and Practice: The Case of South Africa Gold Mining Finance, 1856–1914.* Durham: Duke University Press, 1979.

Kuczynski, R. R. *Demographic Survey of the British Colonial Empire. Volume 111. West Indian and American Territories.* London: Oxford University Press, 1953.

Kwayana, Eusi. *Buxton-Friendship in Print & Memory.* Georgetown: Red Thread Women's Press, 1999.

Kwayana, Eusi and Tchaiko. *Scars of Bondage: A First Study of the Slave Colonial Experience of Africans in Guyana.* Georgetown: Free Press, 2002.

Kwok, Marlene. "Chinese Immigration Into Guyana," *Stabroek News* (February 3, 1994).

Labaton, Stephen. "DeBeers Agrees To Guilty Plea To Re-enter the U.S. Market," *New York Times* (July 10, 2004).

Lachman-White, C. Dennis Adams, and Ulric O. D. Trotz, *A Guide to the Medicinal Plants of Coastal Guyana.* London: Commonwealth Science Council, 1987.

Lamur, Carlo. *The American Takeover: Industrial Emergence and Alcoa's Expansion in Guyana and Suriname, 1914–1921.* Dordrecht, Holland: Floris Publications, 1985.

Lancaster, Alan. "Proposals for Hinterland Settlement and Development of British Guiana, 1884–1900. *History Gazette 59* (1993).

Lanning, George and Marti Mueller. *Africa Undermined: A History of the Mining Companies and the Underdevelopment of Africa.* New York: Penguin Books, 1979.

Larson, Donald and Yair Mundlak. "On the Intersectoral Migration of Agricultural Labor," *Economic Development and Cultural Change 45:2* (1997).

Laurence, K. O. *A Question of Labour: Indentured Immigration Into Trinidad and British Guiana.* New York: St. Martins Press, 1994.

Leechman, Alleyne, ed. *Handbook of British Guiana, 1913.* Georgetown: The Argosy Company Ltd, n.d.

Lowe, Sherwood, A. J. *Historical Analysis of Mining Policy, Law & Administration in Guyana: 1831–2002.* Georgetown: Pavnik Press, 2003.

Madramootoo, Harry. *The Story of Rice in Guyana.* Georgetown: Guyana Rice Corporation, 1973.

Mallon, Florence. "The Promise and Dilemma of Subaltern Studies: Perspectives From Latin America," *The American Historical Review 99:5* (December 1994).

Mandle, Jay. *The Plantation Economy: Population and Economic Change in Guyana, 1838–1960.* Philadelphia: Temple University Press, 1973.

Mangru, Basdeo. *Indenture and Abolition: Sacrifice and Survival On the Guyana Sugar Plantations.* Toronto: Tsar, 1993.

Marsh, Peter T. *Joseph Chamberlain: Enterprise in Politics.* New Haven: Yale University Press, 1994.

McGowan, Winston F, James G. Rose, and David A. Granger. *Themes in African-Guyanese History.* Georgetown: Free Press, 1998.

McNaughton, Patrick. *The Mande Blacksmith: Knowledge, Power, and Art.* Bloomington: Indiana University Press, 1988.

Meredith, Martin. *Diamonds, Gold, and War: The British, The Boers, and the Making of South Africa.* London: Simon and Schuster UK Ltd., 2007.

Midlo Hall, Gwendolyn. *Slavery and African Ethnicities in the Americas: Restoring the Links.* Chapel Hill: University of North Carolina Press, 2005.

Molho, Ian, "Theories of Migration: A Review," *Southern Journal of Political Economy 33:4* (November 1986).

Moohr, Michael. "Patterns of Change in an Export Economy: British Guiana, 1838–1914," Unpublished Ph.D. Diss., Cambridge University, 1971.

———. "The Economic Impact of Slave Emancipation in British Guiana, 1832–1852," *Economic History Review* (1972).

———. "The Discovery of Gold and the Development of Peasant Industries in Guyana, 1884–1914: A Study in the Political Economy of Change," *Caribbean Studies .15:2* (1975).

Moore, Brian L. *Race, Power, and Social Segmentation in Colonial Society: Guyana After Slavery, 1839–1891.* New York: Gordon and Breach Science Publishers, 1987.

———. *Cultural Power, Resistance, and Pluralism: Colonial Guyana, 1838–1900.* Montreal: McGill-Queen's Universities Press, 1995.

———. "The Social Impact of Portuguese Immigration into British Guiana After Emancipation." *Caribbean Freedom: Economy and Society From Emancipation to The Present.* Hilary Beckles and Verene Shepherd, eds. Kingston, Jamaica: Ian Randle Publishers and Princeton: Markus Wiener Publishers,1993.

———. "African Guyanese Political Disempowerment during the Nineteenth Century." Winston F. McGowan, James G. Rose, and David A. Granger, eds. *Themes in African-Guyanese History.* Georgetown: Free Press, 1998.

Morner, Magnus. *Adventurers and Proletariats: The Story of Migrants to Latin America.* Pittsburg: The University of Pittsburg Press, 1985.

Murray, Waveney. "The African-Guyanese Press," *Emancipation: The African Guyanese Achievement* (1993).

Nath, Dwarka. *A History of Indians in British Guiana.* London: By the author, 1950.

Neheusi, Kimani. "The Causes of the Protest of 1905," in Winston F. McGowan, James G. Rose, David A. Granger, *Themes in African-Guyanese History.* Georgetown: Free Press, 1998.

Newbury, Colin W. *The Diamond Rings, Business, Politics, and Precious Stones in South Africa, 1867–1947.* Oxford: Clarendon Press, 1989.

Northrup, David. *Indentured Labor in the Age of Imperialism, 1834–1922.* New York: Cambridge University Press, 1995.

Nugent, Walter. *Crossings: The Great TransAtlantic Migrations, 1870–1914. Bloomington: Indiana University Press,* 1992.

Ochola, Samuel Agonda. *Minerals in African Underdevelopment.* London: Bogle L'Overture Publications, 1975.

Osterhamnel, Jurgen. *Colonialism: A Theoretical Overview.* Translated from the German (1995) Shelly L. Frisch. Princeton: Markus Weiner Publishers, 1997.

Palmer, Annette and Clement London. "Migratory Labor in the English Caribbean (1940–1945) *Journal of Caribbean Studies* (1981).

Palmer, Ransford. *In Search of A Better Life: Perspectives on Migration From the Caribbean.* New York: Praeger Press, 1970.

———. *Pilgrims From the Sun: West Indian Migration to America.* New York: Twayne Publishers, 1995.

Peck, Nathaniel and Thomas Price. *Report of Messrs Peck and Price. 1839 Delegates To British Guiana and Trinidad on Advantages to be Derived From Immigrating to The Colonies.* Boston: Tuttle and Dennett, 1841.

Perkins, H. J. *Notes on British Guiana and Its Gold Industry.* London: Sampson Low, Marston, and Company, 1896.

Phimister, I. R. "The Reconstruction of the Southern Rhodesian Gold Mining Industry, 1903–1910," *Economic History Review 29:3* (1976).

———. "African Labour Conditions and Health in the Southern Rhodesia Mining Industry, 1898–1953" in I. R. Phimister and C. van Onselen, eds. *Studies in the History of African Mine Labour in Colonial Zimbabwe.* Salisbury: Mambo Press, 1978.

Pieterse, Jan Nederveen. *White on Black: Images of Africa and Blacks in Western Culture.* New Haven: Yale University Press, 1992.

Pollydore, Joseph. "A Brief History of Mocha-Arcadia and Its People, 1894–1994," in *Emancipation: African Guyanese Achievement 3* (1995).

Postma, Johannes M. *The Dutch in the Atlantic Ocean Slave Trade, 1600–1815.* Cambridge: Cambridge: Cambridge University Press, 1990.

Prakash, Gyan. "Subaltern Studies as Post Colonial Criticism," *The American Historical Review 99:5* (December 1994).

———. *After Colonialism: Imperial Histories and Post Colonial Development.* Princeton: Princeton University Press, 1995.

Preinvestment Study on the Enhancement of the Jewelry Industry in Guyana. Dublin: Kilkenny Development Partners, 1994.

Purchase and Pride, 170th Anniversary Buxton Village, Guyana, 1840–2010. Celebrating Our Heritage. Commemoration Publication, 2010.

Putnam, Robert D. *Bowling Alone: The Collapse and Revival of American Community.* New York: Simon and Schuster, 2000.

Quamina, Odida T. "The Social Organization of Plantation Mackenzie: An Account of Life in the Guyana Mining Enterprise". *Geneva: U.N. Research Institute for Social Development, 1981.*

———. *The Mineworkers of Guyana: The Making of A Working Class.* Atlantic Highlands, NJ: Zed Books, 1987.

Queenstown 1841 (Westfield, Mocha, Dagaraad) 150 Years Anniversary Publication. New York 1992.

Queenstown 2000—Sankofa "Go Back and Fetch It." New York: The Queenstown, Essequibo Association, 2000.

Queenstown 2004 Anniversary Publication. New York: The Queenstown, Essequibo Association, 2004.

Raleigh, Walter Sir. *The Discovery of the Large, Rich, and Beautiful Empire of Guiana.* London: Hakluyt Society, 1845. Reprint 1596, ed.

Reid Andrews, George. *Afro-Latin America.* Oxford: Oxford University Press, 2004.

———. *Blackness in the White Nation: A History of Afro-Uruguay.* Chapel Hill: University of North Carolina Press, 2010.

Reid, J. M. *British Guiana Commercial Handbook.* Georgetown: The Argosy Printers, 1920.

Roberts, G. W. and J. Byrne. "Summary Statistics on Indentured and Associated Migration Affecting the West Indies, 1834–1918," *Population Studies 20* (1966).

Rodney, Walter A. *How Europe Underdeveloped Africa.* Washington, DC: Howard University Press, 1982. Reprint 1972, ed.

———. "Barbadian Immigration into British Guiana, 1863–1924." Paper presented at *the 9th Annual Conference of Caribbean Historians.* Cave Hill, Barbados, University of the West Indies, April 1977.

———. *Guyanese Sugar Plantations in the Late Nineteenth Century: A Contemporary Description from the Argosy.* Georgetown: Release Publishers, 1979.

———. *A History of the Guyanese Working People, 1881–1905.* Baltimore: The Johns Hopkins University Press, 1981

Rodway, James. *Handbook of British Guiana.* Georgetown: The Committee, 1893.

———. *History of British Guiana from theYear 1668 to the Present Time.* 3 Volumes. Georgetown: J. Thompson, 1891–1894.

Rose, James G. "The Repatriation Controversy and the Beginning of An East Indian Village System," *Guyana Historical Journal 1* (1989).

———. "African-Guyanese Business" and "Africans in Guyanese History" in *Emancipation 2* (1994).

———. "The Strikes of 1842 and 1848," in Winston F. McGowan, James G. Rose, and David A. Granger, eds., *Themes in African Guyianese History.* Georgetown: Free Press, 1998.

Rowland, E. D. "The Census of British Guiana, 1891," *Timehri 4* (1892).

Russell, William. "Rice" *Timehri* (1886).

Schadt, Hermann. *Goldsmiths' Art.* Translated from the German by Ann Potter Schadt. Stuttgart: Arnoldsche Art Publisher, 1996.

Schuler, Monica. "Kru Migration to British and French Guiana, 1841–1857," in Paul Lovejoy, ed. *Africans in Bondage: Studies in Slavery and the Slave Trade.* Madison: University of Wisconsin Press, 1986.

Scobie, Edward. *Black Britannia: A History of Blacks in Britain.* Chicago: Johnson Publishing Company, 1972.

Seecharan, Clem. "Tigers in the Stars:" *The Anatomy of Indian Achievement in British Guiana, 1919–1929.* London: Macmillan Education Ltd., 1997.

———. *Sweetening "Bitter Sugar": Jock Campbell: The Booker Reformer in British Guiana, 1934–1966.* Kingston: Ian Randle Publishers, 2004.

Seymour, Arthur and Elma. *Dictionary of Guyanese Biography,* Georgetown: By the authors, 1986.

Shahabudeen, Mohammed. *Nationalization of Guyana's Bauxite: The Case of Alcan.* Georgetown: Ministry of Information, 1981.

Sharp, William. *Forsaken But For Gold: An Economic Study of Slavery and Mining In the Colombian Choco, 1680–1810.* Chapel Hill: University of North Carolina Press, 1970.

Sheridan, Richard S. *Doctors and Slaves: A Medical and Demographic History of Slavery in the British West Indies, 1680–1834.* Baltimore: The Johns Hopkins University Press, 1984.

Shukla, M. S. *Panorama of Gems and Jewellery in Indian Historical Setting.* Varanasi, India: Kala Prakashan, 2003.

Simmons, Terrence. "Bartica," *Guyana Chronicle* (June 1992).

Simms Hamilton, Ruth, ed. *Routes of Passage: Rethinking the African Diaspora Vol. 1 Part 1.* East Lansing: Michigan State University Press, 2007.

Skerrett, W. Henry. "Gaskin and Jackson, Silver Anniversary 25 Years," *Sunday Chronicle* (May 1991).

———. "Jewelers Dedicated to their Work," *Stabroek News* (January 25 1994).

Smith, Douglas. "The Origins of Porknocking in Guyana," *McGill Studies in Caribbean Anthropology* (1969).

Spackman, Ann. "The Role of Private Companies in the Politics of Empire: A Case Study of Bauxite and Diamond Companies in Guyana in the Early 1920s," *Social and Economic Studies 24:3* (1975).

Spenser, Daniela, "Standing Conventional Cold War History on Its Head," in Gilbert M. Joseph and Daniela Spenser, eds., *In From the Cold War: Latin America's New Encounter With the Cold War.* Durham: Duke University Press, 2008.

Spiro, Jonathan Peter. *Defending the Master Race: Conservation, Eugenics, and the Legacy of Madison Grant.* Burlington: The University of Vermont Press, 2009.

Splenger, Joseph J. "Migration and Socioeconomic Development: Today and Yesterday," in Alan A. Brown and Egon Nueberger, eds. *Internal Migration: A Comparative Perspective.* New York: Academia Press, 1977.

St. Pierre, Maurice. "Industrial Unrest in Mackenzie, Guyana." *McGill Studies in Caribbean Anthropology 5* (1969).

———."The Sociology of Deconstruction: The Case of Guyana Bauxite Ltd." *New World Quarterly 5:4* (1972).

———. "Race, the Political Factor and the Nationalization of the Demerara Bauxite Company, Guyana." *Social and Economic Studies 24* (1975).

———. *Anatomy of Resistance: AntiColonialism in Guyana, 1823–1966.* London: Macmillan Education Ltd., 1999.

Stalin, Black. "More Come," *Rebellion Record Album*, Carnival Performance, Trinidad and Tobago, 1996.

Steele, M. J. "A Philosophy of Fear: The World View of the Jamaican Plantocracy In a Comparative Perspective," *The Journal of Caribbean History* 27:1 (1993).

Stockley, G. M. *The Role of Geological Surveys in Development in British Guiana.* Georgetown: Geological Survey Department, 1953.

———. *The Geology of British Guiana and the Development of its Mineral Resources, Bulletin 25* Georgetown: Geological Survey Department, 1955.

———. "Mineral Development in the Economy of Guiana," *1st. Meeting of the Caribbean Geological Conference, Antigua, 1955.*

Taylor, H. *Handbook of West African Gold Mines.* London: Hutchinson Scientific and Technical Publications, 1946.

Telford, Laurette. "Early Attempts at Gold Mining," *Guyana Chronicle* (August 1992).

The Laws of British Guiana Vol.1, 1774–1865. Oxford: Oxford University Press, 1895.

The Laws of British Guiana Vols. 1–6, 1803–1921. London: Waterlow & Sons Ltd., 1923.

The Mining Laws of the British Empire and Foreign Countries. Imperial Institute. London: H. M. Stationery Office, 1927.

The New Victoria. A Souvenir Magazine 170th Year. Redevelopment Through Revival of Our Traditions & Culture. Vol.1 (Sept 2009).

"The Study on Internal Migration in Three Latin American Countries," in *A Summary Research on Internal Migration in Latin America, Bolivia, Brazil, Peru.* Geneva: Secretariat for Migration, World Council of Churches, 1969.

Thompson, Alvin O. "Gold Mining in British Guiana From Earliest Times To 1939," Unpublished Paper, 1965.

Tobin, Jacqueline L. and Dobard, Raymond G. *Hidden in Plain View: A Secret Story of Quilts and the Underground Railroad.* New York: Doubleday, 1999.

Todaro, Michael. *Internal Migration in Developing Countries: A Review of Theory, Methodology and Research Priorities.* Geneva: International Labor Office, 1976.

Turrell, Rob. "Rhodes, DeBeers, and Monopoly," *The Journal of Imperial and Commonwealth History x:3* (May 1982).

Usman, Aubidsi A. The Nineteenth-Century Black Atlantic." in Toyin Falola and Kevin Roberts, eds. *The Atlantic World, 1450–2000.* Bloomington: Indiana University Press, 2008.

Van de Laan. *The Sierra Leone Diamonds: An Economic Study Covering the Years 1952–1961.* London: Oxford University Press, 1965.

Veiga, Marcello M. "Artisanal Gold Mining Activities in Suriname." *United Nations Industrial Development Organization Report*, December 1997.

Walker, Sheila S. ed. *African Roots/American Cultures: Africa in the Creation of the Americas.* Lanham MD: Rowman & Littlefield Publishers, Inc., 2001.

Walrond, Grantley. "Small Gold Mines: The Production and Declaration Problem—Another Dilemma for a Small Developing Country." *Natural Resources Forum 10: 4* (1986).

Warner-Lewis, Maureen. *Guinea's Other Suns: The African Dynamic in Trinidad Culture.* Dover, MA: The Majority Press, 1991.

Webb Pressler, Margaret. "DeBeers Pleads to Price-Fixing, Firm Pays $10 Million, Can Fully Re-enter US," *The Washington Post* (July 24, 2004).

Webber, A. R. F. *Centenary History and Handbook of British Guiana.* Georgetown: The Argosy Co. Ltd., 1931.

Weinstein, Michael. *The World of Jewel Stones.* NY: Sheridan House Inc., 1958.

Wells, A. F. and D. *Friendly Societies in the West Indies. Report on a Survey.* London: H. M. Stationery Office, 1953.

White, John Sampson. *The Smithsonian Treasury of Minerals and Gems.* Washington DC: The Smithsonian Institute, 1991.

Whitlock, Herbert P. *The Story of Gems.* New York: Lee Furman, 1936.

Who Is Who In British Guiana. Georgetown: The Daily Chronicle, 1940.

Will, H. A. "Colonial Policy and Economic Development in the British West Indies, 1895–1903," *Economic History Review 23:1* (April 1970).

Williams, Dwayne E. "Rethinking the African Diaspora: A Comparative Look at Race and Identity in a Transatlantic Community, 1878–1921" in Darlene Clark Hine and Jacqueline McGowan, eds. *Crossing Boundaries: Comparative History of Black People in Diaspora.* Bloomington: University of Indiana Press, 2007.

Williams, Eric E. *Capitalism and Slavery.* London: Andre Deutsch Ltd., 1991. Reprint 1964 of 1944 edition, University of North Carolina Press.

———. *From Columbus to Castro: The History of the Caribbean, 1492–1969.* New York: Vintage Books, 1984, ed.

Wilson, Andrew R. ed. *The Chinese in the Caribbean.* Princeton: Markus Wiener Publishers, 2004.

Wood, Mboya (Lennox). "Mammy Fiffee: A Tradition of Bush Medicine," *Caribbean Daylight* (May 1995).

Woolford, W. H, "Organising and Development of the Mining Sector of Guyana" Unpublished paper, February 22, 1995.

Yen Ching-Hwang. *Coolies and Mandarins: China's Protection of Overseas Chinese During the Late Ch'ing Period, 1851–1911*. Kent Ridge: Singapore University Press, 1985.

Young, Alan. *The Approaches to Local Self-Government in British Guiana*. London: Longman and Green, 1958.

Yun, Lisa. *The Coolie Speaks: Chinese Indentured Laborers and African Slaves in Cuba*. Philadelphia PA: Temple University Press, 2008.

INDEX

Abolition Act, 1833, 1, 24, 175
Aboriginal Indians, *see* Amerindians
Abraham, Benjamin Victor, businessman, 159, 162
Absconders, gold and diamond mining, 113–16, 118–19
Africa, 1, 6, 11, 21, 24, 161, 164
  see also under African identities, *and individual countries*
African American, businessmen and mining entrepreneurs, 167–8
African American labor migrants in Guyana, 15–16 (n.51)
  in colonizing schemes, 41 (n.9)
  in mining deaths, 57–71
African ancestry, xi, 1, 6, 15, 21, 23, 24, 27, 30, 41, 49, 73, 75, 157
African Association of British Guiana, 14, 25, 72
African Caribbean, *see* Afro-Caribbeans
African continuities, xii, 157–73
African descended, 11, 13, 16, 19, 20, 34, 43, 49, 52–3, 68, 93, 98, 100, 109, 121–2, 141, 146, 148, 164, 176
  see also African Guyanese
African diaspora, 9, 11, 20, 31, 52, 53, 57–9
  contributions to development, xi, 43–5, 54–6, 74–5, 93, 100, 121, 146–7, 171
  mining workers, origins, xi, 12–17, 26–34, 41, 59–70, 93–6
  see also African ancestry, African-descended
African Guyanese, businesspersons and entrepreneurs, 13, 53, 58–70, 86–7, 130–2, 142–3, 148–51, 164, 168–9
African Guyanese, mining workers, 14, 58–9, 68, 100–1, 136–7, 143, 164
  in United States Civil War Navy, 20 (n.81)
African Guyanese jewelers, 163–6
African identities, 15–16, 21, 29, 33, 57–64, 129, 146–7, 157–62
  see also under Krumen
African Methodist Episcopal, Church (AME), 15–16
African rice producers, 29, 75
Africans in Guyana, 1, 6, 9, 10, 12, 13, 15, 16, 21–2, 27–9, 57–70, 75, 122, 130, 145, 159–63
Afro-Caribbeans mining workers, 7, 12, 16
  gold and diamond fields, caualties, 57–70, 164
  see also under African ancesry, African descended, African diaspora, West Indian
Agricola Village, 155
agricultural production, 1, 17, 31, 42, 50, 75, 131–2, 139–41, 159, 162, 168, 171
Air Navigation Ordinance 1925—Air Navigation, 81
air transportation, 9, 72
  provision of, 81–2
Akyma, Demerara River, 53, 94, 97, 103, 107
Aluminum Company of America (ALCOA), 53, 94–5
Aluminium Company of Canada (ALCAN), 94
American Cyanamid Company, 53

Amerindians, 9, 17, 29
and bauxite use, 53
contact, African diaspora interior workers, 141
in mining and forestry, 17, 29 (n.31)
targets of scientific racism, 75
Angola, 12
Apprenticeship Order-in-Council, 25
Apprenticeship scheme, 1, 12, 24
Arakaka Placer and Mining Company, 25
see also mining companies
Arthur, Lionel, Mocha Village, miner, 141
artisans, miners, 163
Asia, Asians, 1, 9, 24, 27–9, 36, 130
see also Chinese, East Indians
Assanah, Joseph, Victoria Village, fisherman, motor mechanic, inventor, 163
Australia, 120, 167
Azores, 33

B. G. Consolidated Goldfields Limited, *see* British Guiana Consolidated Goldfields Co. Ltd.
Bagot, Walter, miner, 164
Bagotstown Village, 131, 155
Bahamas, 12
Baltimore, Maryland, 15–16 (n.55)
Baramani, Essequibo, 24, 120
Barama River, 9
see also Guyana rivers
Barbadian, 26, 27, 33, 62, 65, 66, 136
see also Ince, Fitzgerald, mining migrant to Orinoco, Brazil
Barbados, 15, 34, 53, 61, 64, 121, 123, 137
Barima Gold Mining Company, 25, 26
see also mining companies
Barima River, Essequibo, 9
see also Guyana rivers
Barnard's Syndicate, 43
Barnwell, Octavius, miner, 68
Bartica, entrepot town, 19, 24, 53, 58, 62, 66, 69, 76–7, 79, 80, 82, 85–8, 120
bauxite, xi, xiii, 50, 52–6
bauxite belt, 113, 124–8, 172
bauxite industry, 1, 7, 34, 36, 39, 52, 54–6, 94–5
bauxite infrastructure, 93–108
bauxite miners, workers, 7, 9, 52–6, 95, 99, 106, 108, 112–13, 124–8, 172–3; *compare* South Africa
bauxite mining towns, 2, 53, 86, 97, 100, 124–8
railway in, 108
Berbice, Berbice County, 9, 13, 24, 27, 29, 39, 52, 53, 65, 86, 100, 151, 162
Berbice Mining Company, 127
see also mining companies
Berbice River, 9, 19, 53, 97, 126
see also Guyana rivers
Bernato, Joel and Solomon, international diamond magnates and speculators, 51
Bethany Island, Essequibo, 120
Birth Registers, 15, 161–2
Black Stalin, calypsonian, 34
Board of Commissioners, 19, 26
boat-builders, 167, 172
see also Transportation
Booker's McConnell Co. Ltd, 107
"boom" and "bust," 4
border disputes, with Venezuela, 11, 39, 41, 50
"bottom-house," 100
Bourne, William, jeweler, Virgin Islands, 162
Brazil and Brazilians, 4, 9, 17, 122, 129, 160–1, 171
British Empire, 51, 87, 120
British Guiana
climate of, 3, 9, 20
environmental conditions of, 57–91, 153–4
exports of, xii, 39–56, 93–108, 131
land utilization policies, 3
location, 1–2, 9
plantation conditions, 3, 14, 20
population, 12–18, 41–2
British Guiana Airways, 81–2
British Guiana and Trinidad Mutual Assurance Company, 101
British Guiana Consolidated Gold Concessions Co., 49
British Guiana Consolidated Goldfields Co. Ltd., 52, 56, 83–4, 101, 123
British Guiana Gold, Diamonds, and Mineral Workers Union, 123
British Guiana Gold Company, 11

British Guiana Labour Union (BGLU), 123, 126
British Guiana Miners' Association(BGMA), 52, 123
British Guiana Planters Association, 27
British Medical Association, 66, 86
British Parliament, 14
British West Indies, 1, 15
Brown, Andrew Benjamin, (1857–1939), Honorable, attorney, Den Amstel Village, 6, 25, 135, 169–70
Brown, William Alexander 'Pigeon Island,' entrepreneur and mining millionaire, 169–70
Burnett, F. H., boxing promoter, 46
see also women: women boxers
Burnham, Linden "Forbes" Sampson, (1923–85), attorney, president of Guyana, 124
"bush di-di," 141
bush life, 141–2
see also under British Guiana: environmental conditions
bush-meat or wild meat: flesh of wild pigs, deers, agoutis, iguanas, labbas, 142
bush yaws, 58, 68
"Buxton" and "Plaisance" graveyards, 58
Buxton-Friendship Village, 58, 144, 169
Buxtonians, 72, 143–4
Buxton Village, 13, 131, 135, 139, 142–3, 151, 154, 168–9

Cambridge Academy, 136, 137
Cameron, Norman, E., 67
Cameron Lees, Sir Charles, 43
Camon, Jules, 11
Canada, 43, 54, 86, 94
see also under Mining Regulations and Laws
Cape Verdeans, 15, 27, 29
Cape Verde Islands, 27, 33
Caratal District, Essequibo, 10
Caribbean islands, xi, 11–12, 100, 131, 142, 173
Caribbean plantations, 4
Carrington, Winslow, trade unionist, 124
Carter, Alan, miner, businessman, 143
Carter, Honorable, Sir John P., Guyana's ambassador, Washington, DC, Great Britain, 124
Cayenne, *see* French Guiana
census, 16, 18, 34–5, 41, 46, 59, 97, 161 (n.29), 164
Chamberlain, Houston Stewart, 8
see also pseudeo scientific and stereotypical racial views
Chamberlain, Joseph, colonial secretary, 6, 43–4
Charlestown and LaPenitence, business centers, 89, 120
Cheong, Augustus J., deputy boundary commissioner, 81
Chew, Edward and Richard B., African American labor migrants, 15
China, 29
Chinese, 9, 15, 17–18, 28–9, 68, 120, 175
Choo Kang, A., business enterprise, 107
Christianburg, 2, 53, 56, 97–8, 103, 105–7, 126, 172
Claridge, Charles, mining magnate, international speculator, 51
clergymen, 32
see also Glasgow, Reverend F. C., Reynolds, Reverend D. J.
Clonbrook Village, 155, 162
Cockatara, Linden, 98–100, 106
Cold War, 95, 111, 124, 177
Colombia, 160–1, 162 (n.31)
colonial bureaucrats, 2, 13, 17, 21, 23, 32, 37, 49–51, 86, 94–5, 109, 112, 121–2, 152
Colonial Development and Welfare Act, 82, 90, 104
Colonial Development Corporation, 56
colonial hierarchy, 1, 3, 6, 14, 20, 24, 31–3, 37, 41, 51, 59, 69, 74–5, 95, 111, 121–2, 153, 175, 178
colonialism, 5, 7, 20, 24, 126
Colonial Jewelery Establishment, 165
colonizing schemes, 8, 41 (n.9)
Congo, 12, 122, 124, 143
Constabulary Police Force, 108, 126
contracts, labor, 15, 24, 31–3, 96, 119
Conzius, Robert, diamond miner, 66–7
Cooperative Credit Banks, 145–8, 153, 155
cooperativism, 117, 145–55
Corentyne River, 9, 31, 46, 53
see also Guyana rivers
Corry, James McFarlane, chairman, Den Amstel Village, 135, 153

cost-minimizing
effects of, 71–91
government and private companies, xiii, 71, 125
craftpersons, enslaved in Guyana, 159, 161
Creole Africans (Creoles), 16, 31, 152
see also African descended
Critchlow, Hubert Nathaniel, Order of the British Empire (1884–1958), miner, waterfront worker, trade union organizer, legislator, 6, 117, 123–4
Culley, James, miner, Buxton Village, 58–9
Cuyuni Goldfields Ltd., and a mining town, 84

Danielstown Village, 101, 150–1
Dargan, Patrick, legislator, 123
Death Certificates, 59, 64–5 (n.57)
Death Registers, 15, 75, 87
African diaspora labor migrants in, 59–66, 76
DeBeers Company, 51
de Gobineau, Count, on pseudeo scientific and stereotypical racial views, 8
Demerara Bauxite Company (DEMBA), 47, 52–3, 94–127, 136, 172
Demerara County, 11, 13, 52–3, 66, 94
Demerara Gold Association, 35
Demerara Life Insurance Company, 66
Demerara River, 9, 19, 46, 53, 72, 77, 82, 92–100, 103, 105, 107, 125, 135, 151, 173
see also Guyana rivers
Den Amstel-Fellowship Villages, 13, 153
Den Amstel Village, 69, 135, 155, 168
Denbow, Charles Egerton, Jnr, County Sergeant Major of Police, 143
Denbow, Charles H. A., businessman, 168
Denbow, Claude H. A., dentist, (1911–1979), 124
Denbow, George L. A., clerk, Garnett & Co., 143
Devonish, Lloyd, diamond miner, Fellowship Village, 155
diamonds in Guyana, xi, 39, 42–4, 171–2
miners, xiii, 35–6, 43, 144
smuggling of industrial diamonds, 122
value, diamonds, 44, 54
discrimination in mining, 121
against medical personnel, 65–6, 86–7
diseases
causes, spread, prevention, treatment of malaria, 60, 87–9, 99
clinics and treatment centers, gonorrhoea and syphilis, 89–91
diagnosis and treatment of filaria, 99
infectious, sexually transmitted, 90
dishonesty: mining employees, employers, 113–21
divsified economy, 33, 39
exports, financial returns of, 42–4, 53–6, 73–4, 130–3, 144, 150
"dokpwe," form of mutual aid association, 147
Drakes, Arnold B, mining entrepreneur and businessman, Queenstown Village, 142
drownings, occupational hazard, 57–90, 119 (n.39)
Dutch colonization and mining, 10
Dutch Guiana, *see* Suriname
Dutch West India Company, 10

East Indians, in Guyana, 9, 17–18, 33–4, 64, 75, 87, 120, 154, 175
colonization, repatriation, 21
miners, jewelers, 163–4 (n.41)
stereotypes of, 18, 28
Echo, The, 26, 41, 151
economic conditions and depressions, 4, 11, 21, 29, 33, 65, 84, 152
economic linkages, 7, 67
in bauxite, 172–3
in gold and diamonds, 157–72
Edinburgh University, 86
education, 105–6, 134–7
scholarships, 171
technical schools, 104, 136
Egyptians, 159
El Dorado, fabled city of gold, xiv, 1, 5, 6, 7, 22, 57, 73, 93, 160, 173
El Dorado Company, 39
Eldorado Village, Berbice, 140
Eleazar, Honorable Joseph, attorney, legislator, mayor, 51, 72, 87
Emancipation Act, 1838, 1, 175
commemoration of, 106
Engels, Friedrich, 110
enslaved jewelers, 157–62

entrepreneurial ventures, African Guyanese, 137–43, 162–73
environment, xiv, 1, 19–20, 57–70, 96–7, 124, 150
  environment affecting gold and diamond mining, 3, 41, 143, 150
  environmental reasons, migration, 124, 139
Essequibo County, 10, 13, 19, 24, 42, 61–2, 71–91, 100–1, 151, 167, 169
Essequibo mining districts, 19, 57–91
Essequibo River, 9, 53, 58, 64, 77, 80–2, 170
  see also Guyana rivers
'esusu,' 'susu,' 'ajo,' 'awo,' Credit Associations, Voluntary Credit Associations, Rotating Savings and Credit Associations, 145–8
Ethiopians, 159
Eudell, Demetrius L., 31
Europeans colonists, 22, 34, 68, 120, 122
  mining officials, magnates, elites, 51–2, 110, 161

farmers, as miners, 67, 112, 138–9, 143, 155, 162–3, 169
Felix, Ray, jeweler, businessman, 165
Felix, Samuel, jeweler, businessman, 164
Fellowship Village, 153, 155
Ferdinand, Joseph, St. Lucian, 64
Five Stars District of North West, 49
Forbes, Ernest & Percival, mining businessmen, 26
France, 10, 160
French Guiana, 11, 44, 57–70, 129
  gold exports, value, 44

Garnett, George, 26, 116
Garvey, Marcus, founder of the Universal Negro Improvement Association (UNIA), 135
Gaskin, Herman, jeweler, businessman, 166 (n.52)
geologists, 135
Georgetown, capital city, Guyana, 77, 81, 89, 119–20, 131, 136, 151
Ghana, 5, 160
Ghanaians, *see* Ghana
Gillis, Alexander Barrington, mining entrepreneur, building contractor, 169
Glasgow, Reverend F. C., 35
Golden Grove-Nabaclis Villages, 151
Gold Industry and Crown Lands Commission, 80, 89
gold in Guyana, xi, xiii, 39, 42–4, 64, 71, 133–4, 140, 143, 150, 167, 171–2
  gold, value, 42–3, 54
  gold and diamond fields, 16, 19, 45, 164
  gold diggers, miners, xiii, 35–6, 43, 61, 144; goldsmiths, 89, 120, 166
  gold teeth, 166
Gold Mining Gazette, The, 16, 26, 41, 61, 63–4, 114, 154
Gold Mining Ordinance #16 of 1880, *see* Appendix A
governor Sweetenham, 18
'grape-vine' migration, 97
  see also migration: migration theories
Gravesande, Laurence Storm vans, 10
Gray, Darrell, Queenstown Village, miner, gold, diamonds; bauxite worker, 100–1, 136–7
Great Britain, 2, 11, 28, 41, 54, 71, 94, 124, 128, 145, 149, 176
Greater Senegambia and the Gold Coast, 160
Guiana Mine Workers' Union (GMWU), 126
Guianas, *see* British Guiana, Dutch Guiana, French Guiana
Guinea, 12, 159
Guyana, *see* British Guiana
Guyana Bauxite Company (GUYBAU), 105
Guyana rivers, 9, 53, 71
  see also under individual waterway
Guyanese Miners' Development Association, 123

H. M. Penal Settlement, Essequibo, 86
Hatton, David, Buxton Village, miner, 143
Headley, Josephus, miner and farmer, 143, 169
health care
  in the bauxite regions, 96, 102–5, 107–8, 114, 124–6
  gold and diamond regions, general, 60, 84–9, 95, 102–4, 123–5
Hector, Clifford, jeweler, businessman, 164–6 (n.52)

Herskovits, M. J., 147
Historically Black Colleges and Universities, (HBCUs), 6, 135
Hodge, Christopher, jeweler, Trinidad, 162
Hogg Island, 13
Hope, Aaron, businessman, building contractor, 169
housing, gold and diamond regions, 83–7
  bauxite areas, 97–102, 124–6
  colony-wide, 125–6
Howard University, 6, 65–6, 135
Howell-Jones, B., plantation and mining company owner, 25, 30
hucksters, 1, 5, 21, 101, 142, 154

Ince, Fitzgerald, Barbadian miner, 171
indentured immigration scheme, 15, 23, 27
indentureship, 3, 15, 18
independent prospecting groups, xiv, 1, 5, 31, 43, 51, 68–9, 71, 75, 80
India, 21, 28, 29, 33, 164
industrialization, 3, 16, 95, 177
Institute of Mines and Forests of British Guiana, 24–6, 28, 34, 59, 75, 116, 117
interior infrastructure, gold and diamond regions, 7, 12, 33, 43, 59, 71–91
  interior railroads, 77–8, 81
  interior roads, 75–7
  interior transportation, 78–81
Itanami waterfall, 57, 76, 94, 116–17
Ithaca Village, 13, 155
Ituni, Berbice, 53, 81, 94, 97, 127

Jackson, Eric, jeweler, businessman, 166
Jamaica, Jamaicans, 28, 32, 41, 42, 53, 57–70, 166
Jerrickville, entertainment venue, Victoria Village, 144
jewelers, 162–6 (n.41)
  see also under individual names
jewelry industry and establishments, 56, 89, 163–6 (n.35)
jewels, value, 56
Jones, Iris, woman fatality in mining region, 60
July, Maurice, jeweler, businessman, 164–5

Kanawarook or Konawaruk, 66, 76, 82, 84, 143
Kelly, Robin D. G., 14
Kenrick, W. W., 76
King of Italy, 50
Klondike, 43
knowledge transfer, 145–55
Koriabo, Essequibo, 120
Kraskosky Ltd, 121
Krumen, 15, 28, 29, 64
  see also African identities
Kumakusa, Essequibo, 54, 69, 81, 88
Kwakwani, Berbice, 53

Labor Commission, 1890, evidence, report, 27–34, 59
labor migrants, xi, xv, 1, 9, 15, 16, 23–37, 41, 44, 53, 93, 100, 119
  labor migrants, bauxite, 97
  labor migrants, gold and diamonds, 52, 54, 141–4
  labor migrants, perils, 57–70
  see also Legislation #23, 1890
  labor migrants, shortages, 26–7, 33, 44
labor relations, mining, 109–28
  see also Trade Unions
League of Coloured Peoples (LCP), 124, 135
Legislation #23, 1890, 36
Leguan Island island, 13, 151
Levans, E. J., mining entrepreneur, businessman, 169
Liberty Hall, UNIA, 135
Linden, 53, 81, 101, 107, 126, 136, 173
Linden-Soesdyke Highway, 107
London Missionary Society, 32

Mabaruma, Essequibo, 88, 162
Mackenzie, mining town, 93–108
  excursions to, 106, *see also* Linden
Mackenzie-Christianburg-Wismar, 53, 93–108, 153
Mackenzie High School, 106, 136–7
Mackenzie Primary School, 105
Mahaica, Frederick, Lugard, diamond magnate, millionaire, 170–1
Mahaica, Village, 119
"Mahaica Creek," Essequibo, 171
Mahaicony Village, 28, 131, 151, 171
Majuba Hill at Strong Hope Creek, 49
Manchester Guardian, The, 91
Man Power Citizens' Association (MPCA), 126

Marcus, Elfrieda, 142
Marshall Syndicate, 43
Marx, Karl, 110
Matthews High School, 135
Mazaruni Company, 43
see also mining companies
Mazaruni River, 9, 64
see also Guyana rivers
McD Hackett, George, building contractor, 169
Medical Ordinance #5, 1886, 86
medical practitioners in Guyana
Earle, P. M., 87
Ferguson, J. F., 104
Giglioli, George, 88, 99, 104
Kelly, P. J., 87
King, Duncan Fitzgerald Alfred, 65–6
Nicholson, Charles C., 104
Ozzard, A. S., 62, 86–7
Rohlehr, John Monteith, 62, 72, 84–6, 123
Sharples, Loris, R., 86–7
Singh, J. B., 87
methods of gold and diamond mining, 39, 47–9
migration, 6, 27, 57–70
migration, internal, seasonal, xii, 4, 12, 19–21, 57–70, 96, 129–34
migration, pendulum, xii, 2, 13
migration strategies, 23–56, 112
migration theories, 133–4
migratory movements, 4, 12, 19, 29, 44, 57–70, 93, 133, 162
Miller, Sydney, Magistrate, 46
mineocracy, 2, 7, 21, 23, 27, 70–1, 99, 105, 109–11, 146
mineral, ores, 139
see also bauxite, diamonds, gold
Mingo, Rowena, T., Eldorado Village, 140–1
mining companies, 6, 21, 39, 42–4, 56–70, 83–4, 102
mining industry, 95, 97, 100, 133
mining workers in, 34, 36, 54–6, 74
women, mining, 45–7
see also under individual company's name
mining migrant to Orinoco, Brazil, 171
Mining Regulations and Laws, 113–22
Australia, 120
British Guiana, *see* Appendix A
Canada, 120
South Africa, 119–20
Mocha Village, 13, 66, 132, 141, 155, 164
modernization, xii, 3, 8, 13–15, 125, 145–8, 153–5, 177–8
"Monster petition," 34, 117
Morawhanna, Essequibo, 62, 76, 80, 85, 88, 151, 153
mortality, gold and diamond mining, 40, 57–70, 164

Nabaclis Village, 58, 162
Nassy, Edward, boat-designer, builder, busisnessman, 167
see also Transportation
Negro Progress Convention (NPC), 25, 135
New Amsterdam, 131, 151, 162
New York, 166
New York South American Suriname Goldmining Company, 44
Nikerie, *see* Suriname
Nome, Alaska, 43
North American officials, investors, 2, 52, 110
Not Medically Certified" and "No Medical Attendant, 66
see also Death Certificates

Ogle, Maclean, teacher, attorney, legislator, entrepreneur, Buxton Village, 168
Omai River, 11
see also Guyana rivers
Oppenheimers, Henry, Louis, Otto, diamond magnates, international speculators, 51
Orange Nassau, *see* Buxton Village
"order and progress," xii, 3, 8, 13, 14, 23, 94, 108, 111–12, 125, 152, 177, 178
Ordinances, 15, 39, 52, 143, 147, 148, 149, 150, 154
Organizations in Guyana, 25, 117, 123–4, 135, 146–7
Orinoco, 171
Orna, James Edmund, "Dutchie," boat-builder and operator, businessman, entrepreneur, 173

Pala-Wala, 100
Paramaribo, capital city, Suriname, 44
Parika Village, 19, 77, 64, 81, 135

Paterson, Yasuw, Queenstown Village, 13
Payne and Williams, mining entrepreneurs, 49
Pellew, Trenton Ezekiel, Clonbrook Village, farmer, 155
Peoples' Association, 117
Perkins, H. I., 121
Peter's Hall Village, 155
Peter's Mines (Hatton Garden Dianond Syndicate), 49
Plantation Bauxite Co. Ltd, 53
plantations, 3, 11, 13–15, 18, 20, 27, 54, 132–4, 157
plantocracy and planters, 12, 14, 21, 23, 27, 68, 85, 109, 146
Police Force, 46, 69, 118, 126, 143
Portuguese and Madeirans, 17–18, 21, 121, 148
Post Office Savings Banks, 148, 151–3
Potaro River, 9, 122
see also Guyana rivers
precious metal and gems, 20, 39, 57, 61, 65, 112, 163–4, 177
Primo family, Buxton Village, 58
Proctor, Francis Hector "Frank," tailor, mechanic, musician, Linden, 101–2
pseudeo scientific and stereotypical racial views, 7, 18, 27–30, 49, 50, 75, 120–1
pseudoscience, 27
Puruni River, 52
see also Guyana rivers
"push" and "pull" factors, 20–2, 100, 124
see also under migration: migration theories

Quan brothers, Windsor Forest, in gold and diamond mining, 18
Queenstown Village, Essequibo, 13, 100, 136

R. H. Carr, steamer, Demerara River, 98, 107
race and class, 110–11
race and racism, 18, 27, 49–50, 86–7, 95, 100, 119–22, 171
railroads, 77–8
Raleigh, Sir Walter, 10, 160
Reports of the Commissioner of Lands and Mines, 42
Reports of the Inspector of Villages, 150
Reports of the Institute of Mines and Forests of British Guiana, 16, 27
reverse migration trend, 7, 101–2, 173
Reynolds, Reverend D. J., 33
see also clergymen
Reynolds Metals Company, 53, 126–7
Rockstone, Essequibo, 77, 80–1, 89, 104, 107
Romeo, William, blacksmith, inventor, Clonbrook Village, 162
Rothschild, Consolidated Mines Limited, 51
Royal Agricultural and Commercial Society (RACS), 27, 131
Royal Mail Steam Packet Company, 28

savings bank, 13, 147
Schomburgk, Robert, 10
scientific racism, 75
Semple, Sinclair, miner, Litchfield Village, 65
Shaw University, 6, 135
shopkeepers, 1, 120, 172
Sierra Leone, 15, 29
skills, enslaved, (blacksmiths, coopers, engineers, goldsmiths, mechanics, millwrights), 2, 5, 7, 8, 28, 29, 67, 101, 131, 136, 142, 155, 157–9, 161–7, 173, 177
Slave Registers, British Guiana, occupations of enslaved, 29, 157–9, 161–2
slavery, 3, 13, 17, 29, 31–2, 72, 75, 147, 158, 161
social and economic factors, 4, 21, 57–70, 93–108, 118–19, 129–44
social conditions, 21, 57–70, 106, 109–10
sociopolitical activism, 2, 28, 109–28, 158
South Africa, 5, 43, 65, 98, 120, 122, 124–5
South America, 1, 6, 7, 10, 21–2
countries, 28–9, 34
South and West Africa, 21, 49, 124–5
Sprostons Company Ltd., 49, 79, 89, 103, 106
Stabroek Market, 31, 89
Stephens, Jacob Gorshom, "J.G.,"diamond buyer, businessman, 142, 169

stereotypes and racial views, 7, 27–30, 49–50
of African ancestry, 28, 49–50, 120–1
of East Indians, 18, 28
St. Lucia, 28, 42, 53, 57–70, 173
St. Vincent, 57–70, 173
sugar plantations, 18, 27, 112
see also plantations
Suriname, 4, 9, 17, 20, 31, 44, 57–70, 129, 162, 173
Suriname, enslaved Africans/runaways, Djkus/Maroons in, 44
Suriname, gold, bauxite locations, 44, 53
Suriname, gold exports, value, 44

taxation, fees, 20–1, 30, 50–1, 56, 109, 146, 148–51, 154
on gold and diamonds, 11
Thompson, "Harold, Paris," mining worker, carpenter, Linden, 102
Tikwah Mining Corporation, 84
Tipperary Hall, Buxton Village, 52
Trade Union Ordinance #17, 1921, 123
Trade Unions, 6, 8, 41, 108–28
Trade Unions Ordinance Chapter 57, 127
Transport and Harbours Department, 80, 106, 136
Transportation, 51, 78–81, 89, 107–8, 167
Trinidad and Tobago, 12, 34, 53, 162, 167
Tumatumari, Essequibo, 51, 107
Tuskegee University, 6, 135
Two Friends, Demerara River, 94, 107
Two Friends, East Coast Demerara, 149, 162

Union Navy, 20
United Diamond Fields Co. of B. G. vs. Solomon Bernato, Joel Bernato et al, 51
United Kingdom, 86, 173
see also Great Britain
United States, 94, 129
United States Civil War Navy, *see* Union Navy
United States Marshall Plan, 56
United States wartime base, Atkinson Field, British Guiana, 53

Vagrancy and Labor Laws, 3, 12, 21, 24–5, 30
Venezuela, 1, 6, 9, 10, 11, 17, 27, 29, 41, 44, 50, 64, 76, 122, 129, 162
Victoria Village, 13, 101, 139, 153–4, 163–4
Villages, after emancipation, 13
see also under individual names
Voluntary Subscription and Immigration Society, 15

Wade, Lampley, Berbician, diamond miner, 52
wages, 118, 127
government employees, 56
reduced for gold and diamond miners, 118
skilled and unkilled, in bauxite, 56
sugar workers, 56
Waini River, 9
see also Guyana rivers
Wakenaam Island, 13, 151
watery graves, 57–70
Watooka, Linden, 99–100, 102, 106
Webber, James F., mining businessman, 26
West Bank Demerara villages
Bagotville, 131
Canals #1 and #2, 32, 135, 169
Klien, 1, 131
Middlesex, 135, 169
Pouderoyen, 131
Schoon Ord, 131
Vauxhall, 131
West Coast Demerara villages: Blankenburg; Uitvlugt; Vreed-en-Hoop, 81
West Demerara Railway Company, 19
West Indian, 1, 15, 22, 27, 28, 30, 33, 40–1, 46, 57–70, 84, 100, 103, 121, 147, 168, 170
see also Afro-Caribbeans
West Indian Royal Commission, 1897, 33–4, 59, 62, 65
West Indian Royal Commission, 1938–39, 28, 72, 84, 96
William Fogarty's Ltd., 107
Williams, Arthur James 'Art', aircraft pilot, 81
Williams, Eunice, Linden, 101
Williams, 'Mammy Fiffee' Agatha, 142
Winkle Village, 22
see also Table A2

Winter, James, secretary, Institute of Mines and Forests of British Guiana, 26, 34–5, 116, 117
Wismar, Demerara River, 53, 100, 105–7, 151
women, 5, 45–7, 90, 137–41, 160
  women boxers, 46
  women in farming, 140–1
  women in villages, 137–41
  women mining claim owners, 45–6
  women mining workers, 45–6
  women purchasers, villages, 139, *see also* Table A 1
Wood, Howell, A., Buxton Village, carpenter, miner, 143

Young, A. C., jeweler, businessman, 164
Yukon, 43